The Practice of
Constructivism
in Science Education

OF RELATED INTEREST

The Practice of
Constructivism
in Science Education

AAAS
PRESS

A publishing division of the
American Association for the
Advancement of Science

Kenneth Tobin
editor

Library of Congress Cataloging-in-Publication Data

The Practice of constructivism in science education / edited by
  Kenneth Tobin.
       p. cm.
    Includes bibliographical references (p.) and index.
    ISBN 0-87168-516-7
    1. Science—Study and teaching—United States.
    2. Constructivism (Education)—United States.
    3. Science teachers—Training of—United States.
    I. Tobin, Kenneth George, 1944–   .
    LB1585.3.P73 1993
    507.1—dc20   93-31240
                                  CIP

AAAS Publication: 93-11S
International Standard Book Number: 0-87168-516-7
Printed in the United States of America on acid free recycled paper

Contributors List

Antonio Bettencourt, Science Education Program, University of Saint Louis in Madrid, Spain

Nicholas C. Burbules, Department of Educational Policy Studies, University of Illinois, Urbana-Champaign

William W. Cobern, Education and Human Services, Arizona State University West

Jere Confrey, Mathematics Education Program, Cornell University

Nancy T. Davis, Science Education Program, Florida State University

Thomas M. Dana, Science Education Program, Pennsylvania State University

Mia Lena Etchberger, Lucille Moore Elementary, Panama City, Florida

James J. Gallagher, Institute for Research on Teaching, Michigan State University

Alejandro José Gallard, Science Education Program, Florida State University

Ernst von Glasersfeld, Institute for Scientific Reasoning, University of Massachusetts

Ronald G. Good, Departments of Curriculum and Instruction and Physics, Louisiana State University

Elizabeth Jakubowski, Mathematics Education Program, Florida State University

John St. Julien, Department of Curriculum and Instruction, Louisiana State University

Marcia C. Linn, Mathematics, Science and Technology Education, University of California, Berkeley

Francine P. Peterman, Department of Secondary Education, Ball State University

Lloyd P. Rieber, Educational Technology Program, University of Georgia

Wolff-Michael Roth, Department of Educational Research, Simon Fraser University, Canada

Tom Russell, College of Education, Queen's University, Canada

Kenneth L. Shaw, Mathematics Education Program, Florida State University

Peter C. S. Taylor, Science and Mathematics Education Centre, Curtin University, Australia

Deborah Tippins, Department of Science Education, University of Georgia

Kenneth Tobin, Science Education Program, Florida State University

James H. Wandersee, Department of Curriculum and Instruction, Louisiana State University

Grayson H. Wheatley, Mathematics Education Program, Florida State University

Table of Contents

Preface

Constructivism: A Paradigm for the Practice of Science Education

Currently there is a paradigm war raging in education. Evidence of conflict is seen in nearly every facet of educational practice. At the base of this conflict are the referents used by educators to identify problems, give operational meaning to them, develop a plan to solve them, formulate solutions, and reform policies and procedures. Traditional approaches to educational reform have been built on a platform of objectives that has assumed that problems exist independently of those who perceive them. As a consequence, the traditional approach to problem solving, broadly conceived across the domains of research and educational practices such as testing, administration, and resource development, often is to seek objective solutions and to identify causal relationships among salient variables. Whereas some would argue that traditional practices have served us well and should be maintained, others have argued for a change in epistemology and have endeavored to break away from conventional practices. The resistance to change from those in authority within the educational culture has often been strong. Nevertheless, the revolution has progressed steadily and there is evidence of widespread acceptance of alternatives to objectivism, one of which is constructivism, the focus of this book.

Although constructivism is not a new epistemology, its use has become increasingly popular as a referent for professional actions in the past 10 years. In response to the increasing use of constructivism, a symposium was planned for the 1991 annual meeting of the American Association for the Advancement of Science. This meeting provided the stimulus for the book, which is, in part, based on papers presented at that conference together with chapters from science and mathematics educators who have used constructivism in their professional activities. The book consists of 19 chapters arranged into four sections that focus on the nature of constructivism, constructivism and the teaching and learning of science, constructivist perspectives on teacher education, and conclusions.

To avoid the cumbersome nature of gender-neutral pronouns such as his/her, the female form is used throughout this book.

ABOUT THE BOOK

The first section of the book consists of five chapters on the nature of constructivism. In the introductory chapter Deborah Tippins and I explore the nature of constructivism and explore its applications in terms of science curriculum, assessment of science learning, and undertaking research in science education. The power of constructivism as a referent is portrayed in a context of the limitations of conceptualizing constructivism as a method of teaching and learning. The dialectical relationships between individual sense making and social negotiation of meaning also are discussed.

In the second chapter von Glasersfeld answers 42 questions organized into three sections that relate to epistemology, the role of social factors in knowledge construction, and instruction. The chapter is based on answers to questions generated at two keynote presentations. At the 1990 annual meeting of the National Association for Research in Science Teaching (NARST) in Atlanta, Georgia, von Glasersfeld presented the keynote plenary address in which he discussed implications of constructivism for the learning of science. The latter half of the presentation allowed participants to discuss constructivism in relation to their own practices and to identify questions. Von Glasersfeld then responded to the questions. A similar format was adopted at the Washington, D.C., meeting of the American Association for the Advancement of Science (AAAS). This chapter is based on von Glasersfeld's responses to the questions from both meetings. Accordingly, the chapter makes an invaluable contribution to clarifying constructivism by systematically dealing with most of the critical issues raised by those wanting to better understand constructivism and by those wanting to use challenging questions to demonstrate publicly what they regard to be shortcomings of constructivism as a viable theory of knowledge.

Bettencourt, in chapter 3, discusses radical constructivism in terms of the work of von Glasersfeld, Piaget, and Kitchener. The chapter deals with both the social and the cognitive aspects of constructivism, discusses the role of constraints in relation to knowing, and explores constructivism in relation to the teaching and learning of science. One contribution that Bettencourt makes in his paper is to distinguish three types of constructivism which he terms radical constructivism, hypothetical realism, and pragmatic constructivism. This category system can be used to examine the various types of constructivism identified by Good and his colleagues in chapter 5.

In chapter 4 Cobern begins his discussion of constructivism with an analysis of the construction metaphor and indicates that the cultural context is a significant component. His use of contextual constructivism draws attention to the importance of cultural elements such as social and economic class, religion, and geographical location in the construction of knowledge. Cobern's discussion of constructivism is grounded in a historical context and his interest in world view theory which incorporates an anthropological perspective. Cobern traces the emergence of what he terms *contextual constructivism*, which is founded on two metaphors, the construction site and cognitive ecology. According to Cobern, the term *cultural context* subsumes two ecologies, conceptual ecology and the social and physical milieu of individuals. Neither ecology has meaning without the other. Although constructivists such as von Glasersfeld have built on a Piagetian tradition, Cobern points out that they emphasize Piaget's structural philosophical position and oppose a continued focus on the application of stage theory. According to Cobern, the initial departure from Piagetian research is best termed *personal constructivism* because of the focus on individual learners.

Good, Wandersee, and St. Julien undertake a critique of the various forms of constructivism and urge caution in relation to acceptance of the new "ism" and all of its faces. Chapter 5 commences with a question of whether or not the nature of science is important to constructivists. The authors conclude that the answer to this question, at least for those such as von Glasersfeld who adopt a postepistemological position, is no. However, the meaning they seem to ascribe to postepistemological differs from that which von Glasersfeld provides in his chapter. The next question raised in the chapter concerns whether or not we should agree on the meanings given to constructivism. In this section the authors caution against two all too common trends of reform-without-change and change-without-grounds. The third issue to be addressed concerns the many faces of constructivism. This section addresses the question of diversity among the various constructivist positions. The fourth section is a review of some philosophies that have evolved throughout the past hundred years. The section examines the work of philosophers in relation to their views on the nature of reality. The next two sections examine Piaget's work in relation to radical constructivism and social constructivism respectively. The latter section extends into areas that have become related to the work of those interested in constructivism, namely, semiotics and the sociology of knowledge. Contextual realism is examined in relation to constructivism and then the question is raised as to whether radical constructivism is Aristotelian. Building on a critique of Matthews (1992), the authors seem to agree with Matthews' conclusion that constructivism is a "case of new wine in the old Aristotelian bottle." Such a conclusion appears to be at odds with the radical constructivist position as it emerges in the writings of von Glasersfeld and his answers to the questions raised in chapter 2. The final section of the chapter looks ahead to the future of constructivism, suspending judgment on the extent to which constructivism, in its various forms, will prove to be a viable theory on which progress in science education can be supported.

The second section of the book consists of seven chapters concerned with constructivism and the teaching and learning of science. In chapter 6 Linn and Burbules caution readers against regarding group learning as a panacea that optimizes social interaction and enhances learning. Group learning is not a unitary construct. To illustrate this assertion, Linn and Burbules identify cooperative, collaborative, and tutored group learning situations that are distinctive in terms of the roles of learners. Whereas group learning is advocated for fostering cognitive skills, promoting social skills, and imparting workplace skills, it is apparent from Linn and Burbules that groups might be better for some activities, such as brainstorming and divide-and-conquer tasks, than others such as planning and synthesis tasks. One particular aspect that deserves attention is the importance of an individual's status in group situations. Linn and Burbules assert that the idea selected by a group is often that of the person having greatest status. However, factors other than knowledge of science usually determine a person's status within a group. An additional problem cited by Linn and Burbules concerns stereotyping. Gender-related stereotypes, such as *females are not successful at solving science-related problems* can be reinforced through group activities.

One latent value of this chapter is that it causes the reader to reflect on the meaning of social interaction. Linn and Burbules point out that any use of language is social and that social interaction can occur when an individual reflects on her personal experience. Social interaction, therefore, does not have to involve others in a direct way. The question for teachers thereby becomes one that considers the optimal organization for students. As Roth indicates in chapter 9, there is value in a variety of group arrangements that facilitate social interaction.

In the seventh chapter Wheatley focuses on the process of negotiation in problem-centered learning environments. He discusses a vignette in which two ninth-grade youths discuss possible solutions to a problem. One student is task-oriented and has the goal of constructing mathematical meaning. He has the correct answer but demonstrates a willingness to explain his method to his partner, uses additional examples to assist his partner understand the method and consequent solution, and considers alternative methods and solution. In contrast, the other partner in the group is ego-oriented and is concerned with "getting the right answer" and following rules. He does not use mathematical reasoning in his attempts to persuade his partner of the viability from authority figures. The vignette highlights the importance of mathematical knowledge in the negotiation process.

An important implication of adopting constructivism as a referent for the planning of curriculum is an emergent focus on the learner. From the outset of chapter 8, Jakubowski emphasizes mathematics classrooms as learning places rather than workplaces. The chapter examines learning environments from a constructivist perspective and highlights the need for active learner roles and mediating roles of teachers. Broadly speaking, Jakubowski incorporates a social constructivist perspective in her discussion of problem-centered learning, which is presented as a viable approach to the teaching and learning of mathematics for middle school students. Among the interesting metaphors used to describe learner roles is the need for student presenters to construct a "reader." The interesting implication of this metaphor is that the curriculum can be regarded as a text. Jakubowski cautions that materials should be accessible to all learners in problem-centered environments but that materials might not be needed to solve particular problems as they are framed by students.

When he wrote chapter 9, Roth was a high school physics teacher. He utilized a social constructivist framework to plan and implement the curriculum with an approach that emphasized the importance of language to learning. Four activity types are highlighted to illustrate the vital role of the teacher in facilitating learning of physics with understanding. In individualized settings Roth highlights the importance of students personalizing their experiences through the use of their own language in writing assignments. The significant role of the teacher and more advanced peers as a bridge between personal knowledge and the canonical knowledge of science is illustrated in small group activities and in whole class interactive activities. Roth extends constructivist theory with ideas from Vygotsky to build teaching roles to enable him to attain goals related to students learning science in a way that is consistent with canonical science knowledge. Roth uses the metaphors of cognitive apprenticeship and enculturation to make sense of his teaching as situated modeling, coaching, and fading. From Roth's point of view, collaboration is the key to meaningful learning. Illustrative classroom examples show how students construct meaning, on their own by interacting with their physical environment, in small collaborating peer groups of equal status, and in whole class interactive discussions.

In chapter 10 Gallard draws attention to the critical role of language and culture in learning. Using personal examples from his early childhood to the present time, Gallard makes the case for allowing students to use their own native language to make sense of their experiences and to learn science. The formal language of science can come later. According to Gallard, the early use of formal science terms, in English, can be a significant block to learning with understanding. Gallard also indicates that knowledge is culturally embedded. If students, particularly those for whom English is not the native language, are to use their extant knowledge, it is important for their teachers to provide opportunities and time for them to connect present learning experiences to what they already know. What is needed is not tokenism but substantive efforts to focus on the learning needs of each individual in the class, taking account of their cultural heritage.

Gallagher discusses issues associated with teachers making changes in their science classes. He notes that the dominance of objectivism for so many years rendered alternative ways of thinking, such as constructivism, "strange and often unwelcome." Chapter 11 presents and discusses a model for assisting teachers to change. Gallagher notes that teachers made changes from the traditional paradigm to the alternative slowly, in small increments, in a process in which teachers constructed a vision to guide classroom practices. His chapter presents practical ideas for teachers to assist students not only to acquire knowledge, but to integrate and apply science knowledge. The chapter also describes the significance to the change process of expanding teachers' conceptualizations of assessment to include formative as well as summative goals. Gallagher's model can provide teacher educators and teachers with a tangible way of thinking about their learning with respect to curriculum change while providing concrete examples of activities to pursue.

Rieber adopts a broad definition of instructional technology in a discussion of the enhancement of learning. He views instructional technology as evolving and adaptive, reshaped mainly by advances in human learning and instructional practice. Chapter 12 sets out to show that instructivist and constructivist approaches can be merged in the context of computer microworlds to enhance learning. Although these two approaches traditionally have been viewed as incommensurable, Rieber has adopted an eclectic view to bring the best of each approach into a computer microworld. Through the use of microworlds, a computer environment can be constrained in such a way that particular learning goals are attainable. In this way, Rieber sees it as possible to be clear on the learning goals of a curriculum while allowing students to be engaged in activities that promote active learning and enabling knowledge to be built on prior knowledge through the resolution of problems.

Chapter 13 begins a third section of the book, on applications of constructivism to teacher education. In this chapter I discuss teacher knowledge in terms of metaphors, beliefs, images, and metonymies. Initially it was determined that metaphors can be used by teachers to conceptualize significant teaching roles. Accordingly, metaphors serve to organize sets of beliefs and overt actions that apply to teaching. A metonymy occurs when a part of a concept is used to represent the whole. This idea leads to a situation where some parts of a concept are considered to be more important than others. As an example, a metonymy for teaching might be control, leading to a situation where control of the curriculum is perceived as more important than other teacher roles. Thus, teachers might give greater value to beliefs and overt actions associated with maintaining teacher control over the curriculum. Accordingly, it is important for teachers to identify metonymies and reflect on the manner in which they might constrain the curriculum. From a socio-cultural perspective, it is important to identify the myths of the culture of a school site and associated customs and taboos. I identify three myths that appear to have had a significant role in shaping the traditional curriculum. These myths are objectivism, the teacher as controller of students, and the school as workplace. The issue of curriculum reform can be cast in terms of three alternative myths: constructivism, the student as controller of the curriculum, and the school as a learning place. The question that is then addressed is how to change myths and sustain a new set of customs and taboos.

Peterman also investigates issues associated with teacher change in chapter 14. She indicates that in any change context it is important to take account of the prior beliefs of participants, and the need to elicit and reconstruct beliefs both publicly and privately. Peterman advocates the incorporation in staff development activities of *what* and *how* beliefs of teachers. In addition, Peterman describes the importance of the need to examine the beliefs embedded in staff development practices. Peterman advocates additional study on the interactive nature of beliefs and practice and is somewhat concerned about teachers being able to change their core beliefs. She also recommends the use of constructivism to analyze learning and teaching in relation to the planning and implementation of staff development projects.

Russell commences chapter 15 with a discussion of a number of dilemmas associated with traditional practices. Teachers emphasizing coverage of content rather than understanding science content, an emphasis on preparing students for the next educational level, and preparing students to perform well on tests also were listed as traditional emphases that run counter to recent trends based on constructivism and reflection in action. Russell makes a strong case for learning about science teaching through direct experience with teaching, that is, learning from action. The chapter deals with the relationship between constructivism and reflection-in-action. As has been the case in earlier chapters, Russell has made sense of constructivism in his own way. He defines constructivism as the process by which students deal with content, a method which is guided and mediated by the teacher. Russell advocates learning from experience and in teacher education programs, he recommends less telling and more student action in the practice setting. Russell views practice as the medium through which we come to understand theory. As such, images rather than theory are much more likely to guide teaching.

Shaw and Etchberger make some important points about small group learning in chapter 16. The teacher in the study, Etchberger, wanted to use constructivism as a basis for planning and implementing the curriculum. Initially, she equated constructivism with cooperative learning. However, after initial success, it was apparent that nearly all students had to learn how to learn in groups. The authors conclude that student empowerment is a requirement for learning with understanding. In their discussion of the teacher's learning and change, attention is drawn to the dialectical nature of the relationship between the number of perturbations and commitment to change. The authors highlight the importance to change of constructing a vision, the components of which are presented from a constructivist perspective and include the customs and taboos which apply to the vision.

Taylor deals with the pedagogical change of a mathematics teacher who is teaching out of field. Chapter 17 describes changes that occurred over a period of more than a year and critiques the change process. Some of the obstacles encountered by the teacher include his desire to cling to positivist beliefs of teaching and learning, a view of constructivism that largely ignores the social implications of knowledge construction, and a tendency to retain an objectivist view of knowledge as truth. As the study progressed, Taylor was able to learn about the role of constraints in making significant changes within the culture of schools and the significance of power relationships between the teacher and students.

In chapter 18 Confrey discusses the need for a fresh approach to teacher education within a context of the many calls for reform in mathematics education. In so doing, she examines the roles of others in assisting teachers to implement reform. A vignette presented by Confrey underlines the significance of involving parents in attempts at reform of the curriculum. Unless parents are educated to understand the nature of the reform they might inadvertently subvert the endeavors of teachers to make the changes advocated in a plethora of reports. At the classroom level Confrey raises the interesting question of how purposeful listening and talking can lead to useful knowledge. To respond to this question she turns to constructivism. Confrey then discusses the teacher's role in mathematics classrooms as recognizing students' productive ideas, being a good listener, and using multiple representations of mathematics knowledge. Confrey's thesis is that teachers are not presently educated to do these things, the implication being that reform in teacher education will be a necessary co-requisite for improving the quality of the mathematics curriculum in elementary, middle, and high schools.

The final section of the book consists of chapter 19, written by Dana and Davis, a synthesis of the earlier 18 chapters of the book organized around five themes: knowledge and knowing, teaching and learning, educating teachers, educational research, and future directions.

CONCLUDING REMARKS

Constructivism should not be regarded as a new truth to replace objectivism. On the contrary, constructivism is a way of thinking about knowing, a referent for building models for learning, teaching, and curriculum. These models are hypothetical and they can be put to the test in classrooms. Von Glasersfeld has repeatedly cautioned that constructivism does not tell us what to do, only what not to do. The test for any model, or knowledge construction, is the extent to which it provides an adequate basis for accomplishing goals. Viability, the test of knowledge against experience, is the critical test of models built from constructivism or from any other referent. From a constructivist perspective we might expect different models to be most appropriate in different contexts. It becomes clear that there is no *one* right way to teach and learn, nor *one* right way to think about knowing in all contexts in which one might know.

Within constructivism there is room for different scholars to emphasize different aspects of knowing. For example, if the primary focus of a scholar is to build an understanding of how individuals construct understandings from their experiences, it is predictable that the focus will be on the individual and the sense-making process. In contrast, if the focus is on understanding the teaching and learning process, it is to be expected that socio-cultural processes will be significant from the outset. According to the focus of individual scholars, therefore, constructivism will take on different faces, each with its own emphases. The evolution of constructivist thinking in different ways in different groups is anticipated because of the myriad diverse challenges, idiosyncratic to the situations in which scholarship is embedded, that face those who undertake research and scholarly activities in today's complex world. Those who adhere to a more objectivist way of making sense might object to the many faces of constructivism, expecting to find only one right way of thinking about knowing. However, those who ascribe to the belief that meaning is grounded in socially negotiated processes will anticipate diversity and celebrate the alternative ways of knowing and experiencing that accompany increasing speciation in the genus constructivism.

Kenneth Tobin

Tallahassee, FL
May 1993

PART 1

The Nature of Constructivism

1

Constructivism as a Referent for Teaching and Learning

Kenneth Tobin and Deborah Tippins

THE NATURE OF CONSTRUCTIVISM

V on Glasersfeld (1992, 5) noted that

> from the beginning of the 5th century B.C., the skeptics have shown that it is logically impossible to establish the "truth" of any particular piece of knowledge. The necessary comparison of the piece of knowledge with the "reality" it is supposed to represent cannot be made, because the only rational access to that reality is through yet another act of knowing.

Von Glasersfeld (1992, 6) wrote that constructivists, unlike the skeptics, endeavored to break away from the perennial paradox associated with the requirement that knowledge represent an independent world

> and admit instead that knowledge represents something that is far more important to us, namely what we can do in our experiential world, the successful ways of dealing with the objects we call physical and the unsuccessful ways of thinking with abstract concepts.

Critics of constructivism (e.g., Matthews 1992) often make the accusation that constructivists deny the existence of a reality. Whereas some constructivists might do this, we see constructivism as a form of realism in the sense that the existence of a reality is acknowledged from the outset. What constructivism has to say about that reality, however, is that we can only know about it in a personal and subjective way. To take the example of gravity, a constructivist position is that gravity exists and through our experiences we come to know about gravity. Our knowledge of gravity is both individual and social, and through negotiation, agreement is reached within our social system that the concept of gravity has numerous verifiable properties. We construct a model of gravity that is viable in that the model fits experience, but no matter how elegant, that model cannot claim to be an absolute truth. The model has evolved, as all knowledge evolves, through the processes of negotiation and consensus building. It is a set of assertions about our experiences with gravity. As our experiences have changed, so too has the model been explicated. Each day most of us test our own versions of that model as we experience the world. Things do fall downward and we could not construct a viable theory with things falling upward. Our constructions are constrained by experiences, which comprise subjective interactions with the real world as we have constructed it. Since there is no objective account of what gravity really is, we cannot tell whether our model for gravity gets closer and closer to an absolute reality. Suffice it to say that we can only know gravity in a personal, socially mediated way.

The discussion of gravity provides insights into the way that constructivism explains the development of science. Science does not exist as a body of knowledge separate from knowers. On the contrary, science is viewed as a set of socially negotiated understandings of the events and phenomena that comprise the experienced universe. Knowledge is accepted by the scientific community as viable because of its coherence with other understandings and its fit with experience. Accordingly, scientific knowledge is subject to empirical verification and must be accepted as viable by the scientific community. As Kuhn (1970) and others have pointed out, scientific knowledge changes over time because goals and problems of society change, leading to new experiences; technology provides new ways of experiencing; what is known continues to increase at an exponential rate; and the individuals that comprise the scientific disciplines continually change.

A constructivist perspective acknowledges the existence of an external reality but realizes that cognizing beings can never know what that reality is actually like. This is just one paradox that often is confusing to those wanting to understand constructivism. Von Glasersfeld is clear that radical constructivism does not deny the existence of reality in the presence of cognizing beings to think of it. An absolute reality does exist; however, constructivists point out that an individual can never come to know that reality as a truth. Von Glasersfeld noted that

> a basic misunderstanding of constructivism . . . springs from the resistance or refusal to change the concept of knowing. I have never denied an "absolute reality," I only claim, as the skeptics do, that we have no way of knowing it. And as a constructivist, I go one step further: I claim that we can define the meaning of "to exist" only within the realm of our experiential world and not ontologically. . . . Of course, even as constructivists, we can use the word "reality," but it will be defined differently. It will be made up of the network of things and relationships that we rely on in our living and of which we believe that others rely on, too.

Because of the somewhat different focus of constructivists and epistemologists, von Glasersfeld prefers to think of constructivism as postepistemological. That is, constructivism is not concerned with the question of knowledge as a representation of truth; rather, it focuses on the manner in which knowers construct viable knowledge, that is, knowledge that enables an individual to pursue goals in the multiple contexts in which actions occur.

4

Von Glasersfeld points out that

> viability—quite unlike "truth"—is relative to a context of goals
> and purposes. But these goals and purposes are not limited to
> the concrete or material. In science, for instance, there is,
> beyond the goal of solving specific problems, the goal of con-
> structing as coherent a model of the experiential world as
> possible. (von Glasersfeld 1992, 7)

An often misunderstood aspect of constructivism is that the theory
incorporates a value position that any construction is as viable as another.
Such a position is based on a belief that personal viability is the critical issue
in constructivism. The position ignores the social component of knowledge,
that is, that knowledge must be viable not only personally, but also in the social
contexts in which actions are to occur. Viability is thereby determined with
respect to the actions of an individual and the extent to which those actions
facilitate the attainment of goals in the social contexts of action. If the
subculture of the classroom is considered, therefore, it is not the case that
constructivism suggests that students ought to be able to retain naive theories
or incorrect knowledge. The teacher, representing society, has an obligation to
educate students (i.e., to have students construct knowledge they do not seem
to have, because we think it would be good and useful for them to have it [von
Glasersfeld 1992, 2]), to assist them in learning what is currently regarded by
society as viable knowledge. Accordingly, if a teacher regards the constructions
of any individual to be inviable in the larger set of contexts in which actions are
to occur, it is the teacher's duty as a professional to structure learning
environments to facilitate the process of learning what society regards as
having greatest viability at that particular time. Far too often, however, that
which society regards as having greatest viability dignifies and reinforces a
dominant culture in which certain ways of speaking are legitimate, others are
not, particular forms of knowledge are certified and specific histories taught.
As we structure learning environments to facilitate science learning, we must
keep in mind, as Giroux (1989, 181) suggests, that schools are "contested
cultural sites, not simply places where instruction takes place and bits of
neutral knowledge are transferred."

PERSONAL AND SOCIAL COMPONENTS OF CONSTRUCTIVISM

Over the past decade practitioners in a variety of fields have embraced
constructivism as a theoretical framework on which to base some of their
activities. Not surprisingly, different parts of the theory appeal to different
practitioners. For example, many science educators throughout the world have
studied alternative frameworks in science and conceptual change. These
practitioners focused on the importance of prior knowledge in learning. The
trend for so many practitioners to describe their theoretical rationale in terms
of constructivism led to the emergence of a number of adjectives to characterize the particular brand of constructivism that was used in particular situations. Von Glasersfeld (chapter 2) uses the term *radical constructivism* to
emphasize that knowledge could not be separated from knowing. The term
radical is used as a counterfoil to those who see constructivism mainly or only

in terms of learning being built on prior knowledge. In this weak form, *constructivism* is referred to as trivial constructivism.

An examination of the views of those who have used constructivism (e.g., Bauersfeld 1992; Wood et al. 1992; Saxe 1992; von Glasersfeld 1992) leads us to a synthesis position that knowledge is personally constructed but socially mediated. That is, knowledge only exists in the minds of cognizing beings, but cognizing beings only exist in a socio- cultural sense. From the outset, an organism constructs knowledge in the presence of others who are able to perturb the environment in such a way that a learner's experiences are constrained by the presence of others. The newborn infant learns with the assistance of others to make sense of the signs of the culture, which would not exist without the existence of the culture. Accordingly, the constructions of the individual are constrained by the perturbations that become a part of that individual's experience. A concrete example of this process is the use of language. Language is a tool that facilitates communication between partici-pants in a society. When a learner thinks in terms of language, the thinking is a social process even though it is occurring within the mind of a single individual. Each person constructs her own environment, which includes those with whom she interacts. Thus, individuals construct speakers and listeners and assign roles to those with whom they interact. When we think of knowledge, it is convenient to think in terms of both the individual and the social components. Just as it is sometimes useful to think of an electron as a particle and at other times a wave, so it is sometimes useful to think of knowledge as an individual construct and at other times as a social construct. But at all times knowledge is both social and individual, a dialectical relation-ship existing between the individual's contribution to knowledge and the social contribution. To those who want to give greater emphasis to one than the other this may seem a paradox; however, to those who are comfortable with multiple ways of representing reality it is acceptable for knowledge to be thought of in complementary ways. The recognition that knowledge has both individual and social components that cannot be meaningfully separated enables us to construct science learning environments where multiple ways of knowing (i.e., women's ways of knowing, indigenous people's ways of knowing) are sought and valued.

Evidence of scholars emphasizing the social aspects of constructivism is clearly evident in the work of Cobb and Saxe (e.g., Cobb 1990; Saxe 1992). As Cobb and his colleagues pointed out, it is not helpful to think of the personal and social emphases as an either/or dichotomy. Rather, both have important roles in thinking about knowledge, knowing, and teacher and learner roles in classrooms:

> It is useful to see mathematics as both cognitive activity con-
> strained by social and cultural processes, and as a social and
> cultural phenomenon that is constituted by a community of
> actively cognizing individuals. (Wood et al. 1992, 3)

Saxe (1992) noted that for Piaget the socio-cultural processes were largely unanalyzed. His preference was to build a theoretical framework for knowledge that emphasized the social and cognitive components. Saxe makes a compelling case for a problem-solving approach to the learning of mathemat-ics in which students are involved in structuring their own problem. In his two case studies, the emphasis was on finding a coherent solution to a problem rather than remembering how to apply a recipe to obtain a solution. Indeed, in Saxe's studies, more interest was focused on learning mathematics to be "street wise" and to negotiate effective deals in the commercial sense. In Saxe's subsequent classroom studies, he used a game format to enable students to construct goals that are not directly related to mathematics. To achieve these goals, it is first necessary to learn and apply mathematical knowledge. Saxe concentrated his framework on the notion of goals. Cognitive goals emerge through an individual's daily participation in cultural practices. In attempting to accomplish these emergent goals, children generate new knowledge linked to social and cultural life. According to Saxe, individuals construct novel understandings as they attempt to accomplish goals rooted in both their prior understandings and their socially organized activities.

CONSTRUCTIVISM: METHOD OR REFERENT?

Some authors (e.g., Fosnot 1992) use constructivism to represent a method of teaching whereby the teacher bases what happens on beliefs that are consistent with constructivism. What they mean by this is that constructivism has been used as a referent to build a classroom that maximizes student learning. Typically, the teacher takes account of what students know, maximizes social interaction between learners such that they can negotiate meaning, and provides a variety of sensory experiences from which learning is built. Another example of this practice is seen in chapter 15 where some teaching practices, such as lecturing, are regarded as having little value compared to alternatives such as small group learning or interactive discussions that are "constructivist" in nature. This position, although understandable, reduces constructivism to a set of methods and diminishes its power as a set of intellectual referents for making decisions in relation to actions. Just as constructivism can be used to explain how students make sense of experience in interactive discussions or in small group problem-solving activities, so too can constructivism be used to explain why learning occurs in lectures and how lectures can be adapted to improve the quality of learning.

Wheatley (1991) described approaches to curriculum that have been carefully built with constructivism as a referent. Known as problem-centered learning, students work together in small groups making meaning of tasks and setting out to solve problems that are perplexing. The teacher in such classes has an important mediating role, ascertaining what students know and structuring tasks such that they can build knowledge structures that are commensurate with knowledge of the discipline. Wheatley described how students negotiate meaning in small group situations, and then negotiate consensus in whole class settings. The teacher's role is to monitor student understandings and guide discussions so that all students have opportunities to put language to their understandings and to engage in activities such as clarifying, elaborating, justifying, and evaluating alternative points of view. Such visions of classroom learning environments are exciting and appeal as viable alternatives to those so often reported in studies of learning in traditional classrooms (e.g., Tobin and Gallagher 1987). However, as appealing as these alternative visions of classroom learning might be, to label them as constructivist tends to mask an important application of constructivism.

Constructivism, as a set of beliefs about knowing and knowledge, can be used as a referent to analyze the learning potential of any situation. For example, constructivism can be used to explain why certain students have been successful in learning science in contexts where teachers lecture day after day and students listen and copy notes until their hands ache. To be a viable theory of knowing, constructivism must have explanatory power in all situations where knowledge is constructed or cognizing beings are deemed to know. Similarly, constructivism ought to be useful in predicting how any given set of circumstances might be changed to improve the opportunities of persons who wish to learn in such situations. For example, if a biology department has a policy that all classes will contain at least 200 students, it is probably not feasible to think in terms of Wheatley's problem-centered learning. However, from a constructivist perspective, learning can be thought of as a social process of making sense of experience in terms of what is known. To improve learning, therefore, a teacher might consider how to improve the quality of each of the four components (i.e., social process, making sense, experience, extant knowledge) given the constraint of 200 or more learners. Similarly, in countries such as Taiwan, where class sizes of more than 50 or 60 students are common, teachers can think in terms of improving the quality of social interactions, providing a range of meaningful experiences to each learner, and making it possible for each student to become aware of her relevant prior knowledge and apply it to the process of learning. Constructivism, as a reflective tool, empowers teachers and enables them to fashion learning activities to the circumstances in which they find themselves. Thus, teachers can focus planning and implementation strategies on the needs of learners as they understand them from a constructivist perspective.

In contrast to the use of constructivism as a method is the notion of constructivism as a tool for critical reflection. As such, constructivism can act as a referent for deciding which teacher and learner roles are likely to be more productive in given circumstances. Such use makes it possible to plan and implement activities that are postulated to enhance learning. Constructivism tends to provide different angles on thinking about educational problems. For example, in teacher education a question that needs to be answered is, how can prospective and practicing teachers learn to teach science and mathematics? Traditionally, the question has been addressed in terms of what we know about teaching and learning, as if the knowledge exists as a body out there to be discovered or learned by teachers. Research findings are regarded as contributors to this body of knowledge. Accordingly, solutions to the problem are framed within a set of assumptions that a body of knowledge exists and learners (in this case practicing and/or prospective teachers) can come to know this body of knowledge by interacting with it. The focus is on the interaction between disciplinary knowledge and the learner.

In contrast, two questions are of significance from a constructivist point of view. What does the learner already know about teaching and learning, and how can this knowledge be represented? The associated set of questions relates to the process of making sense of experience. What experiences should teachers have to enable them to build an understanding of teaching and learning? Finally, it seems as if teachers need time to make sense of their experiences. That is, they need time to think things through for themselves and

8

then time for such cognitive activities as clarifying, elaborating, justifying, and considering the merits of alternatives. From a constructivist point of view the emphasis is on the teacher as a learner, a person who will experience teaching and learning situations and give personal meaning to those experiences through reflection, at which time extant knowledge is connected to new understandings as they are built from experience and social interaction with peers and teacher educators.

CONSTRUCTIVISM AND THE CURRICULUM

Thinking of science from a constructivist perspective helps science educators to decide what might comprise a science curriculum. Since all knowledge must be individually constructed, it makes no sense to begin by thinking solely about the disciplines of science in the absence of learners. A learner has to make sense of science through an existing conceptual structure. Whatever science knowledge is constructed will be an interpretation of experience in terms of extant knowledge. Accordingly, two questions are fundamental: first, what experiences should be provided to the learner in order to facilitate learning; and second, how can the learner represent what is known already to give meaning to these experiences? The familiar debate in science education over whether to emphasize concepts or processes has little meaning from a constructivist point of view. Making sense of science is a dialectical process involving both content and process. The two can never be meaningfully separated. The process skills can be thought of as thinking processes, such as using the senses to experience; representing knowledge through language, diagrams, mathematics, and other symbolic modes; clarification; elaboration; comparison; justification; generation of alternatives; and selection of viable solutions to problems.

A curriculum is conceptualized as a set of all learning experiences. Although broad, this definition is consistent with constructivism since it does not regard knowledge as separate from the knower and the culture in which learning is to occur. In the past, objectivist ways of thinking about a body of knowledge existing out there to be learned resulted in curricula that were thought to be transferrable to sites in greatly differing contexts. When teachers and learners adapted the resources to fit local needs, the designers of the materials viewed this as "implementation infidelity." From a constructivist way of thinking, a curriculum is embedded in culture and cannot be separated from culture, which includes other learners, the shared "taken-for-granted" knowledge of the culture, myths, customs, taboos, and history. Also to be considered are the social-political-economical milieu and the influence of others such as parents, administrators, and teachers.

The teacher's role is to mediate the learning of students. This assertion should have an important influence on the way teachers think about teaching. To begin with, the focus ought to be on the learners rather than the discipline. This is in contrast to what Tobin and Gallagher found when they did a case study of a group of Australian teachers to find out what happened when they implemented the science curriculum (Tobin and Gallagher 1987). Teachers in this study focused on content coverage as one of their highest priorities. They planned with this in mind and stuck to the plan to the best of their abilities.

They changed topics at the scheduled time, irrespective of the extent to which students had learned what had been covered. If teachers see themselves as mediating the learning of students, two critical components of their role are to monitor learning and to provide constraints so that student thinking will be channeled in productive directions. To undertake such a role, teachers must interact with students to a greater extent than in traditional classrooms to ascertain what they know and what they are thinking. The interactions can begin with the goals of the curriculum, which ought to take account of the goals of the students. Too often there is a wide discrepancy between the teacher's goals and the students' goals. Many studies have revealed a less than optimal level of student engagement in science classes. In very few classrooms has the majority of students been motivated to learn science for most of the time. With few exceptions the students do not appear to be in class to learn science. Perhaps this problem can be overcome, to a degree at least, if teachers begin to negotiate the goals of the curriculum with students.

As a mediator, the teacher needs to ensure that students are provided opportunities for quality learning experiences that provide a solid base for learning with understanding. A constructivist perspective suggests that teachers can enhance learning in the ways mentioned above, and by constraining experiences, to provide students with a scaffold to build knowledge in directions that would not be possible without the influence of a teacher. (Later in this volume Roth discusses the significance of scaffolding to student learning.) What needs to be clarified here is that the teacher mentioned throughout this section might be a student/learner who already knows about particular content. (Linn and Burbules refer to peer teachers as tutors.) In any classroom there will be the potential of using peers as tutors to enhance the learning of those who do not yet understand specific content. The process benefits those students who can learn through the constraining assistance of peers; it also benefits the tutors because they have an opportunity to clarify and elaborate their knowledge and to represent it in a variety of ways, including the assignment of language to specific science knowledge.

Other metaphors, such as the teacher as provocateur, can bring into focus the variety of productive roles available to teachers. The provocateur may be seen as a master fencer engaged in a struggle with students, thrusting at times, defending at other times, but always making visible the skills that make her a master fencer. Thrust, defend, defend, thrust; the master and student engage in earnest struggle intent on the goal of becoming a more expert fencer. The teacher contributes a great deal and is exhausted by the effort, and in like manner the student is stretched to the limit in a process in which no one is hurt, mutual satisfaction is attained, and the gap between the current skill level and what is needed to be successful is always managed in such a way that it is elusively beyond the student.

What are the essential requisites for student learning? Constructivism suggests that learning is a social process of making sense of experience in terms of what is already known. In that process learners create perturbations that arise from attempts to give meaning to particular experiences through the imaginative use of existing knowledge. The resolution of these perturbations leads to an equilibrium state whereby new knowledge has been constructed to cohere with a particular experience and prior knowledge. Because of the importance of students testing the viability of knowledge claims, teachers must consider how to provide opportunities for such testing through negotiations with students and by providing opportunities for problem solving.

There are other things a teacher can do to promote the learning of students. Planning and implementing tasks is an important role, and teachers should remember that they can only constrain the thinking of students. For example, if the teacher decides to assign activities related to electrochemical processes, then students will spend some time at least thinking about electrochemical processes. To be more specific, if the teacher asks students to discuss the definition of oxidation in small groups, then students will give their thought to oxidation and begin the process of making sense of that concept. The teacher is not transferring knowledge to anyone, but by assigning a particular activity is allowing students to construct certain experiences rather than others. If students do as the teacher wishes, their thinking will be constrained by engaging in the activity, and the chances that they will learn about oxidation are increased.

In the classroom, teachers should provide opportunities for students to represent their knowledge in a variety of ways throughout the lesson by writing, drawing, using symbols, and assigning language to what is known. Student thinking needs to be stimulated by providing time to think: students need time to engage in the processes required to evaluate the adequacy of specific knowledge, make connections, clarify, elaborate, build alternatives, and speculate. Accordingly, teachers should use an average wait time of more than three seconds during explanations and interactions with students (Tobin 1987); during lectures, at least two minutes of every 10 should be provided for interaction with peers (Rowe 1983). For example, after eight minutes of lecture and whole class interaction, the teacher might ask students to discuss linear motion with the person next to them with the purpose of writing three questions for which they do not have answers. This example highlights the time students need to clarify lesson elements and make connections with what they know already; it also shows that an important part of learning is identifying questions that need to be resolved in order to better understand given science content.

In our experience it is most unusual to find teachers who require students to generate questions and seek answers to them. Constructing questions might be one way for students to build conceptual conflict, and seeking answers to them might begin the process of resolving the conflict. The use of groups is discussed in detail in this volume by Jakubowski and also by Linn and Burbules; accordingly, we will not provide a full rationale for group work here. However, group discussions can play a significant role in the learning of students by providing time for interaction with peers to answer student-generated questions, clarify understandings of specific science content, identify and resolve differences in understanding, raise new questions, design investigations, and solve problems. Group interactions also provide a milieu in which students can negotiate differences of opinion and seek consensus.

A most significant role of the teacher, from a constructivist perspective, is to evaluate student learning. In a study of exemplary teachers, Tobin and Fraser found that these teachers routinely monitored students in three distinctive ways: they scanned the class for signs of imminent off-task behavior, closely examined the nature of the engagement of students, and investigated the extent to which students understood what they were learning (Tobin and Fraser 1987). If teachers are to mediate the learning process, it is imperative that they develop ways of assessing what students know and how they can represent what they know.

ALTERNATIVE ASSESSMENT OF LEARNING

Traditional assessment practices seem to be associated with teaching roles akin to judging and rewarding. An interpretive study of science classrooms undertaken by Tobin and Gallagher (1987) revealed that teachers used assessment as a student motivator. Unless an activity was assessed, it was difficult to obtain the active participation and cooperation of students. The effect that appeared to be associated with this trend was an emphasis on completing products for grades. There seemed to be an implicit bargain in the culture that students would work for grades in much the same way that employees work for pay. Students focused on completing tasks and getting the grade, and learning became a by-product of the main activity in the culture. An inadvertent outcome of this process might have been that negotiation between teacher and student led to overspecification of the tasks, making them clear to students, reducing risk, and making it intellectually safe to do science. From a constructivist point of view, such an environment is not conducive to learning science with understanding. To achieve such a goal students must be perplexed at times and struggle to resolve perturbations that are created in the act of experiencing and trying to figure out puzzles associated with engaging in an intellectually rigorous manner. The classes examined in this, and subsequent studies (e.g., Nordland 1990), indicated that the cognitive level of both the academic tasks and the pencil-and-paper tests used to assess learning was low. The assumption underlying many traditional assessment practices is that formal thought, represented by the ingenuity of Newtonian problem solving, is the highest level of thought. Thus, the dialectical interaction between ingenuity and intuition that enabled some scientists (i.e., Einstein) to ask unique questions which moved them outside the boundaries of tradition, is undervalued or ignored in most science classrooms today. The process of education, just like the currency of many countries, appeared to have been inflated.

One study of a beginning middle/high school science teacher graphically describes the problems associated with assessment practices built on an assessment metaphor of fair judge (Tobin and Ulerick 1989). Even though Marsha, the teacher in the study, had gained control of her difficult-to-handle students, they had built very negative attitudes toward her because she was holding the line on what she considered to be legitimate standards of learning—standards built from a consideration of the discipline rather than from a consideration of what students knew. The negative attitude toward the teacher was widespread within the class and manifested itself in a latent hostility toward the teacher and the subject of science. On the basis of the metaphor of bargaining work in exchange for grades, it seemed as if Marsha was trying to negotiate too hard a deal and the workers—the students—were not buying into it. Marsha, like so many teachers we have worked with, had no alternative ways to think about or practice assessment. She had only one conceptualization of assessment, and when it was not working she had no options to fall back on.

Marsha was involved in research on her teaching and had solved some of her management problems by reconceptualizing her role in relation to management by constructing a new metaphor consistent with constructivism. The new metaphor resulted in more focus on the needs of learners and a system of rules that was simple and emphasized student responsibility. Once the new strategies were implemented, a striking transformation of her classroom occurred. Accordingly, Marsha was open to the idea that further improvements would be possible if only she could reconstruct her understandings of assessment. She also had strong beliefs about constructivism as a way of thinking about knowing that opened the possibility of trying different approaches to assessment. Initially, Marsha was inclined to try different approaches advocated by those whom she respected in science education. She tried concept mapping, oral interviews, and problem-centered learning; none seemed to be successful. The essence of the difficulty seemed to be how she regarded science itself and how she viewed assessment. Her beliefs about the nature of science were evolving in relation to her beliefs about constructivism. Since she could not construct a metaphor for assessment, the research team suggested to her that it was like a mirror. Although the metaphor appealed to us (a student looking into a mirror and seeing her knowledge displayed in her mind), it made no sense to Marsha. Subsequently, we suggested a window into the students' minds, an opportunity for them to show the teacher what they knew. Marsha could see how this would work.

Using the window metaphor, Marsha rethought her assessment policies and practices. She implemented a new approach to the class and matters began to improve. The interesting aspect of the metaphor which she built was that it transferred power from the teacher to the students. The students now had the responsibility to make decisions about what they knew, how to represent what they knew, and when to schedule time with the teacher to show that they had learned. To Marsha and the students this seemed like a reasonable approach to assessment. What evolved over a period of time was an approach to assessment that emphasized the autonomy of students and learning with understanding.

Recently, there has been considerable attention given to the idea of creating a portfolio culture as a means of assessing students in alternative ways. For example, Richard Duschl is developing an approach to assessment based on students displaying evidence of what they have learned in a portfolio (Duschl and Gitomer 1991). Angelo Collins also has used portfolios as a way to assess learning (Collins 1991). As we have watched and learned from Duschl and Collins, it seems to us that the use of portfolios can be enhanced by thinking about the process from a constructivist perspective. First, it makes sense to think of a portfolio as a means of enabling students to show what they know. In a sense it is a showcase that provides an interface between the displayer and the assessor. Second, the student can be given autonomy in deciding what goes into the portfolio and what these "artifacts" represent. If students have this responsibility, they can place artifacts in the portfolio and replace them when they have better evidence of learning. At all times students would have the responsibility of describing why an object was in the portfolio and what that object represented in terms of their learning. This description and justification process could take place in writing, orally, or a combination of each. Third, because of the difficulty of understanding what an artifact represents, the portfolio would provide a context for discussion whereby a teacher and a student could negotiate what the artifacts represent and what progress they had made towards meeting the goals of the course. Interactions of this type would undoubtedly require considerable time on the part of the teacher, but it is also the case presently that assessment often consumes a disproportionate amount of a teacher's time. As Duschl and Gitomer imply with the term *portfolio culture*, the use of portfolios might lead to radically different ways of implementing a curriculum and significantly different ways for teachers and students to interact in classrooms.

The availability of portfolios on an on-going basis opens the opportunity for teachers to get in touch with what students know on an on-going basis. Regular perusal of students' portfolios can provide a basis for a learner-focused dialogue between the students and teacher. Also, portfolios provide opportunities for students to interact with one another about what they know. We could foresee a situation where group members spend some time each week examining the portfolios of peers. Some students could assume a tutoring role in respect to those in their group who do not appear to have grasped a particular concept, and all group members can provide formative evaluation of the representativeness of artifacts and the extent to which they accurately represent what has been learned. Much of the learning could be centered on portfolios. The use of portfolios is one tangible way of bringing teaching, learning, and assessment closer together, and from a constructivist perspective, it is easy to see how these interactive processes could result not only in better assessment, but in better learning as well.

CONSTRUCTIVISM AND RESEARCH

Because of the difficulty of changing practices to conform with a given referent, it is not surprising that change takes so much time (Tobin et al. 1992). Actions are not isolated from culture, so changes of an individual might not meet with the approval of others in a culture, and contemplated changes might be regarded as taboo. This possibility points the way to the use of constructivism as a referent for research in science education. This has not been the practice, and because so many of our practices in education, including research practices, have been built on a foundation of objectivism, the change to a constructivist- oriented form of research is difficult: routines have to be deconstructed, much new learning has to occur, and a culture dominated by referents associated with objectivism must change. Within the traditional culture of educational research, many actions contemplated from a constructivist perspective are taboo. In such a context change is difficult.

Doing research is much more than applying methods. Personal theories of researchers can and do influence all aspects of the research process. However, in our experience, the procedures adopted by researchers are very much influenced by the personal history of a researcher and expectations of the profession. That is, the procedures most likely to be used are those utilized in previous studies and mediated by reactions of the profession. Significant changes in our own independent research programs occurred as we embraced qualitative methods of study. These changes were prompted by dissatisfaction with the research questions being addressed. We wanted to identify and investigate the major driving forces on the science curriculum. Qualitative methods were suited to addressing such questions, even though qualitative methods were not embraced by the profession at that time. In opening the doors for qualitative research methods, we had to move beyond the need for certainty in our research and celebrate the existence of ambiguities. The element of risk was not too great, as the larger educational community seemed on the brink of revolution with respect to educational research methods and the theories that best explained educational practices.

In retrospect, it is easy to see that research methods gradually evolved as we became more strongly influenced by constructivism and gained the confidence needed to shun approaches grounded in an objectivist tradition that no longer made sense. The latter course of action has been, at times, hazardous because journal editors and funding agencies have evaluation protocols developed in terms of objectivism. Nevertheless, changes have occurred in our own individual and collective research programs, and now the research methods employed in our studies are consistent with constructivism. Significant changes are evident in all stages of the research process, and in the paragraphs below, implications are presented for the use of constructivism as a referent for conducting research.

Since constructivism acknowledges the impossibility of ever knowing the truth, it is possible to alter the metaphor of researcher as truth seeker to one of researcher as learner. That is, the role of the researcher is to make personal sense of experience and, in a socially mediated way, to build knowledge in a given field. To undertake research in a given area, then, becomes a process of personal learning, ensuring that knowledge is tested for viability in the personal and social settings in which it is to be used. The tests of viability are associated with the evidence for the particular knowledge claims and the extent to which use of the knowledge in particular situations leads to plausible solutions. The standards for judging the quality of research, there-fore, center on the adequacy of data in relation to knowledge claims and the credibility of the assertions, in the sense that use of the knowledge in given circumstances leads to productive outcomes.

Data collection is essentially an objectivist idea that implies that data are out there to be gathered up. As is often the case, the use of the collection metaphor can constrain thinking about actions associated with the process of data creation. From a constructivist perspective data are not collected, but are constructed from experience using personal theoretical frameworks that have greatest salience to the goals of the individual conducting the research. Accordingly, researchers ought to identify the beliefs that have the most significance for a specific study. By identifying such beliefs, they acknowledge that data creation is theory dependent and that a personal framework of beliefs permits some things to be noticed and others to be ignored.

Decisions to categorize data as relevant or nonrelevant can be made at the time data are created, analyzed, or interpreted. Decisions might be deliberative or routine. So many things occur in a given unit of time that no person or group of persons could ever notice them all. The assumption is that the data that are created are representative of the objects, events, and phenomena that occurred in the unit of time. However, the fact remains that the data recorded as relevant conform to the researchers' personal theories of what is relevant in a particular context. For example, in a study of a science teacher of the year (Tobin et al. 1988), the data that were created reflected our unstated theories of what teachers and students ought to be doing in effective science classrooms. To our surprise, the teacher did not see his classroom the same way that we did. His efforts in the classroom were driven by a set of beliefs different from those of the researchers. He was surprised that we chose to record some events as relevant while ignoring others that, from his perspective, were clearly of greater importance. However, we were in the classroom to learn, to explicate an emerging theory of teaching and teacher change. In effect, we saw what we expected to see, and despite efforts to the contrary, we were unable to look outside of our own personal theories. This changed, of course, when the teacher provided his perspectives on our descriptions, interpretations, and theories.

Data creation and interpretation are not separate events, but rather are components of a dialectical process. An important part of the interpretive process is what has been learned from prior studies. When studies are planned, they build on what is known and what was learned from previous studies. This might be expected from a constructivist point of view. Having developed a way of looking into classrooms that is fruitful, a researcher would be reluctant to adopt a new framework. As was mentioned previously, it is not easy to discern when one study ends and another begins. Accordingly, interpretive research progresses by building theory from one study to the next. The assertions of one study become a part of the knowledge base and are incorporated into the plan for the next study. In this way what is known in a given domain becomes explicated through the cases or contexts that are studied.

As we examine the evolution of our research programs we are struck by the way that the implications of one study have provided the framework for the next. It is not clear when one study finishes and another begins. Each time, in the evolution of our research, we seem to arrive at a point where a particular aspect of a paradigm fails to provide meaningful answers, what Kincheloe, Steinberg, and Tippins (1992) have called an "epistemological watershed." In effect what happens during this time of perturbation is that we discover deeper understanding to go beyond to explain what has traditionally eluded us about science teaching and learning. Just how this occurs is seen in the example of a high school science teacher who dramatically switched myriad classroom processes as he changed the metaphor used to guide his practices (Tobin et al. 1990). An understanding of this teacher was grounded in a study of another teacher that began while the former study was being completed (Tobin and Ulerick 1989). The key concepts described for this teacher became most apparent in the subsequent study. Since both studies were proceeding simultaneously, it did not make sense to try to separate out what we had learned in any particular study. Instead, we preferred to illustrate what we had learned in well-chosen vignettes and present the warrants for the viability of our learning. In this way, we began to recognize the holistic nature of research much in the same way that, as Heidegger rotated a glass to observe a particular aspect of it, another aspect of the glass was concealed (Briggs and Peat 1984). Each subsequent study enabled us to consider a previously unconsidered dimension of science teaching and learning.

The issue of voice is one that has become popular recently. For example, when teachers and university faculty collaborate in research, the question of voice is one that soon arises and needs to be considered. Many authors argue that it is important to maintain the voice of the teacher in an interpretive study of teaching. We agree with the sentiment underlying such a position, but we do not think it is always possible or desirable. The goals of a study might well vary for each of the research group members, particularly the teacher, who might have a strong desire to learn more about how to enhance the learning of her students. Accordingly, it might not be possible for the teacher to have a strong voice in the researchers' stories. If voice is provided for each individual, different stories will probably emerge, each story becoming data for the other researcher, which should then become a part of the other's story. In the limit, the stories should converge to a certain extent, each taking account of the other. The significance of the use of the metaphor of teacher as learner, contrasted with researcher as truth seeker, is that differences in the stories of participants are interpreted as differences in what was learned rather than differences in the truth or what really happened. This issue is further illustrated below in the context of a study undertaken by Tobin and Ulerick (1989).

Previous research had provided many insights into the role of teacher knowledge, beliefs, and metaphors in relation to classroom practices. At the beginning of this study, the research goals were related to further learning about teachers, the way they taught, and the way they learned about teaching. Of course the research team had other goals as well; they wanted to assist Marsha to be a more effective teacher and a happier person. Marsha's most significant goals were related to becoming a more successful teacher in the situations in which she found herself, to learn more about herself, and to learn whether teaching was a viable career choice. It is clear that both sets of goals were overlapping, but in an important sense, what was of greatest importance to the research team was of little or no consequence to Marsha. Yet, for the most part, the two sets of goals were sufficiently similar for problems to be minimal. However, throughout the study Marsha was uncomfortable with the interpretations that were formulated; she maintained that our understanding of the way in which she represented herself was not the entire story. If one detects a hint of personal dilemma in this story, one is probably correct. As a member of the research team, the first author would enquire about Marsha's perceptions of the entire story. Depending on the participant's vantage point, the story is seen differently. It was apparent that Marsha's story was about teaching, not about science education; it was an autobiography that focused on why Marsha is what she is today. Drawing upon a foundation and an understanding of constructivism, Marsha's story of what she had learned about herself through participation in the study was one only she could tell. She could tell that story, but other members of the research team could not begin to know even a small part of the essential components and, furthermore, the fine details were not central to the stories others wanted to tell. Although Marsha could editorialize in any story just as much as she wished, and if it was her wish, she could invite us to editorialize in hers, ultimately the stories that emerged would be quite different. This is not surprising, as the meaning that individuals attach to classroom events depend on prior negotiations of meanings, representations, codes, and conventions.

To what extent was Marsha an important part of the story? Marsha was certainly the "principal data source" as she cynically, yet good naturedly, referred to herself throughout the study. However, when the first author observed Marsha in the classroom and listened to her in numerous interviews, he created the data with categories that were in his mind. In this sense, the classification scheme used in the interpretation of data in the on-going data creation-interpretation part of the study was uniquely his, mediated by the voices of others who participated in the research. In effect, the central voice in this study involves the constructions and interpretations of the first author; although Marsha's voice appears throughout many papers, her voice is provided as data to support the learning of others. That Marsha participated in writing the paper suggests that she is content with the interpretation, and perhaps even finds it useful; yet the story that unfolds in the papers remains essentially that of the first author. Marsha still maintains that the story is incomplete. There are other factors that are critical components to the story that are not included, but to include them does not seem possible. Suffice it to say that while we have tried for several years to insert Marsha's missing parts into the story, what is necessary is for Marsha to tell her story, grounding that story in a context of her goals and her theories. Presumably the categories she would choose to tell that story would be quite different from those used by the first author or other members of the research team.

The above interpretation of voice and story takes its form because the personal epistemology of constructivism provides a framework for viewing research as analogous to learning rather than truth seeking. The work of other authors who refer to voice is often couched in terms that suggest that the voice of the researcher dominates over the voice of the subject so that a distorted view of reality emerges. Of course, there are variants on this approach to voice and ensuring that participants in research have their voices preserved in an authentic way. From our perspective, the voices are not alternative views of an accessible reality. Rather, the views of what is happening and why it is happening are dependent on the goals of participants in a study. To some extent the goals will be explicit and complementary, but there always will be personal goals of individuals that are not explicit, even to the individuals who hold them. These covert goals can provide a different frame for categorizing data and experiencing. Accordingly, the different positions that emerge will not necessarily be complementary and will not necessarily address the same issues. It is not essential, or necessarily desirable, to negotiate a consensus on what is the best representation of what has been learned from a study. To maintain procedures that lead to negotiation of a consensus might lead to a power game where the individual with the loudest voice sets the parameters for the consensus. Such a process can mask, or render invisible, stories from which others might learn a considerable amount.

But there is a paradox. The processes of negotiation, with the rich diversity that is fueled by different knowledge and learning goals, actually lead research groups to learn more than they otherwise might. Accordingly, the preservation of voice is not meant to imply that argument and consensus building ought not to occur. What is important is that all participants in a study have opportunities to use their personal theories to form categories that have relevance to them and that the procedures of the research team are such that personal theories become explicated and a part of the stories of what was learned from a study. Thus, if we are to cultivate "voice," we must attend to the multidimensional aspects of collaborative research.

The purpose of reporting is to inform others of what has been learned from a study. In the process it is important to demonstrate that the assertions comprising the grounded theory of a study are plausible in that they fit the data created. However, the focus of a report ought to be on communicating the essence of what was learned, using vignettes to illustrate aspects of the theory in action. The purpose is not to convince readers of the generalizability of what has been learned but to provide sufficient details of the contexts in which the theory is embedded and to enable readers to decide on the extent to which what has been learned can help them meet their goals. Readers will use their personal beliefs and categories to make sense of a story, and will reconstruct the story to reflect what they personally know and value. The knowledge constructed by reading research will reflect the extant knowledge and culture in which each reader has lived her life. Thus, the thick description, so often associated with interpretive research, is necessary to enable readers to understand the context of the study and to maintain the links between the theory and the context in which it is grounded. Readers are entitled to judge the adequacy of the fit of what has been learned not only with what they understand at the present time, but also in terms of the data (i.e., the experiences) of a study.

Journal editors and editorial board members are accustomed to traditional styles of research reporting, styles that focus on convincing readers that a close approximation of the truth has been discovered. If, on the other hand, a researcher regards her research as an example of fresh insights into something that previously has been problematic, the difficulty of reporting is to convince the editor and her editorial board that the paper is one from which others can learn. The generalizability question shifts from the application of findings to a population that approximates the characteristics of a sample to one of providing text from which others can learn. It is for others to decide whether or not what they have learned is applicable to the problems they encounter. What if the editor or her review board do not agree that papers written from the perspective of researcher as learner constitute legitimate research? If the reviewers of papers want to publish research based on the truth-seeker metaphor, researchers who adhere to the learner metaphor might be disadvantaged. Accordingly, changes of the type described above can place researchers in contexts in which they feel their work is reviewed in a process that is not fair. This situation is one that is still to be worked out by educational researchers, editors, and review boards. Prejudice associated with judging the merit of papers written from an alternative paradigm seems to be widespread, and satisfactory solutions are not pervasive in the research community which, at the present time, seems divided.

CONCLUSIONS

Readers of this chapter ought not gain the impression that what we are advocating is a universal truth to replace objectivism, a now invalidated truth. On the contrary, constructivism is conceptualized as a set of beliefs about knowing that has the potential to facilitate different ways of thinking about education, of framing problems, and of formulating answers that extend into areas not considered when objectivism was used as a referent. As we have thought about constructivism, we have come to realize that it is not a unitary construct. Every day we learn something new about constructivism. Like the bird in flight it has an elusive elegance that remains just beyond our grasp. However, we have learned to think of constructivism as a referent, not the only referent for our professional actions, but nonetheless an important one. Over time we have noticed how constructivism has become relevant to one situation after another; however, it is not universally and automatically used as a referent. So many of our actions are based on other referents and they are not all coherent with constructivism. Sometimes the epistemology of action is inconsistent with constructivism, usually because the conscious referents being used are belief sets related to control or constraints.

As our professional goals change, so too do the aspects of constructivism that seem most applicable to the problems that confront us. And as multiple referents become applicable to given problems, additional beliefs are included with constructivism as referents. Accordingly, beliefs about control have become so intertwined with constructivism that we began to use the term *critical constructivism* to include the self-regulation that reveals the psychological, ethical, moral, and political beliefs intertwined with the construction of knowledge. Providing the learner control over her own learning and knowledge began to be more important as we used constructivism as a referent to think about teachers using their knowledge to think about curriculum reform. Rather than having two separate theoretical frameworks, it was clear that there was only one, no longer radical constructivism, but nonetheless a coherent derivative. Similarly, the social component of constructivism has been so important to us that we gave greater emphasis to it, the individual and social components being parts of a dialectical relationship where knowing is seen dualistically as both individual and social, never one alone, but always both.

By way of conclusion, it is our contention that constructivism is an intellectual tool that is useful in many educational contexts. Using constructivism as a referent has led to many changes in our roles as teacher educators and researchers, and as we have watched the evolution of science teaching over the past five years, we have observed that teachers who learn to use constructivism as a referent begin a journey of educational improvement that is comprehensive and on-going (Tobin et al. 1992). We do not claim that use of constructivism as a referent is the only way to initiate changes of such a comprehensive and significant scope, but from our experience we can assert that constructivism can assume a dialectical relationship with almost every other referent in a process that culminates in a coherent world view consisting of compatible referents for action.

REFERENCES

Bauersfeld, H. *The Structuring of Structures*. Paper, alternative epistemologies conference at the University of Georgia, Athens, GA, 1992.

Briggs, J., and F. Peat. *Looking Glass Universe: The Emerging of Science of Wholeness*. New York: Touchstone, 1984.

Cobb, P. "The Tension Between Theories of Learning and Theories of Instruction in Mathematics Education." *Educational Psychologist*, 23, (1990): 87–104.

Collins, A. "Portfolios for Biology Teacher Assessment." *Journal of Personnel Evaluation in Education*, 5, (1991): 147–167.

Duschl, R.A., and D.H. Gitomer. "Epistemological Perspectives on Conceptual Change: Implications for Educational Practice." *Journal of Research in Science Teaching*, 26(9), (1991): 839–858.

Fosnot, C. "Learning to Teach, Teaching to Learn: Center for Constructivist Teaching/Teacher Preparation Project." Paper, annual meeting of the American Educational Research Association, San Francisco, 1992.

Giroux, H. "Educational Reform and Teacher Empowerment." In H. Holtz, et al., *Education and the American Dream.* Granby, MA: Bergin and Garvey, 1989.

Kincheloe, J.L., S.R. Steinberg, and D.J. Tippins (in press). *The Stigma of Genius: Einstein and Beyond Modern Education.* Wakefield, NH: Hollowbrook Communications, Inc., 1992.

Kuhn, T.S. *The Structure of Scientific Revolutions.* Chicago: University of Chicago Press, 1970.

Matthews, M.R. "A Problem with Constructivist Epistemology." Paper, annual meeting of the National Association for Research in Science Teaching, Boston, 1992.

Nordland, F. "The Cognitive Level of Curriculum and Instruction: Teaching for the Four rs." In K. Tobin, J.B. Kahle, and B.J. Fraser, *Windows into Science Classrooms: Problems with High Level Cognitive Learning in Science.* London: Falmer Press, 1990. 135–163.

Rowe, M.B. "Getting Chemistry off the Killer Course List." *Journal of Chemical Education,* 60(11), (1983): 954–956.

Saxe, G. "Cultural Influences on Children's Mathematical Knowledge." Paper, alternative epistemologies conference at the University of Georgia, Athens, GA, 1992.

Tobin, K. "The Role of Wait Time in Higher Cognitive Level Learning." *Review of Educational Research*, 57(1), (1987): 69–95.

Tobin, K., M. Espinet, S.E. Byrd, and D. Adams. "Alternative Perspectives of Effective Science Teaching." *Science Education,* 72(4), (1988): 433–451.

Tobin, K., and B.J. Fraser (Eds.). *Exemplary Practice in Science and Mathematics Education.* Perth, Western Australia: Curtin University of Technology, 1987.

Tobin, K., and J. Gallagher. "What Happens in High School Science Classrooms?" *Journal of Curriculum Studies*, 19, (1987): 549–560.

Tobin, K., J.B. Kahle, and B.J. Fraser. *Windows into Science Classrooms: Problems Associated with High Level Cognitive Learning in Science.* London: Falmer Press, 1990.

Tobin, K., D.J. Tippins, and K.S. Hook. "The Construction and Reconstruction of Teacher Knowledge." Paper, annual meeting of the American Educational Research Association, San Francisco, 1992.

Tobin, K., and S. Ulerick. "An Interpretation of High School Science Teaching Based on Metaphors and Beliefs for Specific Roles." Paper, annual meeting of the American Educational Research Association, San Francisco, 1989.

von Glasersfeld, E. "A Constructivist Approach to Teaching." Paper, alternative epistemologies conference at the University of Georgia, Athens, GA, 1992.

Wheatley, G. "Constructivist Perspectives on Science and Mathematics Learning." *Science Education*, 75(1), (1991): 9–21.

Wood, T., P. Cobb, and E. Yackel. "Reflections on Learning and Teaching Mathematics in Elementary School." Paper, alternative epistemologies conference at the University of Georgia, Athens, GA, 1992.

2

Questions and Answers about Radical Constructivism

Ernst von Glasersfeld

The questions I try to answer in the pages that follow were raised after talks I gave at the National Association for Research in Science Teaching (NARST) meeting in Atlanta, GA (April 1990) and at an American Association for the Advancement of Science (AAAS) symposium in Washington, D.C. (February 1991). Given the time limitation at both occasions, I could address only a small selection. While reviewing the whole collection at a later date, I found that the material could be roughly divided into three subject areas. I begin with the specifically epistemological ones, then consider those that concern the problem of social interaction, and I end with some implications the constructivist orientation might have for teachers and the philosophy of instruction. Because the answers I give are not derived from an established dogma but spring from my subjective point of view, the reader will find a certain amount of overlapping between the three sections. I would like to claim that this is inevitable because, in my experience, once one shifts to the constructivist orientation, everything one thinks and does changes in a way that seems remarkably similar and coherent. Let me emphasize a point I have made in many of my papers: constructivism, as far as I am concerned, is one possible way of thinking. It is a model, and models, no matter how useful they may prove, must never be claimed to be "true."

EPISTEMOLOGY

1. Is constructivism primarily an epistemology or a pedagogy?

Constructivism confronts questions of knowledge—what knowledge is and where it comes from. It can therefore be considered an exercise in epistemology. The idea that knowledge is constructed by the knower is as old as Western philosophy. The pre-Socratics in the sixth and fifth century B.C. were already aware that not everything they knew could be said to come from sensation and that the mind, the specifically human mind, was to a large extent responsible for shaping human knowledge. Some of them also noticed that there was a serious problem regarding the claim that knowledge could and should faithfully reflect a world held to be independent of the knower. They realized that there is no way of checking knowledge against what it was supposed to represent. One can compare knowledge only with other knowledge. The later skeptics have never tired of reiterating this irrefutable argument.

Recently, the reviewer of a paper of mine said that constructivism was "postepistemological," and she was quite right (Noddings 1990). Indeed, the constructivist theory does not fit into the conceptual patterns of traditional epistemology precisely because it posits a different relationship between knowledge and that "real" outside world. Because it changes that important relationship, I prefer to call it a theory of knowing rather than a theory of knowledge. (Please note that constructivism does not deny an outside world; it merely agrees with the skeptics and holds that the only world we can know is the world of our experience.)

2. What is the difference between constructivism and radical constructivism? How do you define radical, trivial, and social constructivism?

A few years ago, when the term *constructivism* began to become fashionable and was adopted by people who had no intention of changing their epistemological orientation, I introduced the term *trivial constructivism*. My intent was to distinguish this fashion from the "radical" movement that broke with the tradition of cognitive representationism.

Social constructionism is a recent development started by some who claim radical constructivism does not take into account the role of social interaction in the construction of knowledge. The claim, as I see it, is partly justified by the fact that neither Piaget nor any more recent constructivist has actually specified a detailed model of how social interaction works from the constructivism point of view. On the other hand, Piaget and all of us who have derived our impetus from him have always maintained that social interaction is a powerful influence in the construction of knowledge. But we were busy devising models for all the elementary constructing that has to be done before a cognitive organism can begin to know and interact with others. As far as I know, the social constructionists tend to take society as a given, and a radical constructivist cannot accept this. From my point of view, "society" must be analyzed as a conceptual construct before its role in the further construction of concepts can be explained and properly assessed.

3. What are the differences between constructivism and idealism? Is the distinction a qualitative one?

Idealism holds that the mind and its constructs are the only reality. Constructivism holds that we can know only what our minds construct but that its constructing is not free (ad lib): just as some ways of acting "work" and others do not, so some of our conceptual constructs and theories work and others do not. For constructivism, the fact that something works (is "viable") does not mean that it therefore is a representation of that "real" world that prevents other things from working. The "real" world remains unknowable no matter how well we manage in the domain of our experience. (This, by the way, also differentiates constructivism from pragmatism with which it has much in common.)

4. Was Descartes revealing his constructivist view when he said, "I think therefore I am"?

Descartes was a very complex thinker and not averse to the notion of construction. Analytic geometry, one of his major achievements, is an explicit example of cognitive construction, but I would not interpret his famous "cogito" phrase as a constructivist statement about the self. The mind was very much an ontological given for Descartes. He called it *"res cogitans"* (the thinking thing), and he saw it as an entity separate from a world that exists in its own right (*res extensa*) and an entity about which the knower can find out some things, not through the senses, but by logical reasoning.

For constructivists the "I," the agent that does the constructing, is a construct like everything we claim to know. This makes the model "circular" and upsets traditional philosophers but not cyberneticians.

5. What is the role of truth in your model?

Constructivism is an attempt to cut loose from the philosophical tradition and specifically from the philosophical tradition that knowledge has to be a representation of reality, where reality is spelled with a capital "R" and what is meant by it is a world prior to having been experienced. Truth in constructivism, as I keep repeating, is replaced by viability. That is a stark step to take. I've given many philosophical arguments (or excuses, if you will) for taking this step in a number of papers. Here let me just mention one thing because it seems the most important in this context: if we want to talk in terms of "truth" and mean by this that what we say and what we think should be a replica of the world as it is, we have to believe that we can visualize or imagine what that "real" world is like. We have to conceive that world in terms of existence, in the sense that it exists by itself prior to our noticing, perceiving, and thinking about it. Now, I claim that we cannot put any meaning to the expression "to exist" outside our sphere of experience. To us, "to exist" means, by and large, to have a location in space and time. But if space and time, as Kant said, are forms of our experiencing and do not actually belong to the ontological reality beyond our experience, we have no way of visualizing things without space and time. For me, this is a good reason for giving up the notion that we have to know or imagine what the world is like before we experience it. Rather, we must try to develop a theory that provides a relatively coherent explanation for how we come to have the kind of knowledge that we use in our living.

6. Is knowledge something to be presented to a learner or is it an activity?

Knowledge is always the result of a constructive activity and, therefore, it cannot be transferred to a passive receiver. It has to be actively built up by each individual knower. A teacher, however, can orient a learner in a general direction, and constraints can be arranged that prevent the learner from constructing in directions that seem unsuitable to the teacher.

7. Does the fact that we can predict physical phenomena with a great deal of accuracy not mean that the picture of reality we have constructed is congruent with the "real" world outside?

No. This question addresses a key element of constructivism. Predictions we make, whether we are physicists or ordinary human beings, predict experiences (things we perceive, feel, and think). What we perceive, feel, and think is essentially the result of our ways and means of carrying out these activities. The results can never be said to be alike, let alone "congruent," with an ontological reality; all we may conclude is that this "real world" allows us to perceive and think certain things. If a prediction turns out to be right, a constructivist can only say that the knowledge from which the prediction was derived proved viable under the particular circumstances of the case.

8. Some concepts seem to be based on a physical stimulus, e.g., "chair"; others seem more abstract, e.g., "I.Q." Is there a continuum from those triggered by some external stimulus to those that are more abstract?

When we say "physical stimulus," we say more than we know. All that we really know is that we have had, or are having, a perceptual, or as Piaget would say, a sensory-motor experience (see preceding answer). To conclude that, because we have a perceptual experience which we call "chair," there must be a chair in the "real" world is to commit the realist fallacy. We have no way of knowing what is or could be beyond our experiential interface. If we can reliably repeat the chair experience, we can only conclude that, under the particular circumstances, it is a viable construct.

Something like "I.Q." is not a sensory-motor experience, but, like any measurement, it is the result of a number of abstractions. In Piaget's terms, it would be an "operative" concept because it is derived from operations we have to carry out. The terms *concrete* and *abstract* are not useful in constructivism because conceptual analysis shows that there are many levels of abstraction. The first level would be abstraction from sensory-motor experiences. Once such an abstraction has been made, its result can immediately be taken as material for a further abstraction, and so on. Thus, there is a continuum of levels of abstraction.

Science largely consists of relational (operative) concepts that are the result of various abstractions having their origin in sensory- motor experience and our own mental operations. This has been beautifully expressed by Einstein (1954) in his essay "Physics and reality." There, he explicitly says that our object concepts are "free creations of the human (or animal) mind."

9. Are there physical scientists who disagree with constructivism? If so, name one or two and indicate where they disagree with what you have said.

Among the great physicists, most have said and written things that imply a constructivist position. Nearly all these statements were anticipated by Hermann von Helmholtz who wrote in 1881, "The principle of causality is in fact nothing but the presupposition of the lawlikeness of all of the appearances of nature" (von Helmholtz 1977, 180). I feel, however, that contemporary physicists are in tacit agreement with Popper, who admits that the skeptics' arguments are irrefutable and that an "instrumentalist" view of knowledge is quite reasonable, but goes on to say that it is better for scientists to continue to believe that they are getting closer to understanding the "real" world. Needless to say, I do not share this double standard.

From the limited experience I have had in the last few years with physics instruction, I would say that most physics teachers have little sympathy for constructivism. They prefer to talk as though they were describing an absolute reality. If pushed a little, they may say, "Well, of course, we don't know for sure," or, "Students don't like uncertainty, they want to know how things really are."

10. Is there a concept of progress in radical constructivism or no change, or is all change random?

Constructivists would be rather foolish to deny that we know more today than did, say, the pre-Socratics or even Newton, but the growth of knowledge concerns knowledge of how to do things. The fact that we can send a man to the moon and can have a phone conversation from our moving car does not mean that the conceptual structures and action programs we have developed are any more a representation of ontological reality. We have learned a great deal about how to set up and manage useful regularities in our experiential world. We know many more things the world allows us to do. We have, after all, learned even how to demolish our experiential world irrevocably. But for constructivism, it is an illusion to believe that what we can do paints a picture of the "real" world. It describes our experiential reality as we happen to be experiencing it, and what we experience is shaped and coordinated according to the concepts and conceptual relationships we are employing at the moment.

11. Is truth, then, an act of faith?

Yes, truth is an act of faith. Constructivism does not deal with the traditional conception of "truth," which would require that one knows, or at least believes, that an idea, a theory, or any conceptual construct is an accurate representation or duplication of something beyond our experiential field. Truth in that sense may be felt (perhaps acquired) by following the mystics, but it is not a rational procedure that leads to it. Note that I am in no way disparaging or discounting what the mystics tell us; I am merely saying, do not try to explain it rationally.

Constructivism replaces the notion of truth with that of viability, which does not refer to anything outside the experiential field. This is analogous to pragmatism, but pragmatists formulated the slogan "true is what works," and their use of the word "true," in my mind, is misleading.

12. If we cannot know reality, can we understand, test, and evaluate others' constructed realities? How can we test another's scientific hypothesis and have valid grounds for rejecting or accepting it?

Scientific hypotheses are "tested" in the experiential world where they either do or do not do what they are claimed to do. Experiential worlds belong to individuals, but, in the course of social interaction, these individual worlds become adapted to one another. They come to form a consensual domain, i.e., an area where the interactors' mutual expectations are more or less regularly realized (see section on the "social"). The consensual domain of the scientific subculture, because of its many specific constraints, is particularly homogeneous, and new hypotheses are "tested" against that (relatively) homogeneous background. If the new hypothesis makes too many new assumptions, it will usually be discarded, regardless of whether it works or not. (For example, Wegener's hypothesis of "continental drift" was proposed at a time when the notion of floating continents was considered preposterous by the entire scientific community. It took a lot of new experiential "evidence" to become acceptable.)

13. Does reality exist aside from one's construction of reality? How do you know?

I don't know. As far as logic is concerned, I could be one of Leibniz' monads (i.e., a closed, impermeable entity whose "experience" is like a movie that automatically runs its preordained course). This is not a particularly useful model. I prefer the model built on the notion of viability within the space left free between constraints whose origin I can never tell.

The main difficulty of the question arises from the word *"exist."* In our human usage, it means to have some location in space, time, or both. But since space and time are our experiential constructs, the words *"to exist"* have no meaning outside the field of our experience, and whatever an independent ontological reality may do, it is not something we can visualize or understand.

14. Shouldn't there be a new label for reality as it exists, to distinguish it from subjective reality?

Yes, but a label is not much use if one can never demonstrate what it stands for. I try to say "experiential reality" as often as I can and, in contrast, put that other mystical "reality" in quotation marks (see Question 11). In German there are two convenient words: *"Wirklichkeit,"* the reality due to one's actions, and *"Realitat,"* which can be used for the other.

15. What harm is there in remaining a realist? What do I have to gain by thinking as a radical constructivist?

No harm at all, as long as you don't tell others that the reality you have constructed is the one they ought to, or, worse, must believe in. In everyday interactions, we all go on as though our experiential reality were the same obligatory reality for everyone. But if one gets into psychology or education, taking account of the subjective nature of experiential realities becomes a matter of honesty.

16. How does constructivism escape the "hermeneutic circle": "Is constructivism an adequate view of knowledge and the world despite the fact that all knowledge is held to be relative?"

Hermeneutics searches for the "true" interpretation, constructivism does not. Constructivism is deliberately circular and never claims "objectivity." We cannot look at our experiential worlds from the outside. We construct them from the inside and have usually lived a good many years in them before we begin to wonder where they came from and what they "really" are. It would be more than a little presumptuous to believe that we could "know" the world that has produced us and our ways and means of thinking. It would require what Hilary Putnam has called "the God's eye view." In contrast, we can always decide whether a way of acting or thinking is "adequate," because it either does what we expect of it, or it does not.

17. Is constructivism a belief?

If you call it "belief" when one has constructed a model and finds it the most adequate one has so far developed, then I would say yes, but it is certainly not belief in an absolute "truth" or revelation. To me, it simply seems to be the most plausible view at the moment.

18. You alluded to the existence of an all-knowing "real" God. How can you reconcile the belief in constructivism with the belief in a real God?

"Alluding to God" is not the same as believing in God. Vico, the first explicit constructivist, said, "A human being can know only what humans have made, God can know the world because He created it." This is not even a hypothesis but merely a conjecture; it might be the case and it might not be the case. A mystic may "know," but that knowledge would not be rational knowledge. This is not a statement of value but a statement of methodology. I have the greatest respect for mystics, although I know nothing about how they operate.

19. How do you hold up constructivism for scrutiny?

By letting people see whether it works, where "to work" means to be as consistent as possible and to cover as much ground as possible. The assessment has to be in terms of viability, not in terms of "truth."

20. Is realism typically a Western philosophy? If so, would you say something about Eastern philosophy?

Our textbooks and television programs (even NOVA programs) are all too often manifestations of Scientism, a pseudo-philosophy that I would call the most recent Western religion because it claims to have found objective "truth." The realism that the scientistic religion entails is certainly characteristic of the West.

The little I have learned about Eastern philosophy does not make me competent to answer the second part of the question. I have the impression that Buddhism and Taoism tend to take rational knowledge and common sense reality less seriously than we do, and the teaching of Zen seems to aim at weaning the student from the frantic attachment to naive realism.

THE SOCIAL ROLE

21. How does constructivism view a body of common knowledge, such as in a textbook or field?

It is an illusion that there is knowledge in textbooks or documents. They contain language, which is strings of words, deposited in them by authors. The words have meaning for the authors and the readers and interpreters, each one of whom has built up her subjective meanings according to her individual experience. Though these individual meanings are constructs that have been through a certain amount of social adaptation (because their users have socially interacted with others), they remain subjective and to some extent idiosyncratic, which never fails to come to the surface in a serious discussion. Texts contain neither meaning nor knowledge; they are a scaffolding on which readers can build their interpretations (see Question 23).

22. Where and how is language taught?

I would say that the foundation of language is not taught at all; children learn that by themselves. You cannot teach language to a two year old or a three year old. You may occasionally be able to show them the use of a new word, but even this often does not work too well. Later, specific interaction can focus on specific linguistic difficulties. That is why teaching language at a later stage by correcting what children say is enormously important. Unfortunately, this seems to have gone by the board because parents no longer have many extensive conversations with their children. Conversation has been replaced by television, and television, being a moneymaking enterprise keyed to the lowest levels of literacy, rarely provides a model for good language.

But, then you might say, a lot of language can be learned from dictionaries. Dictionaries, however, contain definitions that are simply other words. They do not provide the experience from which meanings are abstracted. They do not give us new cognitive elements, but only new combinations of elements that we already possess.

In general, language is learned in the course of interaction with other speakers, because speaking is a form of interacting, and we modify our use of words and utterances when they do not yield the expected results. Insofar as a classroom is interactive, it contributes to the students' linguistic development and also provides them with opportunities to witness the use of words in the context of the experiences to which they refer. That is why it is so important to provide students with experiences that they are able to learn from.

23. What is the role of "social instruction"?

I take it this refers to "group learning," and there is a lot to be said about this:

- Students who work at a problem together have to verbalize how they see the problem and how they propose to solve it. This is one way of generating reflection, which requires awareness of what one is thinking and doing. This, in turn, provides occasions for active abstraction (repeating, writing down, and learning by heart what a teacher says does none of this).

- Explaining something to a peer usually leads a student to see things more clearly, and often to spot inconsistencies in one's own thoughts. When a small group explains its "solution" to the whole class (irrespective of whether it happens to be viable or not), a wonderful opportunity for learning results (cf. the work of Paul Cobb, Ken Tobin, Grayson Wheatley, Jack Lochhead, and others).

- Knowing that those you work with have no ready-made answer increases everyone's courage to try to find an answer.

- If one of the group finds an answer, this usually generates motivation to face a new problem.

- To have an inconsistency or "error" explained by a peer is far less painful than to have the teacher tell you that you are wrong.

24. Please give examples of how social interaction would promote understanding.

Once again, my examples will have to be simplistic. The first kind of example would be the failure to get what one asks for, or what one is trying to get, or to get the wrong thing. I am talking about very young children and their acquisition of linguistic knowledge. Anyone who has attentively observed a child's language acquisition knows the situation where the child uses a word because she wants something, but the word is inappropriate and therefore does not yield what the child expects. The adult misunderstands, gives the wrong thing, and the child gets very irritated. Such moments of failure and perturbation may lead to a change. The child may try another word or the parent may suggest one. The child then may try the new word the next time, and if it works, the child will have learned. It is a very simple inductive procedure. You hang on to the schemes, the gambits, and the methods that have worked in the past. We all do this, not only in the area of language, but in all areas of learning. This principle has an important corollary that I have often mentioned: We have no reason whatsoever to change what we are doing so long as it produces the desired result. This assertion links with the notice of misconceptions in physics and in other teachings. If these conceptions have satisfied what the students demanded of them up to now, the students have no earthly reason to change. The love of truth will not make them change. You have to show them that their conceptions have limitations and that there are situations where those conceptions do not work. That is one way of getting them interested in learning a new conception, because they all want to widen the range of application of whatever schemes or ideas they have.

25. How do we reach shared meanings?

The expression "shared meaning" is misleading. Your meaning and another's are at best compatible; in a given situation, neither reacts in a way that the other could not expect. One only has to get into a discussion of philosophy to realize just how tenuous this compatibility is on the level of abstract concepts. It usually takes a long evening and a good deal of patience before one gets even a vague idea of what the other is trying to say. In a research team, for example, it can take well over a year to establish a workable compatibility with regard to the main terms that are being used. This is what we have called the generation of a consensual domain.

26. How can we teach the tentativeness and personal nature of knowledge, yet not undermine the cultural, social reality we have arrived at?

We cannot, but, as a constructivist, I feel it is one's duty to proclaim the "tentativeness" of everything one teaches and to warn the students that most, if not all, of the "facts" one presents will be seen in a different light two or three decades later. What matters is the process of construction and the fact that this process yields results that seem to be viable at the moment. The cultural and social reality would be a more livable and fruitful one if we could do away with the notion that we have the "truth" and others had better believe it.

27. What is the point of talking or lecturing?

Well, in a way, I believe I have answered this. The point is to foster new combinations of concepts. By talking to an audience I cannot give people any new concepts, but I can prod them to combine in different ways the concepts that they have associated with the words I am using. I can propose combinations that they may not have thought of before or had no occasion to use before. I can perhaps generate some perturbation about conceptual structures that have been used unthinkingly, merely because they felt familiar. If I am not causing some conceptual change, my talking or writing is useless. But in saying this, I am already in the topic of teaching.

INSTRUCTION

28. What are the implications of constructivism for a theory of instruction?

There are many. I can mention only a few here that can be stated briefly:

- If we assume that students have to build up their own knowledge, we have to consider that they are not "blank slates." Even first graders have lived for a few years and found many viable ways of dealing with their experiential environment. The knowledge they have is the only basis on which they can build more. It is therefore crucial for the teacher to get some idea of where they are (what concepts they seem to have and how they relate them).

- Whatever a student says in answer to a question (or "problem") is what makes sense to the student at that moment. It has to be taken seriously as such, regardless of how odd or "wrong" it might seem to the teacher. To be told that it is wrong is most discouraging and inhibiting for the student. Furthermore, given the way that the student interpreted the question, the answer may actually be a good one.

- If teachers want to modify a student's concepts and conceptual structures, they have to try to build a model of the particular student's own thinking. Models of students' thinking can of course be generalized, but before assuming that a student fits the general pattern, one should have some solid evidence that this is a viable assumption in the particular case. It should never be assumed that students' ways of thinking are simple or transparent.

- Asking students how they arrived at their given answer is a good way of discovering something about their thinking; it opens the way to explaining why a particular answer may not be useful under different circumstances.

- If you want to foster students' motivation to delve further into questions that, at first, are of no particular interest (from the students' point of view), you will have to create situations where the students have an opportunity to experience the pleasure inherent in solving a problem. Simply being told that an answer is "correct" does not do anything for the student's conceptual development if she was not interested in the question.

- Successful thinking is more important than "correct" answers. Successful thinking should be rewarded even if it was based on unacceptable premises.

- To understand and appreciate students' thinking, the teacher must have an almost infinitely flexible mind because students sometimes start from premises that seem inconceivable to teachers.

- The constructivist teacher can never justify what she teaches by claiming that it is "true." In mathematics, she can show that the logic derives from certain conventional operations. In science, she cannot say more than that it is the best way of conceiving the situation because, at the moment, it is the most effective way of dealing with it.

29. How do you activate the mind to construct knowledge?

By letting students struggle with problems of their own choice, by helping them (as Maria Montessori said a long time ago) only when they ask for help, we assist them in constructing knowledge. At best, the teacher can orient the students' constructing in a fruitful direction; she can never force it. This is, of course, time consuming, but after students have experienced the pleasure of finding a solution by their own thinking once or twice, they will be quite ready to work on problems suggested by the teacher.

30. What is the role of training interactions? Can episodes of training be used to solicit episodes of teaching?

Making the distinction between training and teaching does not mean that there is no place for training. I am convinced that some training is very useful, but, in itself, it is not conducive to understanding. However, there are many experiences for which one needs to be proficient in certain routines. Routines are necessary. One is not going to be a mathematician unless one has automated some of the elementary operations, but if one has automated them without having grasped their conceptual underpinnings, then one is not going to be a mathematician either. Training and teaching must go hand in hand. If I chiefly emphasize teaching, it is because training is common practice in schools anyway.

31. How does constructivism influence problem solving?

By letting the students discover that it can be fun to solve problems, we influence their problem-solving techniques. It will not be fun if the teacher constantly watches students so that they take the "right" path to the "right" solution. Students often take wholly unexpected and unconventional paths to a solution that is quite viable but perhaps limited in its applicability. If the teacher does not respect this, she merely demonstrates that she is unaware of the history that has led to the present state of the art. That is why, especially in science, I consider it indispensable that students be given a feeling for the long history of magnificent failures.

32. Is knowledge really only built from existing pieces, and does meaning only come from prior knowledge?

I think I have already answered this, at least in my own way. Knowledge is often built up by combining and recombining available concepts or by trying out new conceptual relations. After the child has acquired a certain level of language, language can be used to suggest and to stimulate new combinations. A similar succession of levels can be seen in reflection and abstraction, the processes that generate mathematics and other abstract realms.

33. Are the implications of constructivism the same for mathematics and science education?

Most are, but some are not (see the last bulleted item under Question 28). In mathematics, for instance, the teacher should explain that the step in calculus that leads from infinitesimal bits to continuity is a conceptual decision, not a logical consequence, just as the decimal system, with all its consequences and implications, is a conceptual system, not a God-given characteristic of an external "reality." In science, on the other hand, the first thing to make clear is that scientists do not "unveil" the objective order of a preexisting universe but invent practical ways of coordinating and managing experiences where the range of experiences is always limited by the particular interests of the given period. The history of science is full of examples that show how theories are superseded when new experiences enter the field of exploration. Old theories, as a rule, are not proven to be "wrong"; they merely turn out to be inadequate in an expanded domain of experience or in the pursuit of new goals. Above all, the constructivist teacher should never present a solution as the only solution.

34. How can we structure a teacher education program that allows students to construct an understanding of science, teaching, and learners that is consistent with constructivism?

By making clear that education pursues two goals. First, to foster thinking that does not involve conceptual contradictions and that leads to internally consistent results. Second, to introduce students to the consensual domain that governs the discipline at the moment. The "science" of today is, after all, what the scientists of today believe they agree on. But to a constructivist, "agreement" does not entail that the concepts and conceptual relations used by the people who "agree" are identical; it simply means that, under the given circumstances, they seem compatible (which, to repeat myself, indicates no more than that one is unable to discover discrepancies in one's own and the other's way of speaking and acting).

35. If we adopt the radical constructivism view of knowledge, what changes in the way science teachers talk about the world will be required (i.e., language, terminology)?

Little beyond the fact that what is being talked about is a particular view of the world, not the world as such. It is a way of ordering and coordinating experience that seems useful and consistent at the moment.

36. Technical terms are intensively employed in the teaching and learning of science; what is the status of technical terms (which are rigidly defined) in scientific understanding? Technical terms are memorized, right?

That is a very good question. Technical terms are somewhat different from the ordinary words that children learn in the first three or four years of their lives. The reason that they are different is that there is usually a good operational definition for technical terms, and the student can be shown or guided to do what is necessary to construct the referent. The operations to produce a viable referent may be physical, or they may be conceptual; in the second case, teaching students the term will be more difficult. The physicists' notion of "energy" is a good example. For the last two years, I have been involved with people who do research on physics teaching. As a nonphysicist,

I was, at first, astonished to see that physics teachers and physicists often begin to argue about what energy is and that there does not seem to be a definition that satisfies everyone. Of course, there are definitions in the textbooks, but the physicists tend to say that these definitions reflect a very naive way of thinking about energy. So there are difficulties, and a nonphysicist is not in a position to solve them. All I would say is this: It is no use presenting students with a verbal definition unless they have an opportunity to have some kind of relevant experience. In a laboratory, one can show all sorts of simple experiments in which a construct such as energy becomes, if not obvious, at least quite plausible. So much for technical terms.

37. Does constructivism favor divergent instructional strategies? Or is it consistent with a teacher having an agenda?

The "agenda," I would say, is the attainment of the twofold goal (see Question 34); instructional strategies may be suggested but not prescribed. Every inspired teacher will develop her own strategy. Teaching, after all, is something of an art. It requires not only acquaintance with techniques but also intuition.

38. If understanding is in the individual, does the teacher then abdicate any evaluative function?

My answer is no. In fact, the teacher cannot help evaluating what students do and say. The teacher must show the student what she considers inadequate, and she must show why it does not work. The constructivist teacher is not concerned with "truth" but with successful operating. It is no use simply to tell a student that she is wrong. This only destroys such motivation that she may have. To change a student's way of acting or thinking, she has to see, in her own experience, that what she did was not the most successful way of operating.

39. Whose construction is to be accepted in a school setting?

The first criterion to decide the acceptability of a "construction" is its viability, that is to say, whether it does or does not do what it is supposed to do. If it is viable, one can introduce other criteria such as simplicity, economy, elegance, convention, etc.

40. What are the implications of constructivism for how we assess student learning?

This is the most difficult question. Before it could be answered satisfactorily, one would have to decide what one wants to teach. As long as evaluation is based on "performance," where performance means the ability to reproduce answers that were part of the instructional process, we are not assessing what would be the important "learning" from a constructivist point of view (i.e., the ability to solve new problems). We have to decide whether we want students to develop their ability to think or merely to handle standard procedures and give standard answers in a limited set of situations. Given sufficient authority, the latter can be achieved by rote learning and does not involve the students' conceptual ability. If the students pass this kind of test, they may find the satisfaction of holding a job and will have to discover other ways of attaining intellectual satisfaction. Much research is going on in the area of "assessment of cognitive skills," and some new methods may be found. At the moment, all I can think of is this: present the students with a problem they have not encountered before (in the sense that it is conceptually different), observe (infer) how they conceptualize it, and judge what they do to solve it. It is each student's approach that is more important than the particular solution. By observing the conceptual tools the student is using, one can usually get some inkling as to how far she is on the way towards a workable conceptual network for the particular area. Best of all, give the student a problem to which there is not yet a standard solution. I realize that this is not a satisfactory answer, but then, according to the statistics we read these days, the tests used at present are not satisfactory either.

41. Would a constructivist approach to science cause more students to be "better" scientists?

To answer this question, we should first have to agree as to what it means to be a "better" scientist. From my point of view, I would say yes; to me, a scientist is better if she never forgets that science (1) cannot reveal "objective truth," (2) is forever fallible, and (3) is not the most important thing in the field of human experience. A constructivist attitude is, I believe, a safeguard against the pseudo-religious faith that the search for "truth" justifies all means and that believers can wash their hands of the possible consequences. Constructivists cannot but remain aware of the fact that they are responsible for the way in which they see the world and thus for everything that may follow from their particular way of seeing. Their way of seeing is, in the last analysis, never more than one viable way of seeing.

42. In a largely positivist world, how do you suggest that constructivism can make a bridgehead in teacher education?

We can approach the gap by having current and prospective teachers experience that the constructivist approach works. This is not an easy task, but, where it has been put into practice (at the University of Georgia, at Purdue University, at Florida State in Tallahassee, and also at my own institute at the University of Massachusetts), our initial fears and doubts have been allayed. At the beginning, I was inclined to think that teachers who have taught their particular subject for 15 or 20 years would be extremely reluctant to change their approach to teaching, but practice has proved otherwise. If one succeeds in getting teachers to make a serious effort to apply some of the constructivist methodology, even if they don't believe in it, they become enthralled after five or six weeks. This is not because the results in the children are so wonderful but because the atmosphere in the classroom has changed. The relationship between the children, and between the children and the teacher, becomes radically different; it turns into a productive relationship. It is an overwhelming experience to come into a classroom of six or seven year olds who are all intensely occupied, all doing things that are relevant to what they are supposed to be doing, and it quickly becomes evident that they are enjoying it. Most elementary school teachers find this such a relief that they are very eager to continue.

In other words, it is the success that convinces them. There are difficulties nevertheless; it takes an enormous amount of stamina to bring about a change of mind in school boards, principals, and others in authority. It does not happen in a few weeks. Yet after a year or so, the radical change in the attitudes of the children, and, therefore, in the mood of the school room, has its effect. If then it can be shown, as is the case now, that after two years the understanding generated in the constructivist classrooms is significantly greater than in traditional schools, more than half the battle has been won.

REFERENCES

Einstein, A. *Ideas and Opinions*. New York: Bonanza Books, 1954.

Noddings, N. "Constructivism in Mathematics Education." In R. Davis, C. Maher, & N. Noddings (Eds.), *Constructivist Views on the Teaching and Learning of Mathematics*. Reston, VA: National Council of Teachers of Mathematics, 1990.

von Helmholtz, H. *Epistemological Writings*. Dordrecht, The Netherlands: D. Reidel Publishing Co., 1977.

3

The Construction of Knowledge: A Radical Constructivist View

Antonio Bettencourt

Expressions like "constructivism," "construction of knowledge," "learners construct meaning," and similar ones are starting to become part of the language of science education. We are liable to hear them in professional meetings or inservice workshops and to read them in articles in the professional journals. As the term *constructivism* becomes more widespread, different people tend to use it with slightly different meanings, and some use it in a loose way to designate a complex of different pedagogical, psychological, or philosophical tendencies. (The ideas about constructivism explained in this chapter are in no way to be taken as an attempt to define the "orthodoxy" of constructivism. Consistent with a constructivist view, they are simply a model of what it means to know. The claim of this model is to be a viable view of knowledge. This chapter aims at presenting the model and exploring from there some relations with teaching and learning of science.) These tendencies seem to have in common the central assumption that all we come to know is our own construction.

This position seems, at first, to be fairly common-sensical, given that it is difficult to conceive that the world could enter our minds intact and in one piece. We must, therefore, assume that to know is, in some sense, to transform the object of knowledge. The purposes, processes, and results of that transformation are the subject matter of constructivism.

To begin with, constructivism is a theory of knowledge. This means that it involves a conception of the knower, a conception of the known, and a conception of the relation of knower-known. If we agree that learning has to do with the growth of knowledge and that science is knowledge about certain domains of experience, then constructivism has relations with learning and teaching, and with science.

Different meanings can be given to the expression "construction of knowledge." Some of them take the term *construction* to more extreme consequences than others. This chapter starts by detailing a radical constructivist view of knowing, then explores some of the possible relations between this view and the teaching-learning process in the science classroom.

A RADICAL CONSTRUCTIVIST VIEW

Starting from the assumption that our knowledge is a construction of ours, radical constructivism (hereafter referred to as constructivism) denies the possibility of knowledge transmission, even in principle.

> From that perspective (radical constructivism) there is no way of transferring knowledge—every knower has to build it up for himself. The cognitive organism is first and foremost an organizer who interprets experience and, by interpretation, shapes it into a structured world. That goes for experiencing what we call sensory objects and events, experiencing language and others; and it goes no less for experiencing oneself. "Intelligence organizes the world while organizing itself." (von Glasersfeld 1982, 613)

Knowledge ceases to be a commodity that can be transferred from the mind of a giver (e.g., a teacher) to the mind of a receiver (e.g., a learner) without transformation in the process. Even if the givers intend their communications as transmission, the concepts, ideas, text, etc., transmitted have to be interpreted (and, in that sense, transformed) by the receivers from elements of their experience (von Glasersfeld 1983, 213). We all have had the experience of seeing that what our students made of our teaching bore little resemblance to what we had intended them to learn. That this may be an inherent facet of the process of knowing was pointed out by Ludwik Fleck long ago:

> We forget the simple truth that what we are acquainted with consists rather of what we reach by learning than of what we arrive at by knowing. Yet this is a momentous fact, since in the short span from the teacher's mouth to the student's ear the content of the knowledge transmitted is always slightly distorted. Thus in the course of decades or even centuries and millennia, divergences develop to an extent that it sometimes becomes doubtful whether anything of the original has been preserved at all. In the circumstances the content of knowledge is by and large a free creation of culture. It resembles a traditional myth. (Fleck 1929, 47; translation 1986)

Constructivism, to become a viable alternative to more familiar epistemologies, has to go beyond saying that knowledge is constructed; it has to specify the purposes and processes of knowing.

The first question to be answered is the question of purposes: Why do we construct knowledge at all? The answer seems to be simple and straightforward: We construct knowledge in order to deal with our experience. We should always be suspicious of simple answers. In epistemology (and in life in general), simple answers are false most of the time or hide a great deal of complexity. To deal with our experience means, in this context, to organize it in such a way that our actions bring about desired results and avoid undesirable ones. In order to be able to do this, we have to have some idea (no matter how rough and incomplete) of some of the possible results and how to choose between them (i.e., some valuing scheme). These expectations and preferences come from our previous experience (which includes social and cultural dimensions). Trying to bring about our expectations (i.e., to repeat previous experiences), we construct working hypotheses (no matter how simple) of how entities will behave when acted upon in certain ways. As long as these hypotheses are fulfilled by the results of our action (i.e., by our interpretations of what happens), we will continue to use them. They are our concepts, ideas, theories, models, images, etc., of how the world of our experience behaves. They constitute our understanding of the world and serve as tools to be used in future situations.

This vision of the knowing activity entails a notion of the knower that is based on certain presuppositions about its minimal requirements:

> The *ability* and beyond that the *tendency to establish recur-rences* in the flow of experience; this, in turn, entails at least two capabilities,

> ■ *remembering* and *retrieving* (re-presenting) experiences; [and]

> ■ the *ability to make comparisons and judgments* of similar-ity and difference.

> . . . apart from these, there is the presupposition that the organism likes certain experiences better than others, which is to say *it has some elementary values*. (von Glasersfeld 1989, 128, emphasis added)

The next question to be answered by a constructivist view is the question of processes: How do we come to know what we know? To answer this question, constructivism brings together the insights of Giambattista Vico (1644–1744), Italian philosopher, humanist, and rhetorical theorist, and the research of Jean Piaget (1896–1980), Swiss genetic epistemologist. Vico, besides being a major figure of humanistic thinking, is considered by several scholars to be a remarkable case of foreshadowing some of Piaget's important ideas (Gash 1983; Gash and von Glasersfeld 1978; Mora, 1976a, 1976b; von Glasersfeld 1976a). Radical constructivists consider him to be a forerunner (Gash and von Glasersfeld 1978; von Glasersfeld 1984, 1985, 1989). Piaget, also claimed by radical constructivists as one of them (von Glasersfeld 1974, 1979, 1982), is well known for his essential contributions to the understanding of the ways humans develop their knowledge.

Referring to what it means to know (Latin verb *scire*, root of science), Vico says that "science is knowledge of the genus or mode by which a thing is made; and by this very knowledge the mind makes the thing, because in knowing it puts together the elements of that thing" (Vico, translation 1988, 46). Piaget, two centuries later, expresses a similar conception when he writes that

> knowing consists in constructing or reconstructing the object of knowledge in such a way as to grasp the mechanism of this construction; which is the same as saying (if one prefers to use the terms that positivism has persistently but ineffectually proscribed) that to know is to produce in thought [i.e., in the thinking mode], and the production must be such that it reconstitutes the way in which the phenomena are produced. (Piaget 1961, 441–442; translation in von Glasersfeld 1974)

The research and theorization of Jean Piaget and his collaborators informed by this conception constitutes an extensive and detailed investiga-tion of human ways of knowing. Most of what follows (and some of what came before) is based on Piaget or constructivist readings of his work (von Glasersfeld 1974, 1976a, 1979, 1982, 1984, 1985, 1989).

Confronted with some experience, we tend to use some concept, model, or pattern of action that past experience leads us to expect satisfactory results from, and we act accordingly. If the results of our action are the expected ones (i.e., satisfactory according to our goals), the experience becomes, for all purposes, a case of the concept, model, or pattern of action used. We call the concept, model, or pattern a "scheme" and describe the inclusion in the scheme as "assimilation," borrowing Piaget's terminology. In doing this, we focus more on the similarities we could compare and disregard differences; we "engage in an act of judging, interpreting or bringing the object under a certain category. Hence, assimilation is the equivalent of a judgment: to assimilate a thumb to a sucking scheme is to judge the thumb as something to be sucked" (Kitchener 1987, 53). For all the purposes that matter, the thumb is "a suckable."

As soon as this and other assimilations are successfully made, we tend to use the same scheme over and over again. The repeated use of a scheme has three consequences: (1) the scheme becomes more and more generalized and flexible (e.g., we generate different subsets of "suckables"); (2) the different schemes are integrated with each other; and (3) sooner or later, the use of schemes generates problems (e.g., a spoon may not lend itself to a sucking scheme very easily).

Problems arise when the results of our acting are not the ones we expected from previous experience. We may disregard the differences as unimportant and keep our scheme unaltered, or we may regard the differences as important. In the last case, we have a perturbation in our understanding of experience. This perturbation makes us become aware of our own cognitive activity in order to generate a new solution. The new solution needed is made by stepping back (so to speak) from our concepts, models, or patterns and trying novel combinations (e.g., the spoon may be hit against something) until we are able to produce the expected results, other satisfactory results, or give up for the moment. The stepping back may range from the puzzled look in an infant's face, through the actual physical repetition of the actions to see where we went wrong, to the mental representation of the process and possibilities of modification. The novel combination generated in the process is an altogether different model, concept, or pattern. In the terminology of Piaget, we say that the previous concept, model, or pattern has been "accommodated." The two processes of assimilation and accommodation are never completely independent. In fact, Piaget says that

> assimilation and accommodation are therefore the two poles of an interaction between the organism and the environment, which is the condition for all biological and intellectual operation, and such an interaction presupposes from the point of departure an equilibrium between the two tendencies of opposite poles. (Piaget 1954, 353)

Piaget assumes that, in our search for solutions, we tend to form cognitive structures (i.e., patterns of action, concepts, images, models, theories, etc.) that are in dynamic equilibrium with each other and with our experience. This dynamic equilibrium is a result of structures that are flexible enough to allow disregarding of differences, but not so generalized that anything can be subsumed under them. If we do not disregard differences in some instances, assimilation is not possible and no concept is formed (e.g., if for each dog we see we have to make a different concept, the concept of "dog" as a class cannot arise). If we disregard too many differences, our conceptions are never changed and become ossified, hence useless (von Glasersfeld 1976b).

The two processes of assimilation and accommodation always take us to a state where our structures will be different (hopefully, more useful, interconnected, and elegant) than the previous state. The process of equilibration leads to a new equilibrium, not to the same that there was before. This is why Piaget called it, in French, "equilibration majorante" (1975, 35-44), an expression that was translated as "optimizing equilibration" (1985, 25–32).

The process of equilibration allows our knowledge to change and, at the same time, to remain stable. We preserve and discard schemes in a nonrandom way. Another problem seems to arise here: How are schemes preserved and discarded? The answer to this question makes us introduce another important dimension of a constructivist view: the evolutionary dimension.

A scheme (i.e., a concept, idea, theory, or pattern of action) will be preserved in our repertoire if it helps us deal successfully with our experience. This does not mean that the knowledge in question is the only possible solution to the problem confronting us; it just means that, in using this solution, we have not encountered any obstacles to the obtainment of the results we expect. The solution has passed the constraints of experience and has thus proved viable. In a constructivist epistemology, there is not one right solution, but there are clearly wrong (unviable) ones. A solution that encounters constraints that it cannot pass has to be modified (or remade) to pass them or dropped as useless (von Glasersfeld 1980, 971–972; 1984, 23). The analogy here is with a Neo-Darwinist concept of adaptation.

Adaptation in Neo-Darwinist conceptions of evolution ceased to have teleological connotations. Organisms (or populations) do not have adaptation as a goal. If adaptive modifications are the result of random variations in the genetic pool of a population, we cannot say that organisms (or populations) adapt themselves to the environment. Environmental constraints select out those organisms that do not have the necessary genetic make-up to pass them and leave the others. At any time, the organisms that survive are not those which adapted themselves but those that the constraints of the environment have left. Organisms (and populations) either fit the environmental constraints and survive, or they do not and die (von Glasersfeld 1980, 971–972; 1981, 90–91). Our knowledge can be conceived in an analogous way. The schemes we possess either pass the constraints of experience and prove viable, or they do not, and we discard them (von Glasersfeld and Cobb 1983, 220). The notion of survival applies to our cognitive structures, not to ourselves.

Our knowledge, to use an image of von Glasersfeld (1984, 21), fits our experience as a key fits a lock. The key enables us to open the lock but tells us very little, if anything, about the lock. We all know that different keys can open the same lock. Worse, even, is the fact that burglars use a picklock, a very simple tool, to open different kinds of locks. Our knowledge may have very little relation with the reality underlying our experience, and we cannot assume a one-to-one relationship (i.e., a match) between the two. There is no way of assessing that relationship, given that to compare our knowledge with "reality" we would have to perform another act of knowing and never access reality-in-itself (i.e., before any act of knowing). This paradox, inherent to any act of knowing that purports to know things-in-themselves, was pointed out by the Greek Sophists in the fifth century B.C. and erected into a fundamental objection to the possibility of any certain knowledge by the Skeptics (Sextus Empiricus 1933, 17).

DIFFERENT WAYS OF BEING CONSTRUCTIVIST

The seriousness with which different constructivist epistemologies take this paradox allows us to distinguish three types of constructivism:

■ "Radical constructivism," if the relationship between knowledge and "reality" as a criterion of truth has been abandoned and substituted by a criterion of fitness:

> Radical constructivism, thus, is *radical* because it breaks with convention and develops a theory of knowledge in which knowledge does not reflect an "objective" ontological reality, but exclusively an ordering and organization of a world constituted by our experience. (von Glasersfeld 1984, 22, emphasis his)

Piaget's epistemology, according to some interpretations (von Glasersfeld 1974, 1979, 1982), is compatible with radical constructivism.

■ "Hypothetical realism," if our knowledge (especially scientific knowledge) is considered an hypothesis about the structure of reality and evolving asymptotically to a perfect (though unattainable) knowledge of that reality. (Munevar 1981, 12):

> Nowadays, hypothetical realists are beginning to accept the fact that forms and categories of conceptualization are also functions of central nervous organization, which bear an analogy-relationship to the inherent reality of objects which is just as incomplete as the relationship between the color "red" and the electromagnetic waves of a given range of wavelengths. (Lorenz 1971, 287)

Lorenz, Popper, and most evolutionary epistemologists (e.g., Campbell 1987) can be interpreted as belonging here (Munevar 1981, 4). Piaget, according to Kitchener (1987, 105–109), also belongs here.

■ "Pragmatic constructivism," if our knowledge is considered as a constructed picture of the reality of things-in-themselves (see Cobern, pers. comm.-a).

In its most common form, pragmatic constructivism results from not taking a constructivist position to its extreme consequences.

THROUGH THE NARROW GATE: CONSTRAINTS ON OUR KNOWING ACTIVITY

Evolutionary analogies come into constructivism to answer two questions: the question of purposes and the question of processes. We construct knowledge because that has proved adaptive in the course of our evolution as a species. In the process of construction (which includes use), our schemes are preserved or discarded (analogy with survival), depending on whether they pass the constraints of our experience. Given that the constraints of experience cannot be known in advance, the importance of creativity and multiplicity of ideas cannot be overstressed. Some contemporary philosophers of science apply this idea to science itself and say that the coexistence of different, even competing paradigms, is a desirable feature (e.g., Feyerabend 1975).

The introduction of evolutionary dimensions into constructivism leads to another problem: What are these constraints that our schemes have to fit, and where do they come from? Constraints, in a cognitive sense, can generally be conceived as the elements of experience that prevent us from achieving the results we intend, thereby generating a perturbation. We may distinguish four sources of constraints:

- the elements of our previous constructions

- other cognitive organisms

- the domain of our experience

- the total network of cognitive structures that form our knowledge at any given time

These are the four selecting devices that make us preserve, discard, or modify our cognitive structures. The four sources of constraints are distinguished above for purposes of clarity of analysis, not because they are distinct or act separately. Given their crucial role in the construction of knowledge, they deserve a more detailed look.

The characteristics of our previous constructions impose certain limitations to our future conceptualizations. The elements we abstract from the flow of experience, the ways we combine them, and the regularities we impose on them condition our subsequent ways of knowing and perceiving. As soon as infants achieve some notion of object permanence, they populate their experience with "real objects." The construction of a Euclidean notion of space makes us perceive the world in a certain way. To use another image of von Glasersfeld, building with bricks allows us to make certain kinds of constructions but not others (von Glasersfeld 1984, 37).

Among the permanent objects we populate our experiential world with, there is a class that, very early in our cognitive history, becomes extremely important for subsequent knowing. In our transactions with these "objects," we soon attribute to them properties we have attributed to ourselves. The process is reversible, because we come to attribute to ourselves properties that we had constructed to explain their behavior. This class of "objects" is composed of cognitive organisms (especially humans) that we come to conceive as having feelings, thoughts, and so on, similar to ours (von Glasersfeld 1985, 97–98; 1986; 1989, 130–131). Once these cognitive organisms are abstracted from the flow of experience (i.e., very early), they become an important source of constraints. Piaget goes as far as to say that "others" are the main cause of our assimilation and accommodation:

> The relations of assimilation and accommodation thus constitute, from the time of the sensorimotor level, a formative process analogous to that which, on the plane of verbal and reflective intelligence, is represented by the relations of individual thought and socialization. Just as accommodation to the point of view of others enables individual thought to be located in a totality of perspective that insures its objectivity and

reduces its egocentrism, so also the coordination of sensorimotor assimilation and accommodation leads the subject to go
outside himself to solidify and objectify his universe to the point
where he is able to include himself in it while continuing to
assimilate it to himself. (1954, 356–357)

In our interaction with other humans, their behavior (especially
symbolic) may or may not corroborate our previous conceptions. We have all
sometimes experienced how a disagreement with someone may make us
revise our notions or the concept we have of that person, or both. Another
way in which other persons are crucial for knowledge construction is in
provoking our awareness of inadequacies in our concepts. Called on to
explain what we mean, we may stumble on our own obstacles and generate
the occasion to change.

Another set of constraints seems to come from the domain of our
experience. There is a sense in which our experience, albeit being our
construction, does not bend itself to our imagination, desires, or goals. A
constructivist can never say as the idealist can say, "The world is my dream."
Whenever our experience does not confirm our expectations, we may have
touched on the reality underlying it. "This means that the 'real' world manifests
itself exclusively there where our constructions break down" (von Glasersfeld
1984, 39). The constraints we are encountering may come from that "real"
world. They can only tell us, however, what that reality is not. They leave open
infinite possibilities for what it may be. In this sense, constructivism, like
idealism, maintains that we are cognitively isolated from the nature of reality.
By maintaining this, constructivism does not deny the existence of a real world
(i.e., a world independent of any knower) but merely the possibility of our
knowledge having a matching relation with it. Our knowledge is, at best, a
mapping of transformations allowed by that reality.

The last set of constraints our constructions have to pass comes from the
realization that none of our cognitive structures makes any sense in isolation.
Concepts, ideas, images, theories, and so forth, are related to each other. This
does not mean that our knowledge is, at any time, a completely consistent
system. It is a "conceptual population" (Toulmin 1972, 128). Any new piece of
knowledge must also fit into this conceptual population. Another consequence
of this set of constraints is that we will tend to preserve the stability of the
conceptual population. This means that our assimilations and accommoda-
tions must not cause perturbations that disrupt the whole population.

These constraints may act both as conditions to change our knowledge
or as reasons to preserve existing structures, sometimes in the face of
disconfirming evidence. Concepts that have resisted well to constraints will
tend to be preserved. Scientists will not throw out a theory that has withstood
the tests of experimentation at the first piece of contrary evidence.

Summarizing, we can say that "constructivism is the view that reality
itself is constructed" by the knower (Kitchener 1987, 101). This definition has
two important consequences:

- "knowledge is never a simple copy of reality, but always
  results from a construction of reality through the activities
  of the subject;" (Inhelder 1977, 339) and

- "the subject constructs the cognitive schemes, categories, concepts, and structures necessary for knowledge." (Kitchener 1987, 102)

Abandonment of the notion of correspondence between our knowledge and an independently existing reality does not entail that anything goes. It entails that there is more than one solution to the interaction between the knower and the known. None of the solutions can claim to be the only correct one, but some of them are wrong. The distinction is made in the process of our encountering the constraints of experience (including social and cultural experience) and trying to circumvent them.

CONSTRUCTIVIST EXPLORATIONS IN THE TEACHING AND LEARNING OF SCIENCE

Radical constructivism shows us that the relation between our knowledge and our experience is, at best, underdetermined. The same caveat applies to the relations between theory and practice. Practice is never a simple application of general rules to concrete situations, and theory is never the simple abstraction-generalization from practical situations to general schemes. Practice and theory, like knowledge and experience, stand in a relation of mutual adaptation, of mutual questioning, and of mutual illumination.

Ideas, concepts, theories, and ways of acting are, in constructivism, subject to the constraints of experience. A view of teaching and learning science is required to pass "through the narrow gate." The following reflections illustrate some ways in which constructivism could illuminate the daily practices of teaching and learning science.

Let us take an old monument of the educational landscape: lecture. Constructivism would be naive in saying that students cannot learn from lecture. We all have learned from some lectures and learned valuable things. What constructivism tells us is not to assume that just because we said it, the students learned it as we intended. The students may have learned something totally different from what we intended. This means that constructivism will advocate careful use of lecture and its interspersing with opportunities to encounter constraints. Discussions, images, time for reflection, and opportunities to use the ideas in different situations are needed.

Students, like all human beings, construct knowledge from their experience, including their school experience. In order for a construction to form, there need to be foundations (see Cobern, pers. comm.-b). The foundations are the students' experience and knowledge gathered in the course of their lives. Assuming that students are a blank slate is inviting pedagogical trouble. The students' ideas, conceptions, and views are the very possibility of their learning and knowing. How else could they assimilate experience? And if they do not assimilate, how can they accommodate? Constructivism tells us that these ideas may be very sensible for the students and, hence, resistant to change. It also tells us that these ideas deserve our respect, that is, deserve to be acknowledged, used, challenged, and transformed.

We would be also misled if we said that constructivism advocates the so-called "hands-on science." The important part is not that students manipulate things physically but that they do it for a purpose and engage in discussion about it. That is, prior to the activity, there has to be a question; during the activity, there has to be a reflection (the stepping back) on the difficulties encountered, and, after the activity, the accounting (to oneself, if nothing else) of what happened. Unless hands-on science is embedded in a structure of questioning, reflecting, and re-questioning, probably very little will be learned.

Another idea that constructivism brings us is the appreciation of students' efforts of construction. A student confronted with a situation may come up with an account that seems strange to us. One of the possible ways to respond is to try to understand why such a solution or idea seemed sensible to the student. This will tell us a lot about the student and will also help us to step back from our scientific way of looking at the world. Being able to appreciate the beauty and elegance of a student's construction applies as much in the science classroom as in the art or writing ones. This does not, of course, prevent us from challenging the student about the adequacy of her conception.

Finally, there is the problem of motivation. If the constraints students encounter in the science classroom have to do with spelling, correctness of vocabulary, neatness, task completion, etc., then they may learn to be "good students," but they will not learn about the world of their experience in ways that are useful. The important motivation for knowledge is the satisfaction of seeing how one of our ideas elegantly fits a problem situation. How many students have passed through science classrooms without ever experiencing this?

> Good teachers . . . have practiced much of what is suggested here, without the benefit of an explicit theory of knowing. Their approach was intuitive and successful, and this exposition will not present anything to change their ways. But by supplying a theoretical foundation that seems compatible with what has worked in the past, constructivism may provide the thousands of less intuitive educators an accessible way to improve their methods of instruction (von Glasersfeld 1989, 138).

There seems to be an implication in all this about the kind of classroom where all this will be possible. There has to be discussion, criticism, and time to struggle with problems. The social norms have to be supportive of all this. Can we trust students enough to have all this going on? Can we afford not to trust them?

ACKNOWLEDGMENTS

The author gratefully acknowledges the influence that Ernst von Glasersfeld has had on his thinking about teaching, learning, and knowing. He also expresses the indebtedness of this paper to the many occasions when these ideas were discussed with Roberto Alves Monteiro and Maria Bellini Alves Monteiro. Finally, he is grateful to William Cobern, Wolff-Michael Roth, and Thomas Russell for their comments on an earlier draft. For the faults that still persist, he must, however, claim entire responsibility.

REFERENCES AND NOTES

Campbell, D.T. "Evolutionary Epistemology." In G. Radnitzky and W.W. Bartley III (Eds.), *Evolutionary Epistemology, Rationality and the Sociology of Knowledge*. La Salle, IL: Open Court, 1987. 47–89.

Cobern, W. (pers. comm.-a). Ernst von Glasersfeld (1985, 91–92) calls this "trivial constructivism." This name, besides being pejorative, may lead into the belief that there is a "right" form of constructivism. Such belief is in stark contradiction with constructivism in general and radical constructivism in particular. The name "pragmatic constructivism" was suggested by William Cobern (e-mail, July 16, 1991) in his review of a previous draft of this chapter.)

Cobern, W. (pers. comm.-b). This extension of the metaphor of construction was suggested to me by William Cobern in one of our e-mail exchanges.

Feyerabend, P.K. *Against Method: Outline of an Anarchistic Theory of Knowledge*. London: Verso Editions, 1975.

Fleck, L. "On the Crisis of 'Reality.'" In R. S. Cohen and T. Schnelle (Eds.), *Cognition and Fact: Materials on Ludwik Fleck*. Boston: D. Reidel Publishing Co., 1986. 47–57. (Originally published in German in 1929.)

Gash, H. "Vico's Theory of Knowledge and Some Problems in Genetic Epistemology." *Human Development,* (1983): 26, 110.

Gash, H., and E. von Glasersfeld. "Vico (1668–1744): An Early Anticipator of Radical Constructivism." *The Irish Journal of Psychology*, 4(1), (1978): 22–32.

Inhelder, B. "Genetic Epistemology and Developmental Psychology." In R. W. Rieber and K. Salzinger (Eds.), "The Roots of American Psychology: Historical Influences and Implications for the Future." *Annals of the New York Academy of Sciences*, 291. New York: New York Academy of Sciences, 1977. 332–341.

Kitchener, R.F. *Piaget's Theory of Knowledge: Genetic Epistemology and Scientific Reason*. New Haven, CT: Yale University Press, 1987.

Lorenz, K. *Studies in Animal and Human Behavior* (vol. 2). Cambridge, MA: Harvard University Press, 1971.

Mora, G. "Vico and Piaget: Parallels and Differences." *Social Research*, 43(4), (1976a): 699–712.

Mora, G. "Vico, Piaget, and Genetic Epistemology." In G. Tagliacozzo and D. P. Verene (Eds.), *Giambattista Vico's Science of Humanity*. Baltimore, MD: Johns Hopkins University Press, 1976b. 365–392.

Munevar, G. *Radical Knowledge: A Philosophical Inquiry into the Nature and Limits of Science.* Indianapolis, IN: Hackett Publishing Co., 1981.

Piaget, J. *The Construction of Reality in the Child*. New York: Basic Books, 1954.

Piaget, J. *Les Mecanismes Perceptifs: Modeles Probabilistes, Analyse Genetique, Relations avec l'intelligence* [The Mechanisms of Perception: Probabilistic Models, Genetic Analysis, Relations with Intelligence]. Paris: Presses Universitaires de France, 1961.

Piaget, J. *L'equilibration des Structures Cognitives: Probleme Central du Developpement* [The Equilibration of Cognitive Structures: Central Problem of Development]. Etudes d'Epistemologie Genetique, no. 23. Paris: Presses Universitaires de France, 1975.

Piaget, J. *The Equilibration of Cognitive Structures: The Central Problem of Intellectual Development*. Translated by T. Brown and K.J. Thampy. Chicago: University of Chicago Press, 1985.

Sextus Empiricus. *Outlines of Pyrrhonism*. Loeb Classical Library. New York: G. P. Putnam's Sons, 1933.

Toulmin, S. *Human Understanding: The Collective Use and Evolution of Concepts* (vol. 1). Princeton, NJ: Princeton University Press, 1972.

Vico, G. *On the Most Ancient Wisdom of the Italians Unearthed from the Origins of the Latin Language, including the Disputation with the Giornale de' letterati d'Italia*. Ithaca, NY: Cornell University Press, 1988. (Originally published in Latin in 1710.)

von Glasersfeld, E. "Piaget and the Radical Constructivist Epistemology." In C.D. Smock and E. von Glasersfeld (Eds.), *Epistemology and Education* (Report no. 14). Athens, GA: Follow Through Publications, 1974.

von Glasersfeld, E. *Constructivism in Vico and Piaget*. Contribution to the session on "Vico and Psychology" at the Conference on Vico and Contemporary Thought, New York, January, 1976a.

von Glasersfeld, E. *The Construct of Identity or the Art of Disregarding Difference*. Paper, fourth biennial Southeastern Conference on Human Development, Nashville, TN, 1976b.

von Glasersfeld, E. "Radical Constructivism and Piaget's Concept of Knowledge." In F.B. Murray (Ed.), *The Impact of Piagetian Theory on Education, Philosophy, Psychiatry, and Psychology*. Baltimore, MD: University Park Press, 1979. 109–122.

von Glasersfeld, E. "Adaptation and Viability." *American Psychologist*, 35(11), (1980): 970–974.

von Glasersfeld, E. "The Concepts of Adaptation and Viability in a Radical Constructivist Theory of Knowledge." In I. Siegel, R. Golinkoff, and D. Brodzinski (Eds.), *New Directions in Piagetian Theory and their Application to Education*. Hillsdale, NJ: Erlbaum, 1981. 87–95.

von Glasersfeld, E. "An Interpretation of Piaget's Constructivism." *Revue Internationale de Philosophie*, 36(4), (1982): 612–635. The whole sentence in the text of Piaget (1954, 354–355) reads: "Intelligence thus begins neither with knowledge of the self nor of things as such but with knowledge of their interaction, and it is by orienting itself toward the two poles of that interaction that intelligence organizes the world by organizing itself."

von Glasersfeld, E. "On the Concept of Interpretation." *Poetics*, 12, (1983): 207–218.

von Glasersfeld, E. "An Introduction to Radical Constructivism." In P. Watzlawick (Ed.), *The Invented Reality: How Do We Know What We Believe We Know? Contributions to Constructivism*. New York: Norton, 1984. 17–40.

von Glasersfeld, E. "Reconstructing the Concept of Knowledge." *Archives de Psychologie*, 53, (1985): 91–101.

von Glasersfeld, E. "Steps in the Construction of the Others and Reality." In R. Trappl (Ed.), *Power, Autonomy, Utopia: New Approaches Toward Complex Systems*. New York: Plenum, 1986.

von Glasersfeld, E. "Cognition, Construction of Knowledge, and Teaching." *Synthese*, 80(1), (1989): 121–140.

von Glasersfeld, E., and P. Cobb. "Knowledge as Environmental Fit." *Man-Environment Systems*, 13(5), (1983): 216–224.

4

Contextual Constructivism: The Impact of Culture on the Learning and Teaching of Science

William W. Cobern

Though rooted in neo-Piagetian research, constructivism is an avenue of research that departed from the neo-Piagetian mainstream 20 years ago and has continued on a distinct path of development. The departure was evident by the late seventies, clearly outlined in two publications by Novak (1977) and Driver and Easley (1978). For constructivists, learning is not knowledge written on or transplanted to a person's mind as if the mind were a blank slate waiting to be written on or an empty gallery waiting to be filled. Constructivists use the metaphor of construction because it aptly summarizes the epistemological view that knowledge is *built* by individuals. Since Ausubel et al. (1978), theorists have argued that the construction of new knowledge in science is strongly influenced by prior knowledge, that is, conceptions gained prior to the point of new learning. Learning by construction thus implies a change in prior knowledge, where *change* can mean replacement, addition, or modification of extant knowledge. Learning, by construction, involving change is the basis of the Posner et al. (1982) conceptual change model. In essence, constructivism is an epistemological model of learning, and constructivist teaching is mediation. A constructivist teacher works as the interface between curriculum and student to bring the two together in a way that is meaningful for the learner. Furthermore, if one carries the construction metaphor to its logical conclusion, construction implies a foundation, in addition to the studs and beams of prior knowledge. The construction of new knowledge takes place at a construction site consisting of existing structures standing on a foundation. In other words, construction takes place in a *context*. The purpose of this chapter is, first, to trace the developments in constructivist theory that lead to contextual constructivism, including the types of questions suggested by contextual constructivism. Second, it places those questions in the context of an anthropological world view theory.

THE EVOLUTION OF CONSTRUCTIVISM

Piagetian and neo-Piagetian theory has dominated educational research in the second half of the twentieth century and continues to do so. In science education, the publications of A. E. Lawson provide some of the best examples of Piagetian-based research. In the 1970s, however, a derivative of Piagetian research emerged (see Fig. 1). Researchers in what is now called the constructivist movement maintained Piaget's structural philosophical position but opposed continuing the research focus on mental stages common among the neo-Piagetians of that time period. Instead, they argued for a new epistemological focus on the actual content of student thinking. In science education, the emergence of the constructivist movement as a formidable avenue of research was clearly outlined in two publications by Novak (1977) and Driver and Easley (1978). In the 1980s, the philosophical work of von Glasersfeld (1989) on radical constructivism added to the movement's momentum.

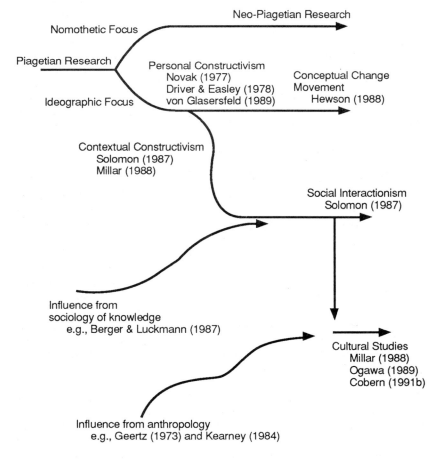

Figure 1. The Emergence of Cultural Studies in Science Education.

At the heart of constructivism is a particular view of knowledge. Knowledge held by an individual is assumed "to have a complex set of referents and meanings" (Magoon 1977, 652) that must be taken into account when a researcher is trying to understand how learning takes place. It is this complex of referents and meanings that led researchers to draw upon David Ausubel's (1963) theory of meaningful learning with its "emphasis on the preexisting conceptual structures to which the students had to relate new items of knowledge" (Solomon 1987, 64). In addition, constructivism involves a distinctive view of the student. From the constructivist perspective, students are knowing beings who construct knowledge that is personally meaningful. Thus, for the researcher to understand learning, the researcher must come to an understanding of the construction process. The focus here is clearly on the student as constructor (Magoon 1977). The early emphasis in constructivism was clearly upon the individual, and, thus, the initial constructivist departure from neo-Piagetian research is best termed *personal constructivism*.

The emergence of personal constructivism in science education had important consequences for science education research. The emergence precipitated a series of very different research questions, pertaining not to mental stages but to the actual content of student conceptions. It is interesting to note that neo-Piagetians such as Lawson maintained a focus on mental stages but replaced Piaget's interview methodology with paper-and-pen instrument methodology. Constructivists did the opposite. While moving away from a focus on mental stages, they kept and developed interview methodologies. Highly structured Piagetian interviews gave way to the less structured interview-about-instances format (West and Pines 1985) and phenomenographic techniques (Marton 1988), which became increasingly qualitative.

STUDENT CONCEPTION RESEARCH

As noted, constructivist research meant a switch in focus from mental stages to the actual content of student conceptions. This was, in part, due to a perceived lack of progress in neo-Piagetian research with respect to improving the efficacy of classroom instruction. Of equal importance, however, was the influence of epistemological studies in the philosophy and history of science. Kuhn's seminal work on scientific revolutions resulted in a lively, enduring debate concerning the development and nature of scientific knowledge (e.g., Shapin 1982). In the 1960s and 1970s, with an eye on these events, science educators around the globe conducted numerous *nature of science* (NOS) studies. This was conceptual research in that researchers wished to know how well student and teacher conceptions of what science is about fit with the researchers' philosophical models. NOS studies ran their course, but research interest in student conceptions continued. Constructivists turned to questions of greater specificity. Rather than asking what students believe science to be about, they asked, What is a student's construction of (say) gravity and how does that construction compare with the epistemological truth of science? Knowledge that compared poorly was called a *misconception* or *alternative conception*. The focus of conceptual research was the student's experience in the natural world and her attempts to make sense of experience vis-à-vis concepts of science. The research quite naturally focused on specific concepts of interest in science education, for example, motion, force, and natural selection.

Conceptual research is important to constructivists because learning is viewed as a process of deconstructing misconceptions and reconstructing valid scientific conceptions in their place. At first, the construction process was considered a personal experience and essentially rational:

> Our central commitment . . . is that learning is a rational activity. That is, learning is fundamentally coming to comprehend and accept ideas because they are seen as intelligible and rational. Learning is thus a kind of inquiry. (Posner et al. 1982, 212)

In the personal constructivist view, conceptual change occurs when a student personally finds that science conceptions are more intelligible, plausible, and fruitful than her own previously held conceptions. Driver (1983) and Driver et al. (1985) are classic presentations of student conception research. That there was a burst of enthusiasm for student conception research, nurtured by studies in epistemology and stimulated by the fascinating research of Driver, among others, was attested to in 1983 and 1987. In those years, Cornell University convened international conferences solely for the purpose of presenting misconception (MC) and alternative conception (AC) research (Helm and Novak 1983; Novak 1987). Personal constructivism continues to be a productive avenue of research, as attested to by the large number of conceptual change and misconception papers read at the 1991 annual meeting of the National Association for Research in Science Teaching.

Nevertheless, the overall result of applying MC- and AC-based conceptual change ideas has been mixed. It has not been found that students *easily* give up their prior ideas. Critics of the conceptual change model consider it excessively rationalistic: "We want to argue that nonrational components are intrinsic to conceptual change in the individual, and that these should not be excluded in investigations of conceptual change" (West and Pines 1983, 37). West and Pines listed power, simplicity in complexity, aesthetics, and personal integrity as potentially significant nonrational components of conceptual change. As early as 1982, Novak raised the Vygotskian point that environment can influence the conceptions that children have about nature events. Sensitive to the critics and perhaps concerned by an apparent lack of sufficient progress in conceptual change research, Strike and Posner (in press) recently revised the conceptual change model. They agree that the original model was indeed excessively rational and in need of broadening. Referring to Toulmin's (1972) notion of "conceptual ecology," Strike and Posner (in press) note that the conceptual change theory "can be stated in a more general form by emphasizing that what it centrally requires is a focus on the learner's conceptual ecology and how that ecology structures learning."

Joan Solomon's work is an excellent example of the more ecological approach to student conceptual research. Solomon looked to sociologists for new ideas that would help her understand the anomalies found in student conceptual research. Drawing from Mead (1934) and Berger and Luckmann (1967), Solomon argued that student ideas about nature stem "not from the logical processes of which science boasts, but from the 'common sense' attitude that relies on being able to interchange perspectives and meanings with others" (1987, 66). The notion of "interchange" introduces a *social* element. Thus,

Solomon made social interactions in the science classroom the focus of her research on student learning. "As students interact with one another, with teachers . . . they develop ideas that, because they are held in common, create a universe of discourse, a common frame of reference in which communication can take place" (quoted in Solomon 1987, 68). The identification of common sense and social interaction as factors in learning brings one to the crux of how Solomon's research differs from personal constructivism. Solomon cogently argued that a crucial component in learning is *context*, that is, learning takes place in context. Solomon's particular interest in context is social interaction. It is clear, however, that context can have a broader definition. Millar's (1989) edited volume contains a series of articles by science education researchers who share the view that research on teaching and learning must be contextualized not only socially but epistemologically and culturally as well.

In summary, the initial break with neo-Piagetian research involved refocusing research on students' actual ideas about nature. This became the basis for a rationalistic, personal constructivist notion of learning. However, a perception of insufficient progress toward improved science instruction led some researchers to doubt the theoretical adequacy of personal constructivism. Drawing on sources in the sociology of knowledge, Solomon (1987), Sutton (1989), and Millar (1989), among others, have moved the science education research field from personal constructivism towards contextual constructivism.

CULTURE AS CONTEXT

What has not significantly changed in the move from neo-Piagetian theory to personal constructivism to a nascent contextual constructivism is the assumption that student views or ideas must ultimately be measured against science conceptions. While not wishing to elevate all student ideas to the level of sacrosanctity, cultural studies indicate that students have other grounds for evaluating ideas. Hills and McAndrews (1987) agree. They noted that student views "provide a different sectioning of experience precisely because the pursuit of scientific knowledge is not the only or even the most important goal they subserve" (216). One can argue that student views are grounded in cultural milieu. As the renowned anthropologist Clifford Geertz wrote, "Man is an animal suspended in webs of significance he himself has spun, I take culture to be those webs, and the analysis of it is not an experimental science in search of law but an interpretive one in search of meaning. . . . " (1973, 5).

Broadly defined, culture is a system of meaning and significance, with the term *system* used advisedly.

> Even the assumption that culture is relatively coherent is under debate. Contemporary critical theory has now posed a major alternative view: culture as multiple discourses that may occasionally come together in large systemic configurations, but that more often exist together within dynamic fields of interaction and conflict. (Outram 1990, 327-328)

Thus, one must not suppose that cultural identification is limited to such conspicuous group identifiers as race, language, or ethnicity. Black America, white America, Spanish-speaking America, English-speaking America—each of these no more identifies a homogeneous cultural group than does the term

American. While there are elements of culture broadly shared in the United States, so that one is justified in speaking of (say) American culture, the United States is composed of many cultural subgroups, most of which are subgroups embedded within other subgroups. In addition to race and language, there are other significant factors that influence the construction of meaning and thus are part of cultural identity. These include economic and education levels, occupation, geographic location, gender, religion, and philosophy. Thus, one can expect to find considerable cultural variation among students. A student constructs knowledge so that the knowledge is meaningful in the student's life situation. Thus, while it is perhaps appropriate, it is neither sufficient to investigate student views as strictly personal views nor to investigate student views vis-à-vis science conceptions. Contextual constructivism carried to its logical conclusion compels the investigation of student views *within* the cultural context that gives meaning to those views. In the colorful phrasing of Hills (1989), this means a change from the study of "domestic affairs" in science education to the study of "foreign affairs."

Foreign affairs in science education is a more familiar idea for educators in non-Western nations, where the difference between students' traditional culture and the culture of Western science is more visible. African educators, for example, understand that science is a "second culture" for students (Ogunniyi 1988). Ogawa believes that

> science, the product of western modernization, should be taught in the context of a foreign culture in school science in a nonwestern society. On the basis of this position, science teachers need not only to know the western science itself but also to be aware of the traditional and scientific ways of thinking, and views of nature. (1989, 247)

In the past, American educators have assumed that scientific explanation is a natural part of our American student culture. But, in the United States, educators face two crucial facts. Firstly, American society is increasingly pluralistic. Secondly, not only is there widespread lack of interest in science among students, there are several cultural subgroups traditionally underrepresented in science, e.g., women, African-Americans, Hispanics, and Native Americans (Vetter and Babco 1987). There is ample reason to consider the possibility that, as in Africa, science is indeed a second culture experience for many American students.

Traditionally, the study of culture is left to the cultural anthropologists. However, in recent years, scientists, historians, and literary critics, among others, have undertaken cultural studies. What unifies these eclectic cultural scholars is that they "take a subject whose working assumptions are considered *natural* and attempt to demonstrate that they are culture-bound" (Heller 1990). Cultural studies in science education turn a series of implicit assumptions into explicit research questions:

- What do students and teachers believe about the world around them, especially the physical world?

- How do students and teachers understand their own place in the world, especially their relationship to the physical world?

- What is the cultural milieu in which these student and teacher beliefs, values, and relationships are grounded and supported?

- What is the culture of science? How is that culture interpreted in the school science classroom?

- What happens when student cultures, teacher culture, and the culture of science meet face to face in the classroom?

In other words, it is important for science educators to understand the fundamental, culturally-based beliefs about the world that students bring to class, and how these beliefs are supported by students' cultures, because science education is successful only to the extent that science can find a niche in the cognitive *and* socio-cultural milieu of students. Thus, the contextual constructivist researcher in science education is led to two of the principal questions of cultural anthropology: *What does a people believe about the world and why?* The *why* is crucial, for this indicates that the question is about culture. In cultural anthropology, these are the questions of world view.

THE WORLD AS OTHERS SEE IT

We live in a rich, experiential world brought to us by our senses, but the data of our senses is an amorphous mass of confusion until interpreted by one's world view. Sociologist Peter Berger (1979) argued that a people's world view provides a special *plausibility structure* of ideas, activities, and values that allows one to gauge the plausibility of any assertion. This shared world view is a fundamental aspect of any cultural group. It may be argued that there is a cultural identity in Western science education and that teachers typically assume students operate within that plausibility structure.

The original notion of a "heretic" was someone who decided things for herself instead of employing society's plausibility structure or world view. In a sense, science classrooms are filled with heretics operating within other plausibility structures. Science educators recognize them by their alleged misconceptions. However,

> talk of misconceptions . . . carries with it the suggestion that
> something has been botched or bungled, or that something has
> gone amiss. . . . And there is often the further implication that
> the student is the culprit: that he or she is the one who has gotten
> something wrong. . . . There is more to error than meets the eye.
> (Hills 1989, 174)

Indeed, there is more than meets the eye. As one begins to study world view, one finds that young children in the classroom are in the process of developing their own world views and plausibility structures. And it is certainly not the case that the school provides the principal, let alone the sole, influence upon this formative process. Older children, on the other hand, may come to class with well developed world views. As a result, these students enter the classroom with culturally validated ideas about the world. But again, the school was not likely the principal formative agent, and thus the students' views are pejoratively labeled as misconceptions. Whether with younger or older children, educators face a difficult instructional problem.

WORLD VIEW THEORY

World view, a concept borrowed from cultural anthropology, refers to the culturally dependent, generally subconscious, fundamental organization of the mind. This conceptual organization manifests itself as a set of presuppositions that predispose one to feel, think, and act in predictable patterns. Kearney referred to world view as "culturally organized macrothought: those dynamically inter-related basic assumptions of a people that determine much of their behavior and decision making, as well as organizing much of their body of symbolic creations . . . and ethnophilosophy in general" (1984, 1). Thus, world view is about the epistemological levels antecedent to the specific views that students hold about physical phenomena, whether one calls those views common sense, alternative conceptions, misconceptions, or valid science. Adapted to science education, world view theory argues first that educators must try to understand the world as students understand it. If the goal in science education is for successful science instruction for all, science educators cannot continue to deny the legitimate variations among people. Only as science educators come to a better understanding of how people view the world and why they hold those views can the structure of science education be changed so that science becomes meaningful for a broader range of people. The issue of science education structure, however, raises a second aspect of world view research: the close examination of the plausibility structures in science classrooms vis-à-vis structures actually necessary for science education. Kilbourn (1984) and Sutton (1989) both found projections in science instruction that are not only unnecessary in science but are also quite sure to conflict with the world views of many students. This is also a criticism frequently made in the feminist literature (e.g., Whatley 1989).

As noted earlier, conceptual change theory argues that students learn science when they see that the scientific explanation is superior to the untutored, common-sense beliefs they brought with them to the classroom. However, that only works when students share the plausibility structure of the science teacher and the science textbook. The documented difficulty in bringing about conceptual change plus the socio-cultural diversity of most classrooms is evidence that many students do not share this plausibility structure. Consider Hawkins' observation:

> The textbook says that heat flows from hot to cold, or that light travels in straight lines, or that the earth goes about the sun; the teacher tries to elucidate . . . . But failure is often imminent. In each case, the intended communication is blocked, more often than not, by a radical mismatch between the presuppositions of the book or the teacher and those of the child. What the book and the teacher obedient to it try to communicate often presupposes (but fails to induce) a radical reorganization, in each case, of some commonsense category of experience. If our early grasp of motion is itself all *geo*graphical, then the earth itself surely does not *go*. (1983, 75)

The mismatch indicates that, for many students, science is a second culture, in much the same way as American educators speak of English as a second language for some students. There is a disjunction between the world view (or views) of students and the world view of the classroom teacher and instructional materials.

The world view of students (and of classrooms), thus, has every appearance of an attractive avenue of research. Unfortunately, investigators have been hampered by a lack of theoretical clarity. For the most part, people who use the term *world view* do not define it or define it vaguely as one's view of the world, our understanding of man and nature, our type of thinking, or how we understand cause and effect. The philosopher W. T. Jones listed 13 different synonyms for world view, commenting that "critics suspect that a concept so variously named is itself somewhat vague, and this suspicion doubtless explains why some students of culture prefer to ignore the notion of world view altogether" (1972, 79). The vagueness of these terms is such that one has done little more than name a hypothetical entity. This ambiguity of definition results in many researchers failing to see how world view can be a useful concept in science education research. As a remedy, Cobern (1991a) has recently adapted for use in science education research a structuralist theory from cultural anthropology—*"logico-structuralism."* Logico-structuralism is the work of Michael Kearney (1984) who, after an extensive review of ethnographies, identified seven universally found, epistemological categories. Logico-structuralism is a composite of these categories: Self, Non-Self, Classification, Relationship, Causality, Time, and Space. Each category is composed of logically related presuppositions. The power of logico-structuralism lies in this composite structure. The seven categories alert the researcher to the complexity of world view while simultaneously providing access to that complexity. Yet, one can still characterize world views by focusing on what are considered to be the salient presuppositions within the seven universal categories.

In summary, world view research concerns fundamental categories antecedent to knowledge as it has been typically investigated by MC and AC researchers. This type of research involves a broad, minimally restricted focus on general beliefs about the world (e.g., nature), rather than a narrow focus on beliefs restricted to specific science topics (e.g., motion). Secondly, world view research is about knowledge in cultural context, rather than about knowledge in isolation. Cultural context subsumes two ecologies. The first is conceptual ecology, consisting of "cognitive artifacts as anomalies, analogies, metaphors, epistemological beliefs, metaphysical beliefs, knowledge from other areas of inquiry and knowledge of competing conceptions" (Strike and Posner, in press). Conceptual ecology is about individuals. The second ecology is about the social and physical milieu in which individuals live. Neither ecology has any meaning without the other. Knowledge of either ecology enables one to understand the other.

DIFFERENT VIEWS OF THE WORLD

It is instructive at this point to look at the issue of world view from a more concrete perspective, lest this discussion become too abstract. Figure 2 is an iconographic visualization of the relation between two hypothetical student world views and a piece of scientific knowledge as presented by a teacher or textbook. "A" and "B" in Figure 2 represent the contours of two different world views, one called the *curved* world view and the other the *angular* world view.

Figure 2. An Iconographic Visualization of World Views and Subject Matter.

A) Curved World View

B) Angular World View

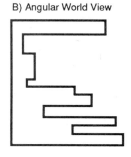

Contextualized Subject Matter

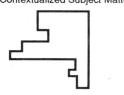

The representation is of fundamental presuppositions. Note that the goodness of fit between the curved world view and the structure of the knowledge to be learned is quite poor; thus, one of two unintended outcomes of instruction is likely. First, students will construct meaningful knowledge. However, because meaning is in reference to the students, the constructed knowledge may be strikingly different from the structure intended by the teacher or textbook author. This is consistent with the findings of misconception research (e.g., Novak 1987). It is also consistent with Chomsky's (1966) description of surface and deep structures and with Chomsky's assertion that surface structure (in this case, the intended learning of instruction) is reconstructed along the lines of an individual's deep structure.

The second possible outcome is that students will simply memorize verbatim what information they can, but, in fact, little or no meaningful learning will have taken place. Hawkins put his finger on the problem when he noted that for a child whose experiences have all been geographical, the earth surely does not go—no matter what the teacher says!

The situation is different for students entering with the angular world view. The angular world view is fundamentally compatible with the structure of the knowledge to be learned. This does not guarantee learning because the student likely holds competing conceptions that must be dealt with. Nevertheless, one has confidence that the intended learning will take place because the new knowledge is fundamentally compatible with the students' world view and thus competitive within students' conceptual ecologies. It is in this situation that one can expect the conceptual change model to be most effective.

And just what are the science teacher and textbook saying? When educators ask questions about science education, the focus is always on education, never science. Thus, Ogawa (1991) observed, "the American approach to multicultural science education is problematic. It seems to me that the movement encourages *universal* science for all Americans without ever considering the possibility of multi-sciences," where multi-science refers to science in various contextualizations. Now this is the point. Students exist in context, and they bring their contexts to the classroom. Science also exists in context, and in the classroom it is contextualized by the teacher and textbook. One should not automatically assume that it is only student context that is of legitimate concern.

To do so is to take the Ptolemaic position that education problems can be solved by adjusting and adding epicycles, when it is quite possible that what is needed is a Copernican reformulation of what science is about.

CONTRASTING CONTEXTUALIZATIONS OF SCIENCE

To further decrease the abstractness of this discussion, Table 1 characterizes science instruction for each of the seven logico-structural, world view categories. This description is based on research that critically examined the cultural form in which Western science is embedded (e.g., Capra 1982; Merchant 1989; Skolimowski 1974; Whatley 1989) and is employed here only as an example. There is research, however, that indicates that this is a relatively accurate description of Western science education (Kilbourn 1984; Proper et al. 1988; Odhiambo 1972; Ogawa 1989; Ogunniyi 1988). Table 1 also shows alternatives for each of the logico-structural categories. The alternatives are drawn from the above citations plus Cobern (1991b). Thus, it can be argued, for example, that the scientific view of the world (i.e., all that is *Other* than one's *Self*) as presented in the classroom is often materialistic, reductionistic, and exploitive. In contrast, students may bring a holistic view of the world with a focus on social and humanistic aspects of the world. Or, the scientist of the classroom is the stereotypical dispassionate, objectively rational man. Some students, on the other hand, may be people who are quite passionate and who blend rationality, emotion, and intuition—and they may not be male. The iconographic portrayal in Figure 2 shows two extremes, total incompatibility and total compatibility. As one considers the two descriptions in Table 1, one can envision how a student's world view can be a mixture of elements represented in the two columns. In a typical classroom setting, reality is likely to be something between total incompatibility and total compatibility with science and science instruction. Research addressing these issues is in its infancy. For examples, see Cobern (1991b) and Ogawa (1989).

This analysis has at least two implications for science instruction. One is that the form science instruction gives to science must be examined. How does the teacher's world view influence the form given to science during instruction? Is the particular form necessary? How well does the particular form fit the students? Can the particular form be adjusted to better accommodate the students? The second is that teachers must consider the possibility of teaching intended to influence students' world views. Of course, it is not uncommon for science educators to claim the inculcation of a scientific world view as a science education objective. However, one of the primary themes in Cobern (1991a) is that this represents a naive understanding of world view. Teaching to influence world view is a far more complicated issue than typically supposed because world view is an inclusive concept referring to fundamental presuppositions (beliefs) affecting all of one's life situation, not just the science classroom. From the literature, it is clear that changing student ideas is difficult, whether an idea is about the nature of science (e.g., Lederman 1986) or about concepts in physics (e.g., Clement 1987). If it is hard to change the conception—the surface structures—surely it is even harder to change the foundations upon which the conceptions are built—the deep structures.

Table 1. Example Descriptors for Logico-Structural Categories.

	Descriptors	
Logico-Structural Categories	Science Instruction	Alternative Views
The Other	materialistic reductionistic exploitive	holistic social/humanistic aesthetic religious
Classification	natural	natural social supernatural
Causality	mechanistic teleonomic	mystical teleological contextual
Relationship	objective nonpersonal	subjective personal
Self	dispassionate independent logical	passionate dependent intuitive
Time & Space	abstract formalism	participatory medium tangible

Influencing student world views is problematic for another reason; it involves changing student beliefs. In one sense, all of teaching intends to change student belief, given that *knowledge* and *belief* are intimately related. Knowledge is thesis, in essence, a belief, supported by lines of evidence. Regardless of what the beliefs are, epistemology is about "whether [one] ought (epistemologically speaking) to hold such beliefs. . . . Epistemology is fundamentally concerned with this normative dimension of belief" (Kitchener, in press). Thus, there is no unambiguous ontological distinction between knowledge and belief, unless one relegates belief to only that which is held blindly. However, from a world view theoretical perspective, the concept of "blind belief" is not helpful because people do not hold beliefs without reason. As Saint Augustine once wrote:

> No one believes anything unless one first thought it to be believable. . . . Everything which is believed should be believed after thought has preceded. . . . Not everyone who thinks believes, since many think in order not to believe; but everyone who believes thinks. (quoted in Neuhaus 1990, 15)

One may question the epistemological soundness of another's reasons but not that the reasons exist. This is, of course, the very question asked of knowledge, what is its epistemological soundness? For all practical purposes, belief and knowledge both represent what one has reason to believe is true.

Science educators do observe an unwritten distinction between *types* of belief (or types of knowledge). The distinction is between belief considered relevant to science, and thus within the purview of science instruction, and irrelevant belief not within the purview. For example, teaching about evolution intends to change student belief about origins but not student belief about the existence of God. What a world view theoretical perspective implies, however, is that this facile distinction is an educator's artifact. What is irrelevant to the educator is perhaps very relevant to the student. This means educators cannot continue to observe the traditional distinctions concerning that which relates to science and that which does not. On the other hand, it would be unethical, and probably counterproductive, for educators to take the presumptuous position that all belief must conform to science. To be sure, one is faced with a rather disconcerting dilemma. However, the analysis presented thus far is not a statement of how things actually are, nor is it intended to drive the development of new teaching strategies; it is intended to raise a good many questions. As Strike and Posner (in press) said of their own theoretical work, this is the presentation of a Lakatosian *"hard core* of a research program . . . that [can] be extended in profitable directions by further work." The methodology of that research program is the next topic.

METHODOLOGICAL CHANGES

As noted earlier, the Novak (1977) and Driver and Easley (1978) articles not only influenced changes in research questions but also stimulated changes in research methodology. Educational research has historically been quantitative. Researchers seek to neutralize the majority of contextual variables and then seek to focus on a very few. Constructivist research, however, has been increasingly influenced by a qualitative research paradigm (Jacob 1987), particularly ethnographic techniques where the first objective is description for understanding. The descriptive accounts are subsequently used in the formulation of hypotheses, which in turn lead to further ethnographic research. The preferred constructivist approach with students has been a Piagetian-styled interview technique called interview-about-instances (West and Pines 1985) or phenomenography (Marton 1988). The interview techniques have become less structured as ethnographic influences have increased and as research moves from a personal constructivist base to a more contextual constructivist base (e.g., Millar 1989).

The reason for the change is quite simply the complexity of people. The constructivist has come to understand that the contextual factors positivist researchers seek to neutralize are in fact factors of considerable significance. World view theory suggests that an appropriate research approach to the complexity of human thought and action is ethnographic research aimed at a Geertzian *thick description*, that is, "an extensive descriptive and interpretive effort at explaining the complexity" (Magoon 1977, 652). This complexity that demands thick description also confounds the researcher's inclination to generalize. Cronbach (1975) noted that the simplest of objectives in educational research are often made inaccessible by the complexity of human thought. He suggested that as a researcher examines an effect in different situations, the first task "is to describe and interpret the effect anew in each locale, perhaps taking into account factors unique to that locale" (125). Generalizing from any single, local study is risky at best. As Cronbach noted, "When we give proper weight to local conditions, any generalization is a working hypothesis, not a conclusion" (125). Rather, joining Cronbach's suggested local description with Geertz' thick description implies that educational research aimed at understanding the complexity of how students learn science or how teachers teach science should comprise a series of local studies where a thick description is pursued in each locality. Generalizations would then be based on the research series, much the same as generalizations in cultural anthropology are often based on several ethnographies rather than one (e.g., Kearney 1984).

There are numerous publications on the methods of the qualitative paradigm (e.g., Brewer and Hunt 1989; Cober 1991a&b; Erickson 1986; Glaser and Strauss 1967; Lincoln and Guba 1985; Lythcott and Duschl 1990; Millar 1989; Smith 1987; Spradley 1979). However, to emphasize the importance of adopting a qualitative paradigm for epistemological research, consider a recently published study involving student beliefs that was done within a traditional quantitative paradigm. Lawson and Weser (1990) attempted to connect certain student beliefs with reasoning skills. The first point to note is that the research involved a neo-Piagetian test of reasoning based on a particular view of rationality. Though this view of rationality is widely held, it has also been criticized for its cultural ethnocentricity (e.g., Cole and Scribner 1974; Modgil and Modgil 1982). The second point is that the research assumed that objective items are a sufficient means of evaluating belief and assumed so without any discussion of sociological or cultural research where the investigation of belief has been a major focus for many years. The research posited belief as discreet, unconnected propositions that can be determined by numerical responses to items such as "the living world is being driven toward greater perfection" (593). The research evaluated the numerical responses for statistical relationships with an equally discreet, unconnected dependent factor. This is a research culture defined by a rather narrow, positivistic view of science. Because many students do not share this culture, it comes as no surprise that the results of the two research instruments were mutually supportive. Ironically, the research supports a conclusion that was not considered by the researchers, i.e., many students do not share significant aspects of the researchers' world view.

In contrast, consider an ethnographic study of college student beliefs concerning the natural world (Cobern 1991b). This study noted that the literatures of sociology, history, theology, and environmental psychology indicate that, historically, Westerners are found to hold naturalistic, aesthetic, and religious views of the natural world. The study used an audio-recorded interview format designed to probe for these specific belief types as well as for belief types to be specified *a posteriori*. The interview transcript analysis showed that the conversation of the informants was dominated by aesthetic and teleological language. This language, isolated from the balance of the transcripts, would provide reason to suspect that, were the Lawson and Weser belief test given to these students, the results would show many responses that Lawson and Weser consider unscientific. However, the balance of the transcripts shows something quite different. The same informants who used aesthetic and teleological language to describe the natural world also indicated that natural science is the appropriate way to explain and understand how the natural world functions. If one were to isolate these comments, one would predict very different results on the Lawson and Weser belief test. However, one does not interpret comments in isolation because a transcript represents a contextual whole. Interpretation of decontextualized data only leads to distortion. Yet, that is exactly the weakness of quantitative research.

This is not to say that anyone's interest in the epistemological issues of vitalism, purpose, origins, and so on is misplaced. However, as Lythcott and Duschl (1990) commented on a similar study, research questions must follow from the research paradigm. In a qualitative paradigm the researcher would begin with the assumption that vitalism, for example, like any other concept, exists in context and thus is rooted in many aspects of a person's conceptual ecology and social-material culture. The researcher would ask what forms the concept takes. How do these various forms operate in the person's thinking and life situation? What other concepts are influenced? Are any of the influenced concepts related to science? What are those science-related concepts? And what is the nature of the influence on them? It is only with this type of thick description that one can avoid the problems of epistemological and cultural solipsism. Vitalism clearly shows the importance of context because it is a relatively extreme example. The function of an extreme example is to demonstrate that one cannot assume *a priori* that contextual factors do not operate in less extreme cases.

CONCLUSION

Epistemological considerations brought science educators to a greater appreciation of the importance of student ideas. Epistemology has also led to an increased awareness that knowledge is connected, a notion that has been profitably exploited with concept maps. Connectedness can be either *intra-*conceptual, the primary focus of constructivism to date, or *inter-*conceptual. Cultural studies, including the sociology of science, draw attention to the importance of inter-conceptual issues, that is, epistemological connectedness beyond the "internal affairs" of science. It is this external dimension, the impact of culture on the learning and teaching of science, that world view theory seeks to address.

The question "How do students make sense of the world?" is an old question placed in a broader arena. Previous science education research touched on the issues of external affairs when it investigated the influence of economic status, gender, and other personal factors on achievement and attitude in science. These factors, however, were always considered discreet entities to be studied individually while simultaneously attempting to control all other factors. Individual factors were never seen as aspects of the seamless web of culture. This is the broader arena, and there is research that can guide contextual constructivism. However, it resides outside of science education. The cultural history of science provides analogies to the notion that science is learned in context. For example, *The Poetic Structure of the World: Copernicus and Kepler* (Hallyn 1990) and *Leviathan and the Air-Pump: Hobbes, Boyle and the Experimental Life* (Shapin and Schaffer 1985) provide excellent discussions of how scientific ideas arise within a cultural and social milieu (also see Latour 1990). In anthropology, there are highly informative accounts, such as "African Traditional Thought and Western Science" (Horton 1967) or *The Domestication of the Savage Mind* (Goody 1977), of how scientific ideas relate to more traditional forms of thought. And, as noted earlier, there are numerous sources on the methods of qualitative research.

This chapter began by attempting to show that contextual constructivism is a natural outgrowth of personal constructivism. These categories are not mutually exclusive but complementary. Figure 1 is thus somewhat misleading. The three avenues of constructivism should be brought back together to show that an adequate approach to teaching and learning research includes all three perspectives. While personal constructivism is the anatomy and physiology of constructivism, contextual constructivism is the ecology. Both categories are needed to achieve greater understanding of how students make sense of the world. In turn, this will enable educators to structure science instruction so that science makes sense to all students. In 1979, Elliot Mishler published a marvelous article with a title that neatly summarizes the case for contextual constructivist research, "Meaning in Context: Is There Any Other Kind?" As constructivist thinking has developed since the late 1970s, its proponents have converged on a "no" response to Mischler's rhetorical question.

REFERENCES

Ausubel, D.P. *The Psychology of Meaningful Verbal Learning*. New York: Grune and Stratton, 1963.

Ausubel, D.P., J.D. Novak, and H. Henesian. *Educational Psychology: A Cognitive View*. New York: Holt, Rinehart and Winston, 1978.

Berger, P.L. *The Heretical Imperative: Contemporary Possibilities of Religious Affirmation*. Garden City, NY: Anchor Press, 1979.

Berger, P.L., and T. Luckmann. *The Social Construction of Reality*. New York: Doubleday, 1967.

Brewer, J., and A. Hunt. *Multimethod Research: A Synthesis of Styles*. Newbury Park, CA: SAGE Publications, 1989.

Capra, F. *The Turning Point: Science, Society, and the Rising Culture*. New York: Simon and Schuster, 1982.

Clement, J. "Overcoming Students' Misconceptions in Physics: The Role of Anchoring Intuitions and Analogical Validity." In J.D. Novak (Ed.), *Proceedings of the Second International Seminar on Misconceptions and Educational Strategies in Science and Mathematics*. Ithaca, NY: Cornell University, 1987.

Chomsky, N. *Cartesian Linguistics*. New York: Harper and Row, 1966.

Cobern, W.W. *World View Theory and Science Education Research*. Cincinnati, OH: National Association for Research in Science Teaching monograph series, vol. 3, 1991a.

Cobern, W.W. "The Natural World as Understood by Selected College Students: A World View Methodological Exploration." Paper, annual meeting of the National Association for Research in Science Teaching, Lake Geneva, WI, 1991b.

Cole, M., and S. Scribner. *Culture and Thought, A Psychological Introduction*. New York: Wiley, 1974.

Cronbach, L.J. "Beyond the Two Disciplines of Scientific Psychology." *American Psychologist*, 30, (1975): 116–127.

Driver, R. *The Pupil as Scientist?* Milton Keynes, UK: Open University Press, 1983.

Driver, R., and J. Easley. "Pupils and Paradigms: A Review of Literature Related to Concept Development in Adolescent Science Students." *Studies in Science Education,* 5, (1978): 61–84.

Driver, R., E. Guesne, and A. Tiberghien (Eds.). *Children's Ideas in Science.* Milton Keynes, UK: Open University Press, 1985.

Erickson, F. "Qualitative Methods in Research on Teaching." In *Handbook of Research on Teaching* (3d ed.). New York: Macmillan, 1986.

Geertz, C. *The Interpretation of Cultures.* New York: Basic Books, 1973.

Glaser, B., and A. Strauss. *The Discovery of Grounded Theory.* New York: Aldine Publishing Co., 1967.

Goody, J. *The Domestication of the Savage Mind.* Cambridge: Cambridge University Press, 1977.

Hallyn, F. *The Poetic Structure of the World: Copernicus and Kepler.* New York: Zone Books, 1990.

Hawkins, D. "Nature Closely Observed." *Deadalus,* 112(2), (1983): 65–89.

Heller, S. "Cultural Studies: Eclectic and Controversial Mix of Research Sparks a Growing Movement." *The Chronicle of Higher Education,* January, 31, 1990.

Helm, H., and J.D. Novak (Eds.). *Proceedings of the First International Seminar on Misconceptions in Science and Mathematics.* Ithaca, NY: Cornell University Press, 1983.

Hills, G.L.C. "Students' 'Untutored' Beliefs about Natural Phenomena: Primitive Science or Commonsense?" *Science Education*, 73(2), (1989): 155–186.

Hills, G.L.C., and B. McAndrews. "David Hawkins, Critical Barriers and the Education of Elementary School Science Teachers." In J.D. Novak (Ed.), *Proceedings of the Second International Seminar on Misconceptions in Science and Mathematics.* Ithaca, NY: Cornell University Press, 1987.

Horton, R. "African Traditional Thought and Western Science." *Africa*, 1, (1967): 155–187.

Jacob, E. "Qualitative Research Traditions: A Review." *Review of Educational Research,* 57(1), (1987): 1–50.

Jones, W.T. "World Views: Their Nature and Their Function." *Current Anthropology*, 13(1), (1972): 79–109.

Kearney, M. *World View.* Novato, CA: Chandler & Sharp Publishers, Inc., 1984.

Kilbourn, B. "World Views and Science Teaching." In H. Munby, G. Orpwood, and T. Russell (Eds.), *Seeing Curriculum in a New Light.* Lanham, MD: University Press of America, Inc., 1984.

Kitchener, R.F. (in press). "Piaget's Genetic Epistemology: Epistemological Implications for Science Education." In R. Duschl and R. Hamilton (Eds.), *Philosophy of Science, Cognitive Psychology, and Educational Theory and Practice.* Albany, NY: SUNY Press.

Latour, B. "Postmodern? No, Simply Amodern! Steps Towards an Anthropology of Science." *Studies in the History and Philosophy of Science,* 21(1), (1990): 145–171.

Lawson, A.E., and J. Weser. "The Rejection of Nonscientific Beliefs About Life: Effects of Instruction and Reasoning Skills." *Journal of Research in Science Teaching,* 27(6), (1990): 589–606.

Lederman, N.G. "Relating Teaching Behavior and Classroom Climate to Changes in Students' Conceptions of the Nature of Science." *Science Education,* 70(1), (1986): 3–19.

Lincoln, Y.S., and E.G. Guba. *Naturalistic Inquiry*. Beverly Hills, CA: SAGE Publications, 1985.

Lythcott, L., and R.A. Duschl. "Qualitative Research: From Methods to Conclusions." *Science Education*, 74(4), (1990): 445–460.

Magoon, A.J. "Constructivist Approaches in Educational Research." *Review of Educational Research*, 47(4), (1977): 651–693.

Marton, F. "Investigating Different Understandings of Reality." In R.R. Sherman and R.B. Webb (Eds.), *Qualitative Research in Education: Focus and Methods*. Philadelphia: Falmer Press, 1988.

Mead, G.H. *Mind, Self and Society*. Chicago: University of Chicago Press, 1934.

Merchant, C. *The Death of Nature: Women, Ecology, and the Scientific Revolution*. San Francisco: Harper and Row, 1989.

Millar, R. (Ed.). *Doing Science: Images of Science in Science Education*. Philadelphia: Falmer Press, 1989.

Mishler, E.G. "Meaning in Context: Is There Any Other Kind?" *Harvard Educational Review*, 49(1), (1979): 1–19.

Modgil, S., and C. Modgil (Eds.). *Jean Piaget: Consensus and Controversy*. New York: Praeger, 1982.

Neuhaus, R.J. "Joshing Richard Rorty." *First Things*, 8 (1990): 14–24.

Novak, J.D. *A Theory of Education*. Ithaca, NY: Cornell University Press, 1977.

Novak, J.D. "Psychological and Epistemological Alternatives to Piagetian Developmental Psychology with Support from Empirical Studies in Science Education." In S. Modgil and C. Modgil (Eds.), *Jean Piaget: Consensus and Controversy*. New York: Praeger, 1982.

Novak, J.D. (Ed.). *Proceedings of the Second International Seminar on Misconceptions in Science and Mathematics*. Ithaca, NY: Cornell University Press, 1987.

Odhiambo, T.R. "Understanding of Science: The Impact of the African View of Nature." In P.G.S. Gilbert and M.N. Lovegreen (Eds.), *Science Education in Africa*. London: Heinemann Educational Books Ltd., 1972.

Ogawa, M. "Beyond the Tacit Framework of 'Science' and 'Science Education' among Science Educators." *International Journal of Science Education*, 11(3), (1989): 247–250.

Ogawa, M. Personal correspondence, 1991.

Ogunniyi, M.B. "Adapting Western Science to Traditional African Culture." *International Journal of Science Education*, 10(1), (1988): 1–9.

Outram, D. A book review of *Cultural History: Between Practices and Representations* by Roger Chartier. *ISIS*, 81(307),(1990): 327–328.

Posner, G.J., P.W. Hewson, and W.A. Gertzog. "Accommodation of Scientific Conception: Towards a Theory of Conceptual Change." *Science Education*, 66(2), (1982): 211–227.

Proper, H., M.F. Wideen, and G. Ivany. "World View Projected by Science Teachers." *Science Education*, 72(5), (1988): 542–560.

Shapin, S. "History of Science and its Sociological Reconstruction." *History of Science*," XX, (1982): 157–211.

Shapin, S., and S. Schaffer. *Leviathan and the Air Pump: Hobbes, Boyle and the Experimental Life*. Princeton, NJ: Princeton University Press, 1985.

Skolimowski, H. "The Scientific World View and the Illusions of Progress." *Social Research*, 41, (1974): 52–82.

Smith, M.L. "Publishing Qualitative Research." *American Educational Research Journal*, 24(2), (1987): 173–183.

Solomon, J. "Social Influences on the Construction of Pupil's Understanding of Science." *Studies in Science Education,* 14, (1987): 63–82.

Spradley, J. *The Ethnographic Interview.* New York: Holt, Rinehart and Winston, Inc., 1979.

Strike, K.A., and G.J. Posner (in press). "A Revisionist Theory of Conceptual Change." In R. Duschl and R. Hamilton (Eds.), *Philosophy of Science, Cognitive Psychology, and Educational Theory and Practice.* Albany, NY: SUNY Press.

Sutton, C. "Writing and Reading in Science: The Hidden Messages." In R. Millar (Ed.), *Doing Science: Images of Science in Science Education.* Philadelphia: Falmer Press, 1989.

Toulmin, S. *Human Understanding: The Collective Use and Evolution of Concepts* (vol. 1). Princeton, NJ: Princeton University Press, 1972.

Vetter, B., and E. Babco. *Professional Women and Minorities: A Manpower Data Resource Service.* Commission on Professionals in Science and Technology, Washington, DC, 1987.

von Glasersfeld, E. "Cognition, Construction of Knowledge, and Teaching." *Synthese,* 80(1), (1989): 121–140.

West, L.H.T., and A.L. Pines. "How 'Rational' Is Rationality?" *Science Education,* 67(1), (1983): 37–39.

West, L.H.T., and A.L. Pines (Eds.). *Cognitive Structure and Conceptual Change.* Orlando, FL: Academic Press, 1985.

Whatley, M.H. "A Feeling for Science: Female Students and Biology Texts." *Women's Studies International Forum,* 12(3), (1989): 355–362.

5

Cautionary Notes on the Appeal of The New "Ism" (Constructivism) in Science Education

Ronald G. Good, James H. Wandersee, and John St. Julien

This chapter is the result of a series of discussions among the authors to explore constructivism, first as individuals and finally in a series of seminars with our faculty and student colleagues. We brought to our conversation differing approaches to questions of learning and knowledge that appeared to be related to constructivism. To the extent that "labels" can capture complex, dynamic positions on these issues, we understood ourselves to be variously a contextual realist (Good), a constructivist (Wandersee), and a pragmatist (St. Julien). We agreed that what was learned was a construction of the active learner. Did this mean that we were all constructivists? The conclusion that we came to as we tried to work out our disagreements was that the term *constructivism* was working more to obscure differences crucial to educational practice than as a guide to practice. Our working out of our differences led us to what we feel are productive questions about constructivism itself and not to a single interpretation of the term. What unifies this work is a common concern that constructivism is currently being accepted as a panacea for educational woes in a way that obscures crucial differences among proponents. It is precisely our own conflicts that make the differences between constructivism's proponents recognizable and make clear to us the depth of the questions which must be answered before constructivism warrants the enthusiastic and, to our understanding, uncritical acclaim that it has received.

Nobel Laureate Bertrand Russell (1872–1970), one of the great social critics, mathematicians, logicians, and philosophers of the twentieth century, warned against uncritical acceptance of the many "isms" that are readily available. Whether the "ism" is part of the name of an organized religion (e.g., Catholicism, Buddhism), a social-political movement (e.g., communism, socialism, capitalism), or an education movement (e.g., associationism, constructivism), Russell would caution against uncritical acceptance of dogma because that reduces the likelihood of reasoned action. It is in this spirit that constructivism is critiqued.

IS THE NATURE OF SCIENCE IMPORTANT TO CONSTRUCTIVISTS?

What seems to be the most basic assumption among scientists within the natural sciences is stated quite simply in *Science for All Americans*: "The world is understandable" (AAAS 1989, 25). It goes on to say, "Science presumes that the things and events in the universe occur in consistent patterns that are comprehensible through careful, systematic study" (25). Without this assumption, it is difficult to understand why anyone would set out to learn more about the world. Firmly set within this scientific world view is the assumption that a foundation of knowledge will develop that can lead us closer to the actual workings of nature. The authors of *Science for All Americans* say it this way:

> Scientists assume that even if there is no way to secure complete and absolute truth, increasingly accurate approximations can be made to account for the world and how it works. . . .The modifications of ideas, rather than their outright rejection is the norm in science, as powerful constructs tend to survive and grow more precise and to become widely accepted. (26)

The nature of science has been debated throughout the twentieth century and before, by scientists, philosophers, and others with interests in the nature and origins of knowledge. Scientists who study the nature of science usually develop positions that can be described as "realistic" in one form or another, as in *Science For All Americans*. The French scientist and philosopher of science, Gaston Bachelard (1884–1962), saw no viable accounts of rationality unless they were based on an understanding of science. In *The New Scientific Spirit* (1934, translated 1984), Bachelard observes, "Sooner or later scientific thought will become the central subject of philosophical controversy; science will show philosophers how to replace intuitive, immediate systems of metaphysics with systems whose principles are debatable and subject to experimental validation" (2-3).

There is reason to believe that constructivism is defined by some in such a way as to eliminate the need to debate the nature of science. This "postepistemological" position has been argued by von Glasersfeld (1990), Noddings (1990) and others. Using the term *postepistemological* suggests that important questions about the nature of science can be bypassed, reducing the domain of study for constructivists to psychological and pedagogical considerations.

If this "radical" constructivist position is intended to eliminate the need to debate the nature of science, it is problematic in a number of ways. Many of the assumptions about science that are stated in *Science For All Americans* would seem to be in serious trouble in the postepistemological world of the constructivist who sees "no point in looking for foundations" (Noddings 1990, 12). How would the many forms of pseudoscience be excluded from the domains of science and science education in the postepistemological world of the radical constructivist?

Perhaps this is a misreading of the constructivists who embrace this postepistemological position. If so, we urge them to discontinue the use of the term *postepistemological*. Its use suggests that important questions about the origin, nature, methods, and limits of scientific knowledge need not be raised as we debate the future of science education.

It would be helpful for the radical constructivists to make public their views on the nature of science as described in *Science For All Americans*. Do they support what appear to us to be statements that are consistent with a form of realism? Achieving higher levels of scientific literacy for all citizens means that reasonable agreement must be reached on the nature of science.

CAN AND SHOULD WE AGREE ON CONSTRUCTIVISM?

A meaningful debate on constructivism (or anything else) assumes that all sides adhere to reasonably similar ground rules, including common use of terms, logic, and rules of evidence.

We take a meaningful debate to be one in which the practice of science education would be challenged in a way that could conceivably lead to its improvement. The Scylla and Charybdis of educational reform debates are the twin rocks of reform-without-change and change-without-grounds. In reform-without-change, we seem to adopt a new or refurbished rhetoric to justify current practice. In change-without-grounds, we have typically instituted sweeping changes without adequate understanding of either the theoretical sources or the implications of these changes. Many failed reform movements have done both. One could offer the behaviorist rejustification of drill-and-practice to exemplify the first and the "new" mathematics of the 1960s to represent the second problem. Because constructivism seems to offer promise to help clear away theoretical blockages and invigorate practice, it is all the more imperative that science educators attempt to chart a passage between these dangers.

THE MANY "FACES" OF CONSTRUCTIVISM

Among the various definitions of "face" are (1) outward appearance, (2) facade, and (3) surface presented to view. The "faces" or modifiers of constructivism found in the literature are listed in Table 1. In some cases, the modifier (e.g., 1. *Contextual*) is replaced by its related part of speech—context. These 15 "faces" of constructivism are used to emphasize the range of implied meanings of this popular movement in education. The appeal of constructivism does indeed seem to be very broad. That knowledge is constructed by people as they try to make sense of their world(s) does not seem to draw much opposition from persons who differ markedly on other issues such as the nature of science, the nature of cognition, and how teaching and learning science should proceed. What are we to make of this apparent agreement within a setting of much disagreement? Is constructivism a platitude that deceives by achieving apparent consensus among otherwise dissenting groups, or is constructivism a way of describing a fundamentally important world view that can really serve as a unifying theme among science educators? Have the various fields of study, such as cognitive psychology, sociology, anthropology, linguistics, philosophy and history of science, and science and mathematics education, reached a point that suggests *constructivism as a metaphor for meaningful activity?*

Table 1. Words used to modify constructivism.

1. *Contextual* (context): 1) The part of a written or spoken statement that surrounds a word or passage and that often specifies its meaning. 2) The circumstances in which a particular event occurs: SITUATION.

2. *Dialectical* (dialectic): 1) The art or practice of arriving at the truth by disclosing the contradictions in an opponent's argument and overwhelming them.

3. *Empirical*: 1) a. Relying upon or gained from experiment or observation <empirical techniques> b. Capable of proof or verification by means of experiment or observation <empirical knowledge>.

4. *Humanistic*: 1) Characterized by kindness, mercy, or compassion.

5. *Information-processing*:
   information: 2. Knowledge derived from study, experience, or instruction.
   processing (process): 1. A system of operations in producing something.

6. *Methodological* (methodology): 1) The system of principles, procedures, and practices applied to a particular branch of knowledge.

7. *Moderate*: 1) Within limits: REASONABLE 2) Opposed to extreme or radical views or measures.

8. *Piagetian*: (not defined)

9. *Postepistemological* (post; epistemology): After; The division of philosophy that investigates the nature and origin of knowledge.

10. *Pragmatic*: 1) Concerned with causes and effects or needs and results rather than ideas or theories.

11. *Radical*: 1) Arising from or going to a root or source: BASIC. 2) Drastic or sweeping: EXTREME.

12. *Rational*: 1) Having or exercising the ability to reason. 2) Of sound mind: SANE.

13. *Realist*: 1) One inclined to factual truth and pragmatism.

14. *Social*: 1) a. Living together in communities. b. Of or relating to society. 2) Of or relating to the upper classes. 3) Enjoying the company of others.

15. *Socio-historical* (society, historical):
   society: 1) a. The totality of social relationships among human beings. b. The institutions and culture of a distinct self-perpetuating group.
   historical: 2) Based on or concerned with events in history.

For the purposes of this chapter, *constructivism* may be defined as an emerging consensus among psychologists, science educators, philosophers of science, and others, that learners (including scientists) must construct and reconstruct their own meaning for ideas about how the world works. According to von Glasersfeld (1989), constructivism is a form of pragmatism that centers on how humans arrive at the knowledge that allows them to cope with the natural world. The word *appeal* used in the title of this chapter is intended to act as double coding: the modern usage meaning "to attract interest" (Meaning 1) and the earlier legal usage meaning "request for review of one's case" (Meaning 2).

FROM POSITIVE FACT TO MEANING IN CONTEXT

Throughout the past one hundred years, most science instruction has centered upon the learning of *facts*, a practice which Joseph Schwab has called memorizing a "rhetoric of conclusions" (Schwab and Brandwein, 24). Although such educational practice derives from logical positivism, an outdated philosophical stance based upon the assumption of an empirically verifiable objective reality, it is instructive to examine the etymology of the word *fact*. That word actually comes from the Latin verb *facere*, which means "to make or do," therefore implying that facts are human constructions and subject to error. In his provocative book, *Writing Biology: Texts in the Social Construction of Scientific Knowledge*, Greg Myers (1990) proposes that biological "facts" *emerge* during the processes of journal article writing and revising; scientific writing doesn't just *communicate* biology, it *produces* biology. From the viewpoint of constructivism, facts have lost their status as snapshots of absolute reality. The assertion "It's a scientific fact!" has lost its persuasive power, a power the history of science shows it never should have had, given its tentativity. Indeed, the history of science is the continuing "saga of people constructing knowledge about natural phenomena" (Wandersee 1990b, 279). Science instruction must change in response to such changing views of the nature of science.

For example, in recent years an increasing number of philosophers of science (e.g., Lakatos, Feyerabend, Kuhn) have come to the conclusion that scientific reality is relative—ultimately, it is what the community of scientists says it is. Relativity, presupposition, content, paradigm shift, and other terms that have entered the vocabulary of science suggest that the observer is no longer considered to be detached from the event observed and that science does not yield absolute truth, nor can we be certain that all of modern science gives us a better approximation of nature than previous science did. While this view might initially appear to damage scientific credibility, it actually strengthens it because the utility of science is not impaired by its tentativity, and science no longer promises more that it can deliver. Latour (1987) points out that scientists are realists when doing normal science but are relativists when it comes to unresolved questions—perhaps helping to explain the paradoxical attraction of science educators to both constructivism and realism.

Why is the memorization of scientific facts an ineffective way of teaching and learning science? There are several reasons: (1) rote learning (memorization without conceptual integration) is less useful and less durable than meaningful learning (Novak 1988); (2) scientific knowledge changes and may even contradict previously learned knowledge; (3) scientific facts are relatively "low-level knowledge"—much less useful than scientific concepts, principles, and theories; (4) teaching is not merely transmitting knowledge, it is a mutually beneficial interaction between teacher and learner that catalyzes knowledge construction and meaning-making; (5) facts are, by their very nature, potentially isolated, disconnected, and misleading; and (6) facts alone do not empower the learner to learn science on her own.

If facts are not the heart of science teaching, what is? Concepts are what all of us (scientists included) think with: concepts are regularities in objects or events designated by a label (e.g., usually a word: glass, wind). Meaningful learning involves the linkage of newly learned concepts to concepts already stored hierarchically in long-term memory. Each linkage unit is called a proposition and consists of two concepts connected by a linking word(s). For example, cells have nuclei, acids taste sour, or temperature affects volume. Constructs are special higher-order concepts such as acceleration (the rate of change of velocity) that involve nested patterns; they are more abstract but also have greater explanatory power. Principles are major propositions (matter consists of atoms) that can often be organized into clusters called theories. Theories have the highest status in science because they describe, predict, and explain large "chunks" of the natural world. Since concepts are patterns and patterns exist only in the eye (and mind) of the beholder, based upon what she "beholds," learning is idiosyncratic. No two learners learn the same thing from a given science lesson nor have they identical cognitive structures. In that sense, *what you perceive constrains what you construct*.

PIAGET AS THE "GREAT PIONEER" OF CONSTRUCTIVISM

Jean Piaget is seen as the "great pioneer" (von Glasersfeld 1990, 22) of constructivism. His theory of knowing is labeled genetic epistemology, a field limited to the problem of how knowledge is increased. In *Psychology and Epistemology* (1970, translation 1971), *Genetic Epistemology* (1968, translation 1970), and *Insights and Illusions of Philosophy* (1965, translation 1971), Piaget explains genetic epistemology and its relationship to psychology and philosophy. Two of his statements from *Psychology and Epistemology* seem particularly relevant here. The second quotation stresses the importance Piaget gave to psychological study (genetic psychology) in explaining the growth of scientific knowledge (genetic epistemology).

> If epistemology does not wish to limit itself to pure speculation, it will always turn more to analysis of the 'stages' of scientific thought and the explanation of intellectual mechanisms used by various branches of science in the conquest of reality. (24)

> Psychological study can render scientific epistemology or the comparative theory of the growth of knowledge exactly the same service. It alone can enlighten us on the true importance and effective connections of fundamental intuitions whose evolution of scientific notion has been either the beneficiary or the victim. (105)

After careful readings of *Psychology and Epistemology, Insights and Illusions of Philosophy*, and other related works by Piaget, one is led to question whether Piaget would consider himself a constructivist in the sense described by von Glasersfeld and other "radical" constructivists. Piaget argued forcefully against speculative *a priori* epistemological considerations and in favor of scientific approaches that limit the scope of the questions and problems posed.

Determined not to charge ahead too hastily, the scientist takes care, on each individual question, to accumulate experimental facts or to delve axiomatically into his reasoning until agreement is reached among all researchers on facts or deductions. Consequently, he forbids, as contrary to his ethics of objectivity, any premature systematization. The fruit of this double sacrifice—delimitation requirement and that of verification—is that in fact science advances, whereas philosophy either keeps harping on itself or benefits from the progress of particular solutions to gain fresh processes of reflection. (Piaget 1970, 94—95)

The constructivist principles offered by von Glasersfeld (1990) seem to be based on what Piaget called "premature systematization" rather than on careful, systematic, scientific study. Apparently in an attempt to "hedge his bets," von Glasersfeld says, "In my view, the following basic principles of radical constructivism emerge quite clearly if one tries to comprise as much as possible of Piaget's writings in one coherent theory—but I would argue for these principles even if they could be shown to diverge from Piaget's thinking" (1990, 22). Does this mean that if Piaget's work is shown to be seriously flawed, the "great pioneer" will become the "not-so-great discard?" Goldin (1990) comments on what he sees as *a priori* aspects of the radical constructivist position:

According to the radical constructivist, these are not conclusions to be derived from controlled empirical studies of learners in which we might imagine one could *distinguish* observationally between constructive and non-constructive learning, and try to ascertain the occasions of their respective occurrence or the degrees of their effectiveness. Instead, it is claimed that they follow from fundamental, *a priori* epistemological considerations, in other words, the nature of human knowledge is that it is *necessarily* "constructed" out of the individual's world of experience, so that learning necessarily entails a process of construction. (38)

Piaget stressed again and again that his genetic epistemology was based on scientific principles and, further, that the progress of philosophy depends on the extent to which persons with a scientific education apply themselves to individual research projects in epistemology that are stated scientifically.

It is ironic that Piaget's work has been criticized as being untestable (cf. Brainerd 1978), even though he stressed the importance of taking a scientific approach to psychology and epistemology. The same criticism seems to be justified on the *a priori* aspects of the radical constructivists.

PIAGET AND THE SOCIAL CONSTRUCTIVISTS

In the category of reform-without-change mentioned earlier in this chapter, there is danger in von Glasersfeld's attempt to reconstruct Piaget as a social constructivist. Unfortunately, this attempt must ignore that Piaget examined social and historical questions only in order to isolate that which, by varying from place to place and time to time, could be demonstrated not to be part of the "genetic" or originating basis (Piaget 1970, translated 1971). It is not simply that Piaget understood science, somewhat archaically, to be a matter of "eternal truths." Rather, it seems that Piaget's cherished goal of establishing a basis for the development of a logical stage in humans was threatened by any notice of the self-referential nature of human learning. Such self-referentiality— as has been noted by those as diverse as Russell (Russell and Whitehead 1910– 1913) and Prigogine (1984)—undermines the necessarily axiomatic basis of logic. The rules (axioms) by which humans learn are themselves "under construction" throughout our development. Unfortunately, a rule that changes self-reflexively can, as Russell complained, lead to self-contradiction. This implication can largely be ignored (even if it is a vital basis of how Piaget's stages were sequentially constructed) when the objects that the student considers are inanimate. But when the objects from which you learn are animate teachers and fellow students, the implications of the mutually reflexive reconstructions of what is being learned cannot be ignored. The danger here is that the recanonization of Piaget will not be sufficiently selective and that teaching practices justified on an asocial Piagetian basis will be imported into a new practice of constructivism. There is a real difference between, say, Vygotsky's "appropriation" and Piaget's "accommodation" that has its roots in just this socially reflexive point. A teaching practice using Cuisenaire rods, for instance, must reflect that, to be successful, such items must necessarily be socially constructed as symbolic for the student; they are not simply concrete and in themselves to be valued.

The differences to be negotiated between Piaget and Vygotsky do not, of course, exhaust the field of possible differences. Within the realm of social ideas, Vygotsky's (Vygotsky 1934, translation 1986) "socio-historical" approach is challenged by formulations that represent different emphases and potentially differing educational practice. A rich stew of approaches seems to be cooking in the field. Various traditions that place a larger emphasis on language, from Bahktin (1986) to Wittgenstein (1958), are adding their emphases and perspectives. A phenomenological position, focusing directly on the experience of learning itself, is visible. Traditionally, a phenomenological analysis has been associated with a sociology-of-knowledge approach, which has roots in sources as various as Mannheim, Durkheim, and Shutz (Whitty 1985).

The pragmatist tradition, though, seems to be the main philosophical backdrop of those who are most visible in the advocacy of constructivism. Even within this renarrowed field of view, large unexplored differences exist. The main reference seems to be to current neo-pragmatists such as Rorty whose position on this question is at best ambiguous (Olson 1989). Surprisingly little attention has been paid to Dewey's work, which is both overtly educational and thoroughly pragmatic. One area in which a reexamination of Dewey would be worthwhile is his discussion of the structuring of the "problematic," from which the problem emerges as a more significant question than solving the eventual problem (Dewey 1949). Similarly, his work on what constitutes a solution to the problem thus generated would benefit the new constructivists. As potentially valuable as Dewey's pragmatic work may be, we can go even deeper into the pool of significant differences among classical pragmatists by pointing out that the "father of pragmatism," C. S. Peirce, developed a semiotic approach to reasoning that has truly radical implications for our understanding of the nature of knowledge itself (Peirce 1940). G. H. Mead, a contemporary of Dewey, was the first to discuss a fully reflexive "mirror" stage in human development (Mead 1934); von Glasersfeld's more one-sided attempt to account for a child's development would benefit from a confrontation with this work (von Glasersfeld 1989). And of course, there is James, whose work in this tradition produced a distinctive psychology with educational implications significantly different from other pragmatists (James 1904). The reader will note that we have followed out merely one strand in the current weave of constructivist ideas and have suggested that, even here, there has been little acknowledgement of differences within the social pragmatism that some advocates espouse. Such a confrontation would be useful to the further development of the adjectival form, pragmatic constructivism. We would like to suggest that a similar exploration could be carried out in any of the other adjectivally designated areas of constructivism and that the result would be similarly useful.

Aside from the social constructivisms that we have discussed, there are also numerous forms that focus on nonsocial factors. Piaget can be cited as one such constructivist, focusing on biological and individual matters. Other constructivisms focus on individual factors such as various cognitive science versions. Differences exist here as well. Schema theoretic approaches, to name one area, contain deep differences. Rumelhart, for instance, as a founding father of modern schema theory, has begat two very different lineages. One is a traditional information processing structural model (Rumelhart 1980) and the other a neurologically-based "connectionist" perspective with very different implications concerning the nature of knowledge and its acquisition (Rumelhart et al. 1986). In contrast to social or individualist models, those that focus strongly on biological or neurological constraints have been somewhat marginalized. Piaget remains a special and anomalous case, an individual whose theoretical base has been largely discarded, but whose perceptual framework remains immensely influential in how educators understand the developmental universe.

CONSTRUCTIVISM AND CONTEXTUAL REALISM IN SCIENCE

Discover is a word commonly used by scientists, historians of science, and many others to describe a process that leads to finding some observable pattern or regularity in nature. In his discussions of the nature of mathematics and the psychology of learning, Goldin (1990) notes that, "it is, however, difficult to talk about *discovering* something, such as a pattern or a structure, if we are unwilling to regard it as 'there,' existing apart from the individual" (45). In the physical world of the natural sciences, when an observer notices a particular pattern of animal behavior, a particular pattern of chemical combination under specified conditions, and so forth, Goldin's assertion seems to be even more compelling. How can these patterns of state, these regularities of process, be discovered if they are not there?

To scientists who "believe that through the use of the intellect, and with the aid of instruments that extend the senses, people can discover patterns in all of nature," (AAAS 1989, 25) it will indeed be surprising to learn from the radical constructivists that the patterns in nature, which science has discovered, are not really there.

Contextual reality is a term used by Schlagel (1986) to label a metaphysical framework for modern science that seems to be more consistent with much of the scientific world view described in *Science For All Americans*. Schlagel's contextual reality recognizes the importance of accounting for developments in modern cognitive science and the limitations of human experience in developing a theory of scientific knowledge. However, he is critical of the followers of Wittgenstein and other language philosophers who believe that knowing is centered in language rather than experience. Schlagel (280–281) quotes David Bohm and attributes to him the description of the physical world as a matrix of inexhaustible qualities and infinite layers of interrelated, though semi-autonomous, levels:

> We assume that the world as a whole is objectively real, and that, as far as we know, it has a [somewhat] precisely describable and analyzable structure of unlimited complexity. This structure must be understood with the aid of a series of progressively more fundamental, more extensive, and more accurate concepts, which will furnish . . . a better and better set of views of the infinite structure of objective reality. We should, however, never expect to obtain a complete theory of this structure, because there are almost certainly more elements in it than we can possibly be aware of at any particular stage of scientific development . . . The point of view described above evidently implies that no theory, or feature of any theory, should ever be regarded as absolute and final. (Bohm 1971, 137)

Thus, Schlagel's contextual reality, unlike von Glasersfeld's radical constructivism, posits reality within certain contexts. Truth or reality is restricted to the context applicable to the event under consideration, recognizing that no theory or knowledge of nature should be understood as final.

Contextual reality, as a metaphysical framework or theory of knowledge in science, seems to be much closer to the scientific world view described in *Science For All Americans* and held by the community of people called *natural* or *basic* scientists. Taking contextual reality in science a bit further, one is required to recognize the importance of the specific nature of a field of inquiry. Philosophy or epistemology of science must be thought of as philosophies/epistemologies of sciences because of the different conceptual frameworks and inquiry methods characteristic of the various natural sciences. Ernst Mayr, one of the leading evolutionary biologists of the twentieth century, begins his *Toward A New Philosophy of Biology* (1988) with the observation that "since the Scientific Revolution, the philosophy of science has been characterized by an almost exclusive reliance on logic, mathematics, and the laws of physics" (v). He notes that *evolutionary* biologists are concerned with such subjects as speciation, adaptation, and macroevolution, and their methods of inquiry are quite different from those of the *functional* biologist, who is interested in functions of a DNA molecule and other structural elements, from molecules to whole individuals. The main inquiry method of the functional biologist is the experiment.

Contextual reality, unlike radical constructivism, does not dismiss the fundamental world view of most scientists that a real world exists and that patterns in nature can be discovered. In this sense, contextual reality is not postepistemological as radical constructivism claims to be. Instead, the contextual realist understands that scientific knowledge is context-bound, dependent upon the constraints of methods of inquiry and applications to various problems and levels of the physical universe.

Abandoning important epistemological questions about the nature of knowledge to concentrate only on psychology and sociology of science (sciences) or mathematics, as some radical constructivists seem to urge, is unlikely to bring many scientists into the efforts to improve science and mathematics education.

IS RADICAL CONSTRUCTIVISM ARISTOTELIAN?

In a recent analysis of radical constructivism *a la* von Glasersfeld, Matthews (1992) notes:

> Von Glasersfeld says often that the orthodox epistemological problem has to be abandoned, and that correspondence as a mark of truth needs to be rejected. This is because on his terms we cannot see reality, we only have our sensations to reflect upon, and so we are never able to judge correspondence between our ideas and the world (a restatement of Berkeley's argument). He sometimes replaces correspondence with pragmatism, and in other places with coherence among experience. My suggestion is that this is a case of new wine in the old Aristotelian bottle. (13)

Matthews accuses von Glasersfeld and other constructivists of failing to distinguish between the theoretical objects of science (e.g., Newton's point mass) from the real objects of science (e.g., the physical objects of study).

Matthews' argument is related to the recent analysis of scientific realism and educational research by House (1991). Using Bhaskar's (1978, 1989) work, House points out the epistemic fallacy of the "standard" view of research; the real world is confused with our sense impressions of it. House (1991) goes on to conclude:

> Events themselves are not the ultimate focus of scientific analysis. Rather events are to be explained by examining the causal structures that produce the events, and events are produced by complex interactions of a multitude of underlying causal entities. Reality consists not only of what we can see but also of the underlying causal entities that are not always directly discernible. Reality then, is stratified. Events are explained by underlying structures, which may be explained eventually by other structures at still deeper levels. (4)

Matthews (1992) and House (1991) raise questions that appear relevant in the debate over constructivism in its many forms. It is likely that some "faces" of constructivism are influenced by a confusion between the theoretical objects of science and the objects themselves.

CONSTRUCTING THE FUTURE OF CONSTRUCTIVISM FOR SCIENCE EDUCATION

As the "great pioneer" of constructivism, Piaget stated many of his views on education in a brief book with the title *To Understand Is To Invent* (1973). In the first section under "methodological trends," Piaget identifies empirical associationism (e.g., Skinner's work), innateness (e.g., Chomsky's work), and constructivism as the trends in education. Applying his constructivist ideas to science teaching, Piaget (1973) says:

> It is obvious that the teacher as organizer remains indispensable in order to create the situations and construct the initial devices which present useful problems to the child. Secondly, he is needed to provide counter-examples that compel reflection and reconsideration of over-hasty solutions. What is desired is that the teacher cease being a lecturer, satisfied with transmitting ready-made solutions; his role should rather be that of a mentor stimulating initiative and research. Considering that it took centuries to arrive at the so-called new mathematics and modern, even macroscopic, physics, it would be ridiculous to think that without guidance toward awareness of the central problems the child could ever succeed in formulating them himself. But, conversely, the teacher-organizer should know not only his own science, but also be well versed in the details of the development of the child's or adolescent's mind. (16–17)

Piaget goes on to underline the importance of experimentation in science with considerable freedom of initiative on the part of students:

> In short, the basic principle of active methods will have to draw its inspiration from the history of science and may be expressed as follows: to understand is to discover, or reconstruct by rediscovery, and such conditions must be complied with if in the future individuals are to be formed who are capable of production and creativity and not simply repetition. (20)

In a recent article on constructivism in science and mathematics learning, Wheatley (1991) emphasizes the shift to knowledge as a process in saying "viewing mathematical and scientific knowledge as a learner activity rather than an independent body of 'knowns' leads to quite different educational considerations" (12). Wheatley's use of the words "rather than" are an indication of the focus of attention for some constructivists. Whereas behaviorists focus on the influence of the environment, constructivists focus on the learner. This lack of attention to the nature of scientific or mathematical knowledge seems to be consistent with the postepistemological nature of constructivism. Matthews (1992) criticizes Wheatley's position for "failing to distinguish the theoretical objects of science, which do not lie around, from the real objects of science, which do lie around and fall on people's heads" (12). In science education circles, the process-product controversy has been debated for years, with many persons concluding that science and science education are both process and product (Wandersee 1991). It seems to us that it is not reasonable to separate process from product as Wheatley tries to do.

Early work by Allen Newell, Herbert Simon, and others who saw learning and related things from an information-processing perspective showed that emphasizing process over product (i.e., general problem-solving heuristics over extensive declarative knowledge bases) led to a dead end in the effort to construct computer systems that display expertise. Although some constructivists seem to feel that the information-processing paradigm has little, if any, relevance for their work, it would be prudent to look carefully at all knowledge bases as we search for ways to improve science and mathematics education. Cognitive science perspectives on induction (Holland et al. 1986), prediction (Good 1989), scientific discovery (Langley et al. 1987), and problem solving (Smith and Good 1984) offer science educators ideas on learning and teaching science that could turn out to be very useful. Lessons learned should not be ignored simply because they have resulted from work in a different field. Both Piaget's genetic epistemology and cognitive science require that specialists in various fields contribute solutions to specific problems of interest to them in order for overall progress to occur.

More than any other principle, *knowledge as active construction* by the individual is promoted by constructivists. How do people come to have their ideas about how things work in nature? Goldin (1990) raises interesting questions in his critique of radical constructivism. He asserts, "one need not have recourse to arguments from radical constructivist epistemology in order to justify, on empirical grounds, a distinction between 'constructive' and 'nonconstructive' models for learning" (41). Goldin goes on to say that one can advocate discovery learning and creative thinking without claiming all the baggage of radical constructivism.

How does one make progress from less scientific to more scientific ideas? How does the novice become more expert? How can the level of scientific literacy, as proposed in *Science For All Americans,* be achieved? Let us assume that knowledge *is* a personal construction. How can we as science educators help students construct their knowledge about nature so that they become more scientifically literate? Does becoming more scientifically literate mean progressing along the novice-to-expert continuum? In short, what is it that we are trying to construct? The authors of *Science For All Americans* say, "Students must be encouraged to develop new views by seeing how such views help them make better sense of the world" (146). We must assume that "make better sense of the world" means to move closer to a current scientific view of some part of nature. Constructing knowledge that makes better sense of the world assumes that we can decide when this conceptual change occurs.

Carey (1985) has provided a very thoughtful account of changes in young children's biological concepts during years 4 to 10. She does not posit *a priori* ideas about the mechanisms of conceptual change, preferring instead to describe the changes that seem to occur as children develop ideas in closer agreement with scientists' ideas.

Focusing on the description of conceptual change rather than on its possible mechanisms seems a sensible approach for a field of study in its infancy. The mechanisms of mind posited by various theoretical camps are referred to as structures, schemas, scripts, and so forth. Depending on the particular context, conceptual change can be understood from a variety of viewpoints, as Goldin (1990) argues. Cognitive science involves many disciplines that study various aspects of cognition, yielding a "patchwork quilt" of knowledge that many believe can provide a good foundation for understanding knowledge growth in humans as well as machines. To reject this knowledge base because of its information-processing emphasis or because it does not accept *a priori* constructivist assumptions would be a mistake at this early stage of exploring conceptual change in the science education of students.

Practically, constructivism may be gaining influence because it appears to confirm what teachers have come to know via experience. It acknowledges the importance of the world that the child brings to the classroom in constructing the child's understanding of the classroom task; it places the teacher back into an active structuring role without implying that such a structuring is under the sole control of that teacher. Some forms of constructivism begin to balance the traditional assumption that learning takes place within individuals with a perspective that points to the importance of the social situation in which the learning takes place. That is, the class itself can be understood as constituting a learning entity; this is also a long-held teacher perception which has had little theoretical support.

Constructivism, then, may prove a useful and even unifying force in the theorizing and practice of science education, but such a happy outcome can result only from a confrontation with the real differences that exist among different constructivisms.

As with any formulation, constructivism offers not only possibilities but also problems. One example has to be the penchant of some constructivists to fly the banner of a postepistemological or purely "methodological" constructivism (von Glasersfeld 1990; Noddings 1990). While granting that the particular constructivisms espoused by these workers does not need to focus on traditionally-posed ontological questions or on the epistemological questions that grew out of these ontic debates, it is doubtful that they can escape the "onus" of positing a theory of knowledge. Considering the history of logical positivism's attempt to evade any necessity to confront ethical responsibility for their assumptions by using just this ruse ("we are only advocating a method," e.g., Ayer 1959), the very attempt is rhetorically questionable. This move to place constructivism as a method serves to shut down the conversation with our colleagues who hold to a realist epistemology by making the misleading claim that constructivism has no implications about the nature of knowledge. Certainly it does, and this is arguably one of its strongest points. Another weakness in current constructivist formulations is the apparent lack of awareness of valuable insights from related fields. One example is cognitive science whose findings are seldom discussed; another is any perceptible attempt to ground a constructivist metaphor in the actual physiological body. Connectionism, a hybrid cognitive science/neurology research program, offers some very significant insights into what may limit the forms thinking, memory, and perception can take. The ability to demonstrate that certain constructivisms can exist within the limits such research demonstrates, while traditional formulations cannot, may be very powerful in convincing others of the power of a constructivist position. Constructivism also needs to deal with the alogical component implied by the idea that the rules governing constructivist development change. This self-referentiality is at the heart of the possibility of self-organization and growth that make the theory so potentially powerful, but these characteristics cannot be easily approached unless one is willing to confront the alogical implications in a way that, for instance, Piaget was unwilling or unable to do. In this regard, we may be in an advantageous position vis-à-vis Piaget since we have before us models of science that admit self-referentiality as an organizing principle rather than as a source of illogical disorder. This self-referential approach is the one made famous by Prigogine's dissapative structures (e.g., hurricanes, stable eddies in a stream, or Barnard cells) and Lorenz's strange attractors.

Broadly, all these "talking points" are encompassed by the larger problem of acknowledging real differences between various constructivisms and their implications. In calling for and trying to indicate the desirability of a vigorous debate on this topic, we hope to make clear that we do not desire that those of different constructivist colorations cease research and practice in order to "talk about" the best possible approach to constructing constructivism. Rather, we anticipate that a truly fruitful conversation will include both challenging research data and convincing practice. Our conviction is that, in enriching our conversation, we will find ourselves enriching our practice and extending our knowledge in ways which we suspect will be difficult to envision from our current vantage point.

The metaphor of "construction" is, at once, plausible and appealing. It seems to fit the data gathered in studies of students' conceptions of nature, and it falls under the fashionable umbrella of relativism. However, the notion of "knowledge construction" may actually be more prescriptive than descriptive, more platitude than explanation, and more transitory than permanent. As we learn more about how human memory works, for example, we will continue to revise our views of how our minds work. Therefore, the best strategy may be to read—widely and deeply—about the emerging philosophy of constructivism, to reserve judgment about its potential to improve science education, and to check its congruence with modern learning theory and the findings of cognitive science. Learning may be more than just "carpentry" and teaching may be more than just "negotiation" and "building inspection." We must always be on guard for the imprecise thinking to which all of us can fall prey (Wandersee 1990a, 95-96). John Stuart Mill once wrote, "The grand achievement of the present age is the diffusion of superficial knowledge" (Seldes 1985, 288). We do not have to be irritatingly skeptical or passionately gullible, just reasonably cautious.

The intellectual "appeal" (Meaning 1) of constructivism for explaining the events of science education is undeniable; the philosophical "appeal" (Meaning 2) of constructivism is still "under advisement."

REFERENCES

AAAS. *Science for all Americans*. Washington, DC: American Association for the Advancement of Science, 1989.

Ayer, A. Editor's Introduction. In A. J. Ayer (Ed.), *Logical Positivism*. New York: Free Press, 1959. 3–28.

Bachelard, G. *The New Scientific Spirit*. Boston: Beacon, 1934 (translated 1984).

Bakhtin, M. *Speech Genres and Other Late Essays*. (V. W. McGee, Trans.). C. Emerson and M. Holquist (Eds.). Austin, TX: University of Texas Press, 1986.

Bhaskar, R. *A Realist Theory of Science*. Sussex, GB: Harvester Press, 1978.

Bhaskar, R. *Reclaiming Reality*. London: Verso, 1989.

Bohm, D. *Causality and Chance in Modern Physics*. Philadelphia: University of Pennsylvania Press, 1971.

Brainerd, C. "Learning Research and Piagetian Theory." In L. Siegel and C. Brainerd (Eds.), *Alternatives to Piaget: Critical Essays on the Theory*. New York: Academic Press, 1978.

Carey, S. *Conceptual Change in Childhood*. Cambridge, MA: MIT Press, 1985.

Dewey, J., and A. Bently. *Knowing and the Known*. Boston: Beacon, 1949.

Goldin, G. "Epistemology, Constructivism, and Discovery Learning in Mathematics." In R. Davis, C. Maher, and N. Noddings (Eds.), *Constructivist Views on the Teaching and Learning of Mathematics*. Reston, VA: National Council of Teachers of Mathematics, 1990.

Good, R. "Toward a Unified Conception of Thinking: Prediction Within a Cognitive Science Perspective." Paper, 62d annual meeting of the National Association for Research in Science Teaching, San Francisco, 1989.

Holland, J., K. Holyoak, R. Nisbett, and P. Thagard. *Induction: Processes of Inference, Learning, and Discovery*. Cambridge, MA: MIT Press, 1986.

House, E. "Realism in Research." *Educational Researcher*, 20(6), (1991): 2–9, 25.

James, W. "Does Consciousness Exist?" In *Essays in Radical Empiricism*. Cambridge, MA: Harvard University Press, 1904, 1976.

Langley, P., H. Simon, G. Bradshaw, and J. Zytkow. *Scientific Discovery: Computational Explorations of the Creative Process*. Cambridge, MA: MIT Press, 1987.

Latour, B. *Science in Action*. Cambridge, MA: Harvard University Press, 1987.

Matthews, M. "A Problem with Constructivist Epistemology." Paper, 65th annual meeting of the National Association for Research in Science Teaching, Boston, 1992.

Mayr, E. *Toward a New Philosophy of Biology: Observations of an Evolutionist*. Cambridge, MA: Belknap/Harvard University Press, 1988.

Mead, G. *Mind, Self, and Society*. (C.W. Morris, Ed.). Chicago: University of Chicago Press, 1934.

Myers, G. *Writing Biology: Texts in the Social Construction of Scientific Knowledge*. Madison, WI: University of Wisconsin Press, 1990.

Noddings, N. "Constructivism in Mathematics Education." In R. Davis, C. Maher, and N. Noddings (Eds.), *Constructivist Views on the Teaching and Learning of Mathematics*. Reston, VA: National Council of Teachers of Mathematics, 1990.

Novak, J. "The Role of Content and Process in the Education of Science Teachers." In P. Brandewein and A. Passow (Eds.), *Gifted Young in Science: Potential Through Performances*. Washington, DC: National Science Teachers Association, 1988.

Olson, G. "Social Construction and Composition Theory: A Conversation with Richard Rorty." *Journal of Advanced Composition*, 9, (1989): 1–9.

Peirce, C. *Collected Papers*. Cambridge, MA: Belknap Press, 1940.

Piaget, J. *Insights and Illusions of Philosophy*. New York: New American Library, 1965 (translation 1971).

Piaget, J. *Genetic Epistemology*. New York: Columbia University Press, 1968 (translation 1970).

Piaget, J. *Psychology and Epistemology*. New York: Viking, 1970 (translation 1971).

Piaget, J. *To Understand is to Invent*. New York: Viking, 1973.

Prigogine, I., and I. Stengers. *Order out of Chaos: Man's New Dialogue with Nature*. New York: Bantam, 1984.

Rumelhart, D. "Schemata: The Building Blocks of Cognition." In R. Spiro, B. Bruce, and W. Brewer (Eds.), *Theoretical Issues in Reading Comprehension*. Hillsdale, NJ: Erlbaum, 1980. 33–58.

Rumelhart, D., P. Smolensky, J. McClelland, and G. Hinton. "Schemata and Sequential Thought Process in PDP Models." In J. McClelland, D. Rumelhart, and the PDP Research Group (Eds.), *Parallel Distributed Processing:*

Explorations in the Microstructure of Cognition, Vol. 2. Cambridge, MA: MIT Press, 1986. 7–57.

Russell, B., and A. Whitehead. *Principia Mathematica.* (Vols. 1–3). Cambridge: Cambridge University Press, 1910–1913.

Schlagel, R. *Contextual Realism: A Meta-physical Framework for Modern Science.* New York: Paragon House, 1986.

Schwab, J., and P. Brandwein. *The Teaching of Science.* Cambridge, MA: Harvard University Press, 1962.

Seldes, G. *The Great Thoughts.* New York: Ballantine Books, 1985.

Smith, M., and R. Good. "Problem Solving and Classical Genetics: Successful vs. Unsuccessful Performance." *Journal of Research in Science Teaching,* 21, (1984): 895–912.

von Glasersfeld, E. "Cognition, Construction of Knowledge, and Teaching." *Synthese,* 80, (1989): 121–140.

von Glasersfeld, E. "An Exposition of Constructivism: Why Some Like it Radical." In R. Davis, C. Maher, and N. Noddings (Eds.), *Constructivist Views on the Teaching and Learning of Mathematics.* Reston, VA: National Council of Teachers of Mathematics, 1990.

Vygotsky, L. *Thought and Language* (rev. ed.) (A. Kozulin, Trans.). Cambridge, MA: MIT Press, 1986. (Originally published in Russian in 1934.)

Wandersee, J. "Imprecise Thinking: A Baconian Checklist." *Journal of Research in Science Teaching,* 27(2), (1990a): 95–96.

Wandersee, J. "On the Value and Use of the History of Science in Teaching Today's Science: Constructing Historical Vignettes." In D. Herget (Ed.), *More History and Philosophy of Science in Science Teaching.* Tallahassee, FL: Florida State University, Department of Philosophy, (1990b). 277–283.

Wandersee, J. "Guest Editorial: Mantras, False Dichotomies, and Science Education Research." *Journal of Research in Science Teaching,* 28, (1991): 211–212.

Wheatley, G. "Constructivist Perspectives on Science and Mathematics Learning." *Science Education,* 75, (1991): 9–21.

Whitty, G. *Sociology and School Knowledge.* London: Methuen, 1985.

Wittgenstein, L. *Philosophical Investigations.* (G.E.M. Anscombe, Trans.). New York: Macmillan, 1958.

PART 2

Teaching and Learning of Science and Mathematics

6

Construction of Knowledge and Group Learning

Marcia C. Linn and Nicholas C. Burbules

Enthusiasm for group learning is so widespread that one takes the risk of seeming retrogressive by even raising the possibility that it may not be the best mode of learning for all educational aims, for all subject areas, or for all students. Yet historical experience, if nothing else, should make us cautious about regarding educational ideas like group learning as panaceas. Certain ideas have a cyclic nature and return again and again in slightly different guises, despite mixed or even nonexistent evidence of success. Such proposals are often couched in language that is broad and ambiguous, taking on as many different meanings as there are advocates for them. Rhetoric praising such approaches often replaces serious attention to more endemic problems of education. For this reason, we approach the idea of group learning with some skepticism.

Arguments for group learning are usually buttressed by the claim that students learning together co-construct more powerful understandings than they could construct alone. For example, a recent issue of *The Professional Teacher* acknowledges that group learning has a long history in education but then asserts that the new twist is "cooperative learning" and concludes, sweepingly, "The process of learning cooperatively actually improves the acquisition and retention of content and skills throughout the curriculum. Kids learn better when they learn cooperatively" (Dockterman 1990, 8).

In this chapter, we argue that advocacy of group learning as a mechanism for knowledge construction oversimplifies important issues concerning the social structure of groups, the goals of individuals in groups, and the diverse nature of knowledge construction. The weight of evidence supports the conclusion that knowledge is constructed (e.g., Eylon and Linn 1988), but the evidence on the process of co-construction is much less definitive. Although most research groups start with the premise that social interaction facilitates cognitive development (Azmitia and Perlmutter 1989; Brown and Palincsar 1989; Doise and Mugny 1984; Slavin 1983), they disagree about how and when group learning fosters effective knowledge construction (Cohen 1986; Damon and Phelps 1989; Elshout, in press; Fraser 1989; Kulik and Kulik 1989; Salomon and Globerson 1989; Schoenfeld 1989; Webb 1989).

Our goal here is to define the merits of group learning more carefully. We will examine specific student needs and learning goals in terms of group learning. We will point out that co-construction of knowledge in group learning is but one of many constructive mechanisms. We will identify when group learning may be less effective than autonomous learning for certain students or for specific educational goals. Furthermore, we will argue that, considering the diverse goals we pursue in education, group learning may be helpful in attaining some of them but counterproductive in attaining others.

"Group learning" comprises a number of quite different activities. They all assume that learning requires participants to discuss a task with each other before completing it. This way of framing the concept stresses the *communicative* aspect of group interaction. In the course of this communication, students jointly negotiate understanding, plan complex tasks, explain things to each other, direct activities, contribute ideas, and coordinate actions with one another. However, this broad definition encompasses several distinct types of group learning, including *cooperative learning*, *collaborative learning*, and *tutored learning*. *Cooperative learning* involves dividing a task into parts and having each group member complete one of the parts. In *collaborative learning,* two or more students jointly work out a single solution to a problem. *Tutored learning* occurs when one student helps another gain expertise; usually, the person doing the tutoring is more expert or uses a method for fostering learning in the student being tutored. It is, of course, possible to design activities that combine more than one of these types of group learning.

Furthermore, this definition relates to a range of learning goals. Group learning has been postulated to benefit students in at least three ways: by fostering cognitive skills, by promoting social skills, and by imparting workplace skills. Researchers have suggested that group learning might foster cognitive skills such as the ability to solve complex problems, to communicate experimental results, or to understand difficult material (e.g., Johnson and Johnson 1979; Pea, in press; Phelps and Damon 1989; Webb 1989). Others have stressed that group learning is ideal for imparting the social skills and dispositions necessary in a democracy, such as respect for others, reliance on others for assistance, and sensitivity to group expectations (e.g., Cohen 1986; Sharan and Sharan 1976). Recently, some researchers promoting group learning have pointed out that such experiences can also prepare students for the workplace by helping them learn to work in teams, to take the role of leader or follower depending on the situation, to deal with "difficult" associates, and to learn the norms and practices of a group (e.g., Cohen 1986; Lave and Wenger 1991; Schoenfeld 1989; Slavin 1983). Naturally, group activities may yield more than one of these benefits.

These distinctions frame the issues surrounding group learning in a helpful way. Rather than an all-or-nothing endorsement of the concept, we believe the key questions ought to be: What is best learned in groups? When will one or another group activity foster knowledge construction? For what additional goals is group learning helpful or detrimental? Who benefits from group learning? And how, in light of these reflections, can group learning be made an effective part of a repertoire of teaching and learning activities?

In this chapter, we explore the benefits that have been claimed for group learning, examine the processes that have been hypothesized to foster co-

construction of knowledge, and analyze how these factors might work together. We assess research supporting these interactions and draw some conclusions about what is currently known and what can be reliably recommended about cooperative, collaborative, and tutored group learning. We will focus here specifically on examples in science learning, including the learning of computer science.

GROUP LEARNING AND COGNITIVE SKILL

Claims that group learning can foster cognitive skill take two forms: some argue that groups perform certain cognitive skills such as problem solving or decision making better than individuals; others assert that group learning helps students develop certain cognitive skills. We examine these claims, look at the theoretical perspectives from which they arise, and consider the mechanisms of group interaction that might foster such cognitive outcomes.

Much of the enthusiasm for group learning stems from the success of groups in one aspect of problem solving, namely, brainstorming. Research suggests that groups are effective at brainstorming and generating ideas (e.g., de Bono 1973) because participants can build on ideas suggested by others, thereby engaging in co-construction of knowledge. The benefits of brainstorming were illustrated vividly in an account by Von Fange, who described two engineers who labored for over a month to generate 27 alternate solutions to a control-device problem. When they gathered together a group of 11 young engineers having no particular background with the problem, the group was able to generate, in 25 minutes, all 27 of those solutions, and several more solutions that the original engineers had not thought of (Von Fange 1959). Brainstorming does not occur spontaneously but, rather, results from training in idea generation and group interaction: groups are trained to generate ideas and to accept and elaborate the ideas generated by others but not to criticize them. Research suggests that trained individuals generate more and better ideas than untrained individuals (Meadow and Parnes 1959).

Once the brainstorming has yielded promising ideas, however, research and observation suggests that planning and synthesis are often best performed by individuals. For example, virtually every software designer interviewed by Lammers (1986) reported that brainstorming can yield exciting ideas but that the most successful plans are created by one individual. Anyone who has served on a committee can attest to the pitfalls of group planning and point to group plans that were worse than any solution suggested by an individual.

Why is brainstorming successful as a group activity while planning seems difficult? With training, individuals can allocate all of their processing capacity to generating ideas in a brainstorming session, and every idea generated becomes a product of the session. In contrast, planners must coordinate several ideas and present them persuasively to the rest of the group. Working out the details of a software design plan requires keeping many pieces of information in mind, which may be difficult in a group.

After the plan is created, a group might again contribute by speeding up the process of implementation with a divide-and-conquer strategy. A planner can dole out tasks to group members, and when these subtasks can be performed simultaneously, this approach will reduce the total time required for the implementation.

Thus, it appears that certain cognitive skills are better matched to particular group strategies and that individuals might be better at some aspects of problem solving than groups are. These conjectures are supported by research on decision making as well as our experience with the Computer as Lab Partner project.

Research on decision making illustrates both the advantages of training in brainstorming and the problems of group planning. Janis (1972, 1982a, 1982b, 1982c) studies foreign-policy decision making and reports that when decisions have major implications, the common strategy is to make groups rather than individuals responsible. He finds that untrained groups may make more avoidable errors than individuals because group processes produce shared miscalculations. Janis argues that often the norms and expectations in a decision-making group can limit the alternatives considered, reinforce excessive risk taking, prevent reanalysis of a planned course of action, reduce motivation for getting information from experts, lead to selective bias in evaluating evidence, and support hasty, unreflective decision making. Janis demonstrates how cohesive groups develop symptoms of what he calls "groupthink" because the expectation of concurrence-seeking overshadows other group goals. These problems are not limited to groups of foreign policy makers.

Similar effects of group norms on decision making arise when students solve science problems. The Computer as Lab Partner (CLP) project has examined group problem solving in middle school physical science classes over a five-year period (Linn and Songer, in press-a; Linn and Songer, in press-b). The shortage of hardware in precollege classrooms forces teachers to use group learning in technological environments. Recently, the CLP project has been exploring the strengths and limitations of group learning and investigating effective procedures for helping students work in groups. Our studies of group learning involve observations of students working together in two-, three-, and four-person groups as well as interventions to improve group interaction. Students are required to predict the outcomes for complex everyday problems, such as whether wool or aluminum foil is the best for wrapping a drink to keep it cold. After making predictions, students conduct experiments either in simulation or using real time data collection (see Fig. 1). They then reach group conclusions and record them in an electronic notebook (for details, see Linn 1991; Linn, et al., in press-c).

Group behavioral norms interfere with productive problem solving at many points in this process. In generating a prediction, students often prefer to accept the first idea generated, instead of coming up with several ideas and then choosing the best. Our interviews with students who did not contribute often reveal that the student felt pressure from the group to move along rather than consider alternatives. Often, the idea generated by the student with greatest status is the one selected. Yet status in science class may result from success in other areas (student government or sports) rather than mastery of science concepts. In addition, students often misinterpret the results of their experiments rather than seriously considering the evidence. Thus, some students are surprised that wool turns out to be better than foil for keeping a drink cold and simply assume that the results were the opposite. When asked, they indicate that wool is for keeping people warm, not for keeping a drink cool. Thus, they draw on everyday knowledge rather than attempting to reconcile

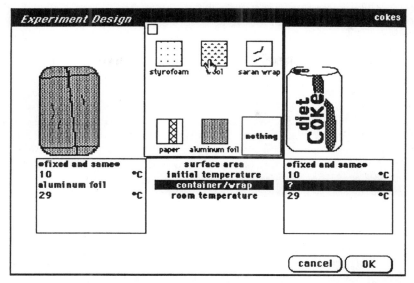

Experiment Design Screen: Students select experimental conditions

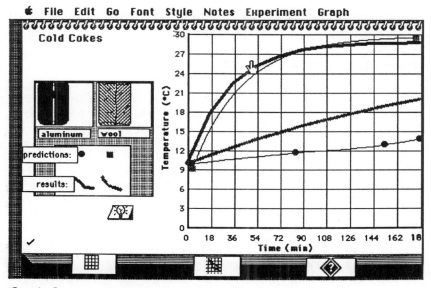

Graph Screen: Predictions and results are shown

Figure 1. Screen displays showing experiment design and graphs of predictions and results.

conflicting information. Even if a student voices dissent, the group often embarrasses the individual into agreement by sarcasm or by invoking status differences. Finally, when things turn out differently than expected, the group often invokes stereotypes, justifying invalid predictions by saying such things as "We were wrong because we listened to Jessica," or even, "That proves girls cannot do science."

Implementing group learning in the classroom may intensify students' propensities to be cognitive economists and reinforce stereotypes about who has the best ideas. Furthermore, since group learning is usually unfamiliar to students, they may lack discourse strategies effective for discussion of scientific ideas. In particular, students' everyday discourse strategies are inappropriate for making inferences about conflicting evidence in scientific domains, and their models of academic discourse are often incomplete. The rules of everyday discourse help individuals make inferences about what people mean by what they say, based on the expectation that it will "make sense" or conform to group expectations (Grice 1978; Levinson 1983). Students follow a set of implicit principles that result in few controversies.

In contrast, academic discourse tends to follow the rules of formal logic and regularly results in controversy. However, students' models of academic discourse are generally taken either from textbooks or from their teachers, neither of which is likely to model the use of academic discourse for identifying or resolving controversy. Rather, teachers and textbooks generally model the process of asserting information on the basis of authority. As Forman (in press) demonstrates, this authority-based form of discourse leads to dysfunctional scientific discussions where students assert ideas but do not provide explanations. If several group members adopt this strategy, the interaction reduces to a shouting match.

Thus, students have two unhelpful models for co-construction of knowledge in a group setting: either they can use the model of everyday discourse and remain silent about disagreements or they can assert ideas with authority. We added *agreement bars* to our software to help groups recognize when a mem-

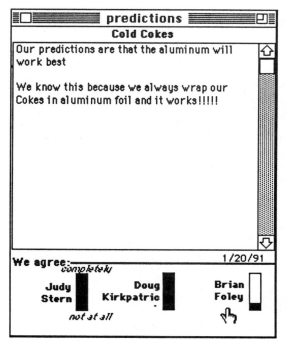

Figure 2. Screen display showing agreement bars.

96

ber was silent yet disagreed (see Fig. 2). Often groups reported surprise when a bar was set in disagreement. When students assert ideas with authority, the person who is listened to may possess status as a result of accomplishments unrelated to the understanding of science. Thus, we must ask not only whether groups can construct knowledge but also what knowledge the group is likely to construct. Both the CLP experience and the research on decision making suggest that groups may construct knowledge to meet group norms and that these norms may differ from the norms of natural scientists.

Aside from efficiency or productiveness in generating a particular outcome, these interactive processes, in school contexts, suggest that we must also ask which group activities are most likely to benefit learners. Group members may contribute components of a good solution, yet the participants themselves may be no more knowlegeable after making their contributions than they were before. Such considerations force us to ask when group work leads to better learning and when students might learn more by working autonomously. This, in turn, requires us to clarify which goals we want group learning to serve.

RATIONALE FOR GROUP LEARNING IN FOSTERING COGNITIVE GOALS

A commitment to group learning is often inspired by the work of Dewey (1920), Piaget (1970), and Vygotsky (1962). These authors do emphasize group learning as a factor in fostering knowledge construction because, for them, all learning takes place in a social context, and group learning *per se* is only one influence on the social construction of knowledge. Even autonomous learning is, for them, "social" since it requires symbol systems, modes of reflection and inquiry, and motivations that inevitably depend on participation within a broader social sphere. Vygotsky, for example, does not discriminate between cognitive activity and social activity because he views these as entwined (Vygotsky 1987; Wertsch 1985). These theorists broadly advocate taking advantage of the social nature of learning to foster knowledge construction, but they do not always specify how this might happen.

Vygotsky argued that the social context of learning may be used to extend what he called the "zone of proximal development," such that learners can be more effective than if they were learning alone. The role of the social context is to scaffold the learner, providing hints and help that make it possible for the learner to progress further than would be possible without this support. Of course, the social context in this case could be a group of learners, a teacher available to provide appropriate hints, a tutor offering helpful comments, or other features of the learning environment. Newman (1990) translates this idea into what he calls the "construction zone." He argues that the contributions of teachers, peers, and the technological environment can all help or hinder student progress.

Piaget focuses on the development of the capacity to understand and respond to the views of others. He argues that young children are too egocentric to benefit from social interactions. Piaget asserts that, as they get older, students become capable of recognizing disputes and controversies. They can then use these conflicts to spur construction of more sophisticated ideas. Thus, for Piaget, group learning benefits students when controversy is identified and resolved. As we have already seen, however, many factors stand in the way of students' ability to recognize conflict. Piaget recognized the role of status as potentially preventing conflict and argued that groups of equals were more likely to help each other learn than were groups with status differences. For Piaget, then, the mechanism of group learning is controversy, and many factors need to be present in order for controversy to be effectively identified and resolved. Thus, the case for group learning in fostering cognitive goals is complex. On the one hand, we see that all learning takes place in a social context, but, on the other hand, we learn that individuals often pay little attention to the context in which they are learning. Such issues lead us to examine the mechanisms of group learning that have been hypothesized to foster cognitive skill.

MECHANISMS OF GROUP LEARNING AND THEIR RELATION TO COGNITIVE SKILLS

How can group learning help students construct understanding? Researchers have investigated mechanisms that contribute to effective learning in groups, including (1) motivation to meet the expectations of the group, (2) appropriation of the ideas of group members, (3) access to knowledge distributed among group members, (4) negotiation of understanding with group members, (5) monitoring of progress provided by group members, (6) hints or feedback provided by group members, and (7) dividing the task among group members. We consider how these mechanisms work with cooperative, collaborative, and tutored learning.

Motivation in Cooperative Learning, Collaborative Learning, and Tutoring

Many postulate that group learning motivates students to persist at a task (Cohen 1986; Phelps and Damon 1989). Some see this as intrinsic motivation while others stress competition between groups (Slavin 1987a, 1987b). Group learning could be used to encourage students to persist at tasks that are known to be beneficial, tasks such as reflection and knowledge integration (Linn and Songer, in press-a). Both reflection and knowledge integration require students to persist as well as to marshall intellectual resources. Group learning may motivate some students to identify the needed information, integrate it, and reflect on it. On the other hand, unless integration activities are an integral part of the curriculum, they are not likely to be a part of group learning. As numerous analyses of the curriculum reveal, knowledge integration is rarely taught or assessed (Eylon and Linn 1988). Students regularly remark, "I don't have to understand it, I just have to answer these questions" (e.g., Songer 1989). Much science instruction emphasizes memorization rather than knowledge construction. Thus, to be effective, the motivation of group learning must be channeled into appropriate intellectual activity.

Appropriation in Collaborative Learning

Appropriation occurs when students build on someone else's idea to create an idea that they could not have generated alone, as in brainstorming. For example, Newman (1990) reports that students in an interactive earth science course came up with innovative group ideas by building their ideas upon those of other students. He argues that the scaffolding of the group extends the zone of proximal development for students and enhances knowledge construction. Under the correct circumstances, the opportunity for instruction to make appropriation effective.

Does appropriation require group participation? Perhaps the same process is at work when an individual is stimulated by a written work, a video, or a clever problem. For example, Chi and Van Lehn (in press) asked students to comment on their learning of physics and found that successful students spontaneously generate what they call "self-explanations" that sound very much like appropriations of the ideas in the textbook. Thus, there is no reason to assume that appropriation is limited to group learning. Rather, it is a characteristic of productive learners.

Distributed Knowledge in Collaborative Learning

Groups can draw on the knowledge of all participants to locate ideas that help construct knowledge. Success depends on the knowledge available. If the group lacks sufficient information, this process will fail. For example, the Computer as Lab Partner program is concerned with helping students construct knowledge of thermodynamics (Linn 1991). On a pretest about thermal events, middle school students are essentially without knowledge of these processes. Putting them in groups to come up with problem solutions would not succeed since they would lack distributed knowledge. When these students try to generate explanations jointly, they often reinforce ideas that do not predict outcomes. Group efforts to draw on distributed knowledge are, therefore, counterproductive. Instead, the students first gain knowledge by conducting a variety of experiments in groups. They take advantage of cooperative learning by dividing the task into parts and performing different activities simultaneously. Then, students can compare their results to those of other groups to take advantage of distributed knowledge.

In contrast, if two students have a small amount of knowledge between them, they might pool it and behave differently from an individual. For example, Burbules and Reese (1984) observed both individual and paired students learning about logic from an instructional computer game called *Rocky's Boots* (Robinette 1982). Students had considerable latitude in choosing activities. Instead of using the tutorial intended to explain the nature of the game, paired students frequently chose to explore the game by drawing on their distributed knowledge. Some pairs discovered the role of each element in the game by trial and error, learning both what was possible and what was not possible. Although individuals and groups were equally successful at solving the problems in the game, the pairs were more willing to try unusual combinations and to experiment with the materials. This willingness to test the limits of the situation occurred spontaneously in the cooperative groups set up to work with *Rocky's Boots*, perhaps because the students collectively had a range of strategies for exploring software. The *Rocky's Boots* software had the advantage that it provided a significant amount of feedback, while, at the same time, it allowed students to try different solutions to each problem. In a sense, the computer feedback became part of the distributed knowledge available to the group.

Of course, there are other ways to access distributed knowledge; a group is far from necessary. Students might use a networked environment to access a database, read a manual, or go to the library to get the needed information. Nevertheless, in a group another student might supply information that the learner did not realize would be relevant, thus combining distributed knowledge with the opportunity for appropriation of an idea. Such opportunities could also arise with databases or books but are probably more likely in groups.

Negotiation of Understanding in Collaborative Learning

Students in groups have the opportunity to compare ideas and construct a common point of view. Negotiation of meaning is the crux of the argument for co-construction of knowledge. In writing this chapter, for example, the authors co-constructed a view of group learning by joining knowledge of philosophy and cognitive science (see also Schoenfeld 1989). While promising in many contexts, such mechanisms for knowledge construction are difficult to harness. Several research groups report examples of how groups negotiate an unpredictive view of a scientific event (Brown et al., in press; Linn and Songer, in press-a). Leonard (in press) reports on a group of students designing a computer program; they negotiate a poor compromise by combining the worst features of each of their ideas. This is another example of what Janis (1972, 1982a, 1982b, 1982c) called "groupthink," or a faulty group decision-making process.

As before, negotiation of understanding is not limited to group learning. This process also occurs when one student attempts to integrate two sources of information such as two experiments or two essays. In this case as well as in the group case, there are many examples of negotiation gone awry.

Monitoring of Progress in Tutoring

Students often benefit from tutoring precisely because the tutor helps them monitor their progress and models the process of self-monitoring. The tutor might cue students to check their work, compare one solution to another, generate self-explanations, or break a problem into subparts. Furthermore, tutors often reduce memory demands for individuals by keeping track of progress, supplying details that otherwise would need looking up, and prompting helpful behaviors. Typically, a tutor knows the material being taught and can therefore monitor the constructive process in another learner. This process has also been used successfully by training students to assume the role of tutor. Following careful analysis of the process of reading comprehension, Brown and Palincsar (Brown and Campione 1990; Brown et al., in press) devised a group-learning procedure called "Reciprocal Teaching" that allowed students to monitor the behavior of other students. In Reciprocal Teaching, one student serves as leader, following a well-specified role. This student guides other group members to construct a summary of a text by determining what question each paragraph answers. The group leader monitored by the teacher gradually assumes more and more responsibility for monitoring group activity. All students eventually become group leaders. These researchers trained students to implement a component of effective tutoring without acquiring the usual expertise of a tutor.

Are tutors necessary for students to learn self-monitoring? At least for some students, self-monitoring can be induced from autonomous learning activities. For example, we find that when students learning a new programming language are required to keep a "bug journal" (recording their errors), they often reach insights about their own learning that allow them to anticipate and avoid certain mistakes in the future (Linn and Clancy, in press).

Providing Hints or Feedback in Tutoring

Another aspect of tutoring that can be implemented in group learning concerns providing hints or feedback. Vygotsky argued that appropriate hints can expand the zone of proximal development and scaffold students as they learn. To provide hints, the tutor needs more expertise than the learner. This approach to group learning is described in the apprenticeship approach (e.g., Collins et al. 1989).

Some investigators argue that accomplishments, such as first language learning, are the result of group learning more than organized schooling (e.g., Bruner 1976). For example, Miller (1977) refers to language learning as a "spontaneous apprenticeship" where learners observe words in context and get feedback when they try out words they have partially understood. First language learning primarily involves tutoring by more accomplished practitioners, not groups of first language learners jointly figuring out how the language works. In general, language acquisition is more rapid for children with adult tutors than it is for children learning primarily from siblings and same-age friends. Leaving first language learners to rely primarily on each other leads, at best, to the "special languages" of twins and, at worst, to delayed language acquisition.

Tutors may or may not construct a more complete understanding of the material being taught. However, they are likely to construct a more complete understanding of how students learn. Many successful approaches to teaching science have emerged from analyzing successful tutoring. For example, Clement developed the "anchoring conceptions" that help introductory physics students from work on tutoring (Clement et al. 1989). Similarly, the "benchmark experiments" of Minstrell were first used in tutoring sessions and then expanded to whole class activities (Minstrell 1982).

Divide and Conquer

Although the process of dividing a task into components and parceling them out does not by itself foster knowledge construction, it is efficient and may, therefore, free time or attention for construction of knowledge. The "divide-and-conquer" approach reduces cognitive load for the group and allows the group to accomplish a more complex task. In many cases, a single individual (the teacher, the group leader) divides up the task. In these circumstances, participants gain little direct experience with planning. For example, students in the Computer as Lab Partner project learn how to divide up the task from the teacher. A potential benefit is that the group could model the process determined by the teacher in subsequent work.

In summary, creating a community that takes advantage of these cognitive processes to improve knowledge construction will involve judicious use of groups, of informal communication, and of individual work. Ideally, those with greater expertise will be able to help others who wish to or need to develop this expertise. At the same time, individuals should not engage in joint tasks that could be better carried out by one person. Furthermore, there is a danger that the most expert people will be unable to perform their main tasks because they are too much in demand as tutors. Developing an effective learning community requires considerable planning and negotiation and involves trade offs that may be difficult to anticipate.

Taken together, these arguments suggest that in learning cognitive skills, the primary benefits of group learning involve accessing distributed knowledge. Groups may be more motivated, efficient, and expedient at this process. We now analyze the relationships between group learning and social skills.

How Does Group Learning Impart Social Skills?

Arguments for group learning derive primarily from two separate sets of claims: those concerning the benefits of learning more subject matter, developing richer understandings, or fostering complex cognitive skills; and those concerning the benefits of learning to work effectively in groups. Society has a general interest in promoting the skills and dispositions of cooperation, negotiation, collective problem solving and decision making, completely aside from their role in promoting learning per se. However, as we have already discussed, broad enthusiasm for these goals has sometimes obscured the ways in which they are best pursued. Implementing certain group learning activities in schools may, in fact, be counterproductive of such ends. In this context, whether or not group learning activities are implemented in ways that promote learning may have much to do with whether students associate such activities with success or frustration, and, hence, with whether they see any value in acquiring the skills and dispositions of working cooperatively with others.

Advocates for group learning argue that by learning in groups, students will gain respect for others and recognize their reliance on society for survival (Cohen 1986; Dewey 1915, 1916; Slavin 1983). These discussions often have assumed that group members contribute willingly and effectively to the group project. Such assumptions may be false. Moreover, the social norms of the group and the school strongly influence how groups interact. We share in the hope that participation in group learning activities will broaden students' understanding of and tolerance for others, as well as promote cooperative dispositions; group learning is necessary for students to practice working with and communicating with others. However, the composition and organization of schools and classrooms (the contexts within which group learning activities are implemented) frequently work against such results. Furthermore, students already spend much time in extracurricular group activities, including organized sports and clubs. They also find numerous social activities to occupy their time outside of school. Yet these social activities are not always organized with learning goals in mind. In these settings, students may acquire harmful as well as beneficial social skills and attitudes: they may learn antagonism or suspicion as well as more productive forms of group interaction. Students can also develop unproductive learning habits as the result of group activities. Peer influence can be negative, as we know from teenage smoking, pregnancy, and drug abuse. Group activities are not intrinsically educative.

Finally, working in environments where cooperation is a requirement imposed on students by a well-meaning teacher can make students resent the activity. When group learning is difficult, frustrating, or inefficient (as it frequently is), students might lose respect for others and come to distrust social problem solving. They might actually become less disposed toward sharing, cooperating with, or helping others. Thus, the question to be considered is, When and how can we encourage group learning in schools so as to increase respect for, and the ability to work effectively with, others? What are educationally productive forms of group interaction?

RATIONALE FOR GROUP LEARNING IN FOSTERING SOCIAL SKILLS

Dewey (1916) argued that the formation of a democratic public sphere depends on people being able to identify common concerns and to extrapolate the consequences of actions they may take, as a community, to address them. He is also well known for arguing that the classroom should be a microcosm of the kind of society we want to create. This background helps us consider the relation of group learning to alternative visions of social interaction. As we have argued, group learning can mean several things, including pooling separate sources of information, negotiating different understandings to achieve some consensual understandings, team modes of investigation, peer tutoring, or "divide-and-conquer" strategies of subdividing complex tasks into components that can be carried out largely separately, then merged into an integrated solution to the problem.

From a democratic standpoint, these different styles of group learning reflect different models of cooperation in the social sphere, but they also reflect different avenues of access to that sphere, avenues that will have varying appeal to, and tolerance for, different sorts of people. As discussed previously, not every student takes equally well to the rough-and-tumble of cooperative investigation, debate, negotiated understandings, and so on. A democratic approach to group learning needs to respond to such diversity by finding alternative routes of access to knowledge, while still maintaining the value of communicative interchange among the different participants in that process.

However, changing classroom norms to accommodate such democratic values raises considerable difficulties (Cohen 1986). For example, traditional classroom norms, including methods of evaluation, mitigate against group work; these norms do more to encourage listening to the teacher and working alone. Successful group learning projects generally emphasize the process of investigation over "getting the right answer," yet most of the messages surrounding school learning socialize students into ends-oriented motivations.

Group learning projects often involve tasks in which there is no right answer or where multiple answers might be acceptable, yet we do very little in schools to prepare students for such problems. School problems are typically less complex and ambiguous than the problems generally encountered by society.

Some educators have addressed these school problems. Mann (1991), in her programming classes, teaches students questioning strategies and emphasizes that group members have the responsibility to answer the questions of others. Her approach allows all students to ask questions and obtain answers. In this sense, she builds on research by Webb (1989) that demonstrated the effect of questioning in groups. Only those students who asked questions and obtained answers benefited from group work.

One of the primary arguments for group interaction is that it helps students develop social skills, but, without guidance, students may adopt restrictive and status-based norms of classroom interaction. When students choose their collaborators, they tend to select friends and follow agreed-upon social behaviors. When groups are assigned, teachers report serious problems with dysfunctional social interaction (e.g., Cohen 1986). As a result, most advocates of group learning recommend explicitly teaching students social interaction skills (Cohen 1986; Johnson and Johnson 1979; Slavin 1987a). To reduce conflict, educators recommend group arrangements that focus primarily on social skills or that capitalize on acceptable social behavior. Unfortunately, if a teacher wants to implement group work, her efforts to impart social skills for group interaction may overshadow the advantages of group work. Teachers may spend all their available time just imparting social skills and have no time left for cognitive activities.

As Cohen (1986) suggests, often the goal of respecting others conflicts with the goal of accomplishing the best project. To come up with the best approach, the ideas of some students will be ignored while others will be promoted in importance. Talented leaders can ensure that the strengths of each group member are acknowledged, but students rarely have these leadership skills. Instead, good student leaders often go for compromise, thereby making sure that all the ideas are represented but not reaching the most ideal conclusion. Poor student leaders may permit group members to reinforce stereotypes and insult students of low status. For example, the CLP project found that students in four-person groups engaged in status-seeking behavior at least 30% of the time. Students regularly made comments such as, "Well, that's because you're stupid," or, "Girls can't do science anyway, so why bother?" Other group members condoned these comments, thereby reinforcing these stereotypes. As a result, Cohen recommends focusing group interaction on learning to respect each other rather than on accomplishing cognitive goals (but this creates the awkward trade off described earlier). Furthermore, such activities are most successful when addressed school-wide rather than limited to one course or one group project. Thus, the Child Development Project (1991) reports tremendous success with elementary schools that adopt a cooperative approach but require all of the teachers in a school to agree to participate.

Others, wishing to take advantage of group interaction without devoting extensive time to teaching social skills, have sought to minimize the need for social skills while maximizing the advantages of group interaction. Brown and Campione (Brown et al., in press) recommend "reciprocal teaching," a form of student tutoring to capitalize on group work, which we described earlier. Nix (1990) reports success when students volunteer for group work, engage in highly motivating tasks such as making their own videos, and choose their own collaborators. Linn (in press) advocates two-person groups and suggests that the CLP software can help monitor interaction among group members and therefore free the teacher to address cognitive issues.

MECHANICS OF GROUP LEARNING AND THEIR RELATION TO SOCIAL SKILLS

In another essay (Burbules and Linn, in press), we argue that science educators can benefit from studying the actual processes of scientific investigation and the view of science knowledge that such an understanding yields. Specifically, we argue that the history and sociology of science reveal processes of inquiry that are contextual, provisional, and imperfect; scientific discovery and theoretical development happen, as Kuhn, Lakatos, and others have shown, largely via the social interactions of communities of scientific investigation, not through bursts of brilliance by isolated "great individuals." There are lessons here for science education, and especially for our present purposes, for rethinking the social character of group learning in science.

Motivation in Cooperative Learning, Collaborative Learning, and Tutoring

We do know that for most students of all ages, opportunities to interact with peers is a powerful interest both inside and outside the classroom. This is true for practicing scientists; why not for science learners? One of the primary arguments for group learning is that by drawing on this intrinsic motivation, students will be attracted to learning activities because they provide an opportunity for social interaction, and they value the learning activity itself. In such contexts, as Dewey was famous for arguing, contexts of "work" and "play" become indistinguishable. On the negative side, competition as a motivation in group contexts can have ambivalent or even counterproductive effects.

Appropriation in Collaborative Learning

The process of hearing and responding to the ideas of others, as well as building additively to a collective understanding, provides more value than simply that of the knowledge gained. Students learn to appreciate the social processes by which knowledge claims can be enhanced through the contributions of others. Students can learn the limits of their own capacities and appreciate the strengths and insights of others. Unfortunately, as we have discussed, students may also find their ideas reduced and their contribution marginalized when status concerns override intellectual considerations.

Distributed Knowledge in Collaborative Learning

Similarly, when groups engage collaboratively in problem solving, a high degree of diversity in their outlooks and approaches to a problem can be beneficial. From the standpoint of social skills, this means not only that students acquire new information and perspectives but that they can also gain a respect for diversity and for the potential benefits of even highly unconventional vantage points on a problem. However, such interactions require the kind of mutual respect that has been difficult to achieve in classrooms.

Negotiation of Understanding in Collaborative Learning

If there is one conclusion supported from historical and social studies of science, it is that the mechanisms by which scientific claims are postulated, criticized, and modified are fundamental to what "counts" as knowledge at any given time, and that, in general, the extent to which scientific investigators submit themselves to such processes, the better the condition of understanding that results. Students who come to participate in and appreciate such processes, therefore, benefit not only by having more sound beliefs but by having a more accurate understanding of how scientific knowledge actually emerges. Furthermore, by participating in and appreciating such processes, students also acquire a mode of discursive interaction that is more reasonable and open-minded (Siegel 1988). Currently our research shows students generally believe that science knowledge is "static," or established, rather than "dynamic," or progressing and open to revision (Songer and Linn, in press), so they may not gain insight into the nature of science knowledge from group processes.

Research in group decision making, however, reveals that groups often engage in unproductive collaboration because of subtle social constraints (Janis 1972, 1982a, 1982b, 1982c). In particular, highly cohesive groups often come to prefer the camaraderie of the group to rigorous analysis of each other's ideas; as a result, they may fail to criticize alternatives and avoid presenting controversial issues.

Monitoring of Progress and Providing Hints or Feedback in Tutoring

There are reciprocal social benefits in tutoring situations that arise when students monitor each other's behavior. Levin (Levin et al. 1983) reports that this is common when one student controls the keyboard, and another keeps track of group contributions. When this succeeds, the processing demands of a situation are split among group members.

Divide and Conquer

One of the important outcomes of group activities, in any area, is the potential for specialization and efficiency by separating tasks that can be carried out simultaneously. Often the biggest problem in this process is identifying and coordinating the activities that can work together to produce creative solutions; this planning and coordination often requires leadership. Learning when to step forward with one's own ideas (which can be counterproductive when everyone tries to do it at once) and when to let some other individual within the group take direction are important social skills to master, apart from their particular learning benefits. Similarly, once the pieces of an investigative project are developed, the group often needs help in integrating them and making sense of the information. Frequently, as in examples discussed earlier, this task can best be carried out by an individual who formulates a way of synthesizing the information, then brings it back to the group for consideration.

Thus, the potential for exercising social skills arise regularly in group learning. Without guidance, however, these opportunities may foster the construction of authoritarian, arbitrary, competitive, or stereotyped social interaction patterns. To take advantage of group learning to impart productive societal skills requires carefully designed instruction; however, such instruction, valuable as it is in itself, can siphon away time from the subject matter at hand. Is group work worth the risk and effort? The relative balance of advantages and disadvantages is far from clear.

How Can Group Learning Impart Workplace Skill?

Workplace success involves the social and cognitive skills already discussed, including working effectively with others, taking the appropriate role of leader or follower depending on the situation, dealing with "difficult" people, and learning the norms and expectations of the group. Learning such workplace skills is certainly a goal of schooling. However, we believe that in many discussions of the relation of schooling to workplace skills, the cart had pulled the horse; by focusing on what current work conditions require, the previous two sets of goals (namely, the cognitive and social skills we generally desire for students) have been neglected, especially for broad student populations. When discussion of cognitive and social educational aims takes place only in the context of whether they serve workplace needs, in our view, this reverses the priority of things.

Education for the workplace, aside from some narrow vocational skills (which are less and less important to many jobs), primarily refers to a set of cognitive and social characteristics that presumably are of general benefit, *aside* from whether they specifically serve the needs of the workplace. Justifying them in terms of workplace utility deflects attention from more important aims and priorities in schooling. Instead, we should be asking what constitutes an educated person; we should be concerned about the qualities necessary for democratic citizenship, for a rich cultural and aesthetic life, for an appreciation of the natural environment, for good friendships and family relations, for an active sense of participation in a broader human community, and so on. Preparation for the workplace, if conceived too narrowly and vocationally, actually interferes with the attainment of these broader cognitive and social goals. Finally, given the rapidly changing nature of the workplace, any present set of objectives is likely to be, in many ways, obsolete by the time students actually enter the adult workplace (and many students at the high school level are already gaining plenty of vocational training in their after-hours jobs). For these reasons, as many have argued, schools are better off aiming toward a broadly conceived set of cognitive and social dispositions that are *educationally* justified, in the expectation that these constitute the most flexible and adaptable workplace preparation that schools can provide.

There are obvious drawbacks to letting the methods and aims of schools be established entirely by demands of the workplace. The typical work environment hardly offers a model for schools to emulate. Preparing students for the workplace requires a focus on what should become functional group activities rather than a focus on current behaviors. In fact, studies in the workplace reveal a broad range of dysfunctional groups (Janis 1972, 1982a, 1982b, 1982c). Authoritarian styles of interaction, subtle coercion to adopt group norms, and complacency about risk taking are all elements of the current workplace. Rather than supporting these practices, schooling should seek to overcome them.

The question, then, is how to design group and individual learning experiences with an awareness of the needs of the workplace, without compromising other, more valuable, educational purposes in the process.

Rationale for Group Learning in Fostering Workplace Skills

Recent interest in the teamwork and group processes found in successful Japanese companies have spurred educators to emphasize these skills in schools. Reformers have called on educators to prepare students for the workplace by imparting skills in working cooperatively (National Commission on Excellence in Education 1983). It is a shame, perhaps, that it has taken our current anxieties about international competitiveness to draw attention to reforms that should have been considered long ago, but, for whatever reason, modes of group learning that promote cooperation are now given greater prominence in school planning.

A profound difference between American and Japanese schools concerns the emphasis on respect for others and a sense of responsibility for group work. When these values are inculcated from preschool, students and adults agree on the standards for group work and status-seeking behavior is minimized (e.g., Stevenson 1990). Considerable evidence suggests that few U.S. communities establish these desirable working relationships involving group learning. Students need practice in working in a group and in completing the group activity. For example, if the activity is a science experiment, the group needs ideas for the design, implementation, and interpretation of the results. To work as a group rather than leaving the whole task to one individual, students also need skill in communication, leadership, respect for each other, and responsibility for joint tasks. Aligning all of these skills in one group is a considerable challenge.

In addition to the vocational aspect of group learning, proponents of the apprenticeship model of learning (e.g., Collins et al. 1989) argue that group learning is more "authentic" because it more closely approximates workplace activities and adult learning. They argue that enculturation into a community with shared goals, skills, and values is more motivating and instructive than the "contrived" activities in traditional classrooms (Dewey 1915; Lave and Wenger 1991). Similarly, learning about the models that students observe adults using also has some intrinsic motivational appeal. Cohen (1986) refers to group learning as a more "adult form of learning" that is neglected in precollege instruction.

WHAT ARE THE "AUTHENTIC PROCESSES" WE INTEND FOR STUDENTS TO IMITATE IN GROUP LEARNING?

A careful examination of actual work environments, if these are to provide a model for classrooms, reveals ambiguous evidence on just how "cooperative" group processes actually are. Earlier, we reviewed our arguments about whether the investigative practices of real scientists in working environments might offer a good model for the redesign of a science classroom (Burbules and Linn, in press). Such activities include a range of investigative methods. Scientists look things up, approximate when necessary, copy results from others, jointly compose chapters, consult colleagues on technical details, and so on (aside from active investigative work in the laboratory). Much of the work of a research group occurs after a coordinator has planned the activities and doled out tasks; after this, much of the work consists of individuals autonomously completing parts of the project. The group may have participated in the initial brainstorming and may review progress and comment periodically; but joint problem solving is rare, and individuals may not be aware of all aspects of the undertaking.

Examining the comments of successful software developers, the book *Programmers at Work* (Lammers 1986) reveals that they have mixed views about working on teams or in groups. For example, Jonathan Sachs (the author of *Lotus 1-2-3*) remarks:

> I like to do the whole project, the design and implementation. I like to have control over all the pieces. I'm not really much of a team player. (169)

Charles Simonyi (author of Bravo for the Alto computer) believes that having more than one programmer work on a project does not necessarily lead to efficiency. He says:

> The actual amount of code produced per person is smaller, the more people there are writing the program. Consequently, the total code produced is greater for awhile and then it may actually decrease. With two people, you might get 50% more code per unit of time. By the way, the efficiency of the code also decreases with an increase in the number of people working on the program. The most efficient programs are written by one person. (17)

Bill Gates (CEO of Microsoft Corp. and the author of *Microsoft BASIC*) says:

> It's kind of painful sometimes if you have somebody else working on the project. They never code stuff exactly the way you like to see it coded. I remember when we were working on BASIC, I'd go back and recode other people's sections of code, without making any dramatic improvement. That bothers people when you go in and do that, but sometimes you just feel like you have to do it. (76)

Gates goes on to say:

> It's nice to have someone who's up to speed to talk to when you're debugging code or you aren't sure about some particular tradeoff. In a sense, it's a way of taking a break, relieving the intensity without having to switch topics, just going in and

discussing it with somebody . . . . Paul and I learned how to work together in an effective way. You don't find too many partnerships like that. (84)

John Page (author of *PFS:FILE*), comments:

If the product requires four of five people, then you go about it differently than if you write it by yourself. I believe very strongly that one person, certainly no more than two, should do the design and high-level structure. You get a consistency and elegance when it springs from one mind. You can be misled by trying to make it fun for everybody and have a committee design the thing. That is absolutely fatal. (98)

As these quotes suggest, group work in "authentic" software design environments is primarily of the "divide-and-conquer" form. Group work can slow the process. Of concern is the size of the project carried out by an individual. There are projects that most believe can only be done by one person working incessantly, as reported in *The Soul of the New Machine* (Kidder 1981). One person does the high-level plan for a new computer with reviews by others. Individuals do their part but do not take much interest in what everyone else does. After the "Soul" team divided up the design of a new machine, large parts were done by individuals working for months with little interaction with others. Those implementing the hardware and software are particularly oblivious to issues of compatibility with the product line and marketing, even though these are crucial to the success of their venture.

Nevertheless, there can be advantages to modeling science classrooms on certain aspects of the organization and activities of such real scientific work settings. Enculturation into a scientific research community involves learning such group-constructed norms as standards of discourse, rules of evidence, stances toward knowledge, and the appropriateness of methodologies, including modeling, experimenting, and theorizing. However, such enculturation into the processes and norms of the scientific research community does not happen spontaneously in group activities; it involves extensive planning in a structured classroom setting. Otherwise, as noted earlier, negative and counterproductive aspects of group dynamics can take hold in the classroom (as they do in other work settings): favoritism, discrimination, plagiarism, secrecy, and inefficiency (e.g., Treweek 1988).

One significant difference between workplace experiences and classrooms is that, in the workplace, individuals often experience only certain aspects of project activities, while, in science classes, the goal is typically to impart an understanding of the whole process. How might this be accomplished? Rotating roles so all students experience each activity is one approach. The Computer as Lab Partner project tried this approach by assigning students to roles such as group leader, experiment coordinator, and the like. As expected, students differed in their enthusiasm for each role and in their success in implementing each role. Some students were experienced in leadership and respected by their peers. They succeeded as leaders. Others found it impossible to lead a group because their efforts were impeded by others. These students reported being criticized and insulted by group members. Stereotypical beliefs about who could succeed in science were apparent, and those with less status had greater difficulty performing the role of leader.

Thus, the classroom situation provided an authentic workplace experience, but part of this "authenticity" was that some students became convinced that they were unwelcome in science. Here, again, attention to developing skills and dispositions of social interaction must accompany a concern with "getting the right answer" or "covering the material."

Another issue concerns the judgments of quantity of information, quality of information, and efficiency in gaining information quickly that need to be made in real contexts of investigation. While, in a learning environment, it is easy to think that more information equals more learning and that more is always better, in real-world situations, this is rarely the case. More often the purpose at hand, time and resource constraints, and even the feelings one has toward one's co-workers constrain group learning activities. In what ways are such factors appropriate to "authentic" group learning activities in the classroom? Specifically, are schools prepared to set aside quantitative standards of how much students learn in order to move other educational objectives to the forefront? Current trends here are not encouraging.

As a result of these considerations, we believe, all of the educationally significant aims of workplace preparation already stand under the heading of broader cognitive and social aims. It is an error to set a separate set of vocational goals over and apart from these; when this has been done, it is invariably the more educationally significant goals that have suffered. While there are aspects of group learning that do anticipate and prepare students for aspects of their future adult work environments, there are also drawbacks to such experiences (just as there are drawbacks to those work environments themselves). Status conflicts, competitiveness, recalcitrant colleagues, and so on, are part of the "real world" of work, but this is not a good reason for endorsing their role in the school classroom.

"Authenticity," in the sense of mirroring the actual processes of work, therefore, is not a good reason in itself for interfering with educational priorities. The account we have provided here, of identifying the cognitive and social goals we mean to serve primarily through group learning, is meant to show how certain group learning activities often conflict with authentic practices. The educational choices we face should address alternative goals and incorporate styles of group learning activities in light of a broader and more inclusive range of priorities.

CONCLUSIONS

In summary, each of the three common arguments for group learning (as scaffolding the construction of knowledge, as a means of fostering certain values important to democratic living, and as a mode of socialization into the habits and dispositions of adulthood, including those of work) represent ambiguous justifications at best.

We need to ask how different group learning approaches succeed, what knowledge they help students construct, and for whom they are appropriate. Furthermore, whatever the merits of group learning, this approach should not drive other instructional strategies off the agenda. Rather than choosing between explicit instruction, guided construction, or autonomous learning, we argue that all are components of effective teaching. In light of these considerations, what conclusions can be drawn about the merits of group learning, particularly for the goal of constructed knowledge? We gather our comments under several broad questions that recap the issues discussed in this chapter.

What are the different types of group learning?

First, we need to remember that group learning comprises at least three very different sorts of activities, each with its own merits and limitations. Of the three forms of group interaction identified earlier, tutoring is the most compatible with current classroom norms for behavior and, therefore, the most commonly implemented. *Tutoring* establishes status differences on the basis of expertise or authority. In many classrooms, tutors are selected from those students having high test scores and are accorded status in the classroom based on expertise. The situation is more equitable in reciprocal teaching where students take turns being the tutor and acquire tutoring skills by modeling the behavior of an adult teacher. Since all the students are trained to be tutors, each student can monitor the behavior of the others.

Cooperative learning can also be partially implemented within current classroom norms. For example, in the CLP classroom, the teacher determines how tasks will be divided so that student leaders are not needed to assign responsibilities. There are still difficulties in four-person groups, however, when students fail to perform group tasks. In two-person groups, the strong influence of one's partner seems to increase the likelihood that each student will perform the assigned tasks.

Collaborative learning requires the most well-developed social skills for success, as well as the most substantial change in classroom norms for success. In collaborative situations, students have the most opportunity to show disrespect for the ideas of others. Furthermore, when students fail to respect the ideas of others, they may ignore or reject data or other information that might change their conclusions. Collaboration succeeds when students are effective at communicating their ideas and able to help other group members see why their idea contributes to the group goal. It also depends on group adherence to a form of discourse that values argument, reliance on evidence, and explanation. Large collaborations require someone to make sure that all participants are included.

Who benefits from group learning?

In assessing the advantages of group learning, we must also ask whether all students benefit equally. Some students eschew group interaction and autonomously master complex skills, while others find considerable satisfaction in traditional assignments and are uncomfortable in the open-ended situations common in groups. Not all students stand to benefit from group learning activities, which can foster anxiety and fear of failure as well as motivating students to work for peer approval (Shneiderman 1980).

Groups often reinforce status differences and stereotyped behavior rather than alleviating these impediments to effective learning (Cohen 1986). In a discussion of the role of group activities in altering stereotypes, DeVillar and Faltis (1991) conclude:

> It is therefore imperative to underscore that physical proximity,
> a common curriculum, and cooperatively based learning meth-

ods, while necessary, are in themselves not sufficient to change exclusionary attitudes based on race or ethnicity to inclusionary ones. (60)

In the context of gender, Sadker et al. (1991) argue similarly:

The different and contradictory findings of the relatively few studies analyzing cross-gender performance in cooperative learning organizations suggest that, by itself, the implementation of cooperative learning groups does not necessarily lead to a more equitable and effective learning environment for females. (307)

Evidence shows that in science, and perhaps in other areas of the curriculum as well, group learning may be *less* equitable than autonomous learning for females. Group learning may reinforce stereotypes, biases, and views of science as a male domain. Male students may discredit females, and the classroom may become a microcosm of the "old boys" network that has frequently discouraged women and minorities from participating in science in general.

More subtle group processes may also determine who benefits from working with other students to solve problems. Webb (1989) reports that students who ask questions *that are answered* benefit from group learning. What of those students who are ignored by the group and do not get their questions answered? Similarly, Webb indicates that those who are able to answer questions learn in group situations. What happens to students who lack the knowledge to answer questions asked by others? It is surprising that, in basing a model of "authentic" learning on the dynamics of group processes in other segments of society, we seem to neglect the aversive aspects of group activities that we all know from personal experience.

Besides asking *who* stands to benefit from group learning situations, we must also consider *when* a student may be ready to engage in particular group learning activities and derive benefits from them. Both Vygotsky's "zone of proximal development" and Dewey's "teachable moments" assume that certain challenges foster learning only when they fall within a conflux of motivation, developmental readiness, interest, and attention for the learner. Group learning activities are as subject to these sorts of considerations as are any other sort of learning activity (and like the others, imposition before a student is prepared to benefit from them can be sharply counterproductive).

Thus, we note that the benefits of group processes are anything but uniform from student to student. To appreciate this form of learning properly, we must look at it from the standpoint of the individual learner as well as from the standpoint of the group. The norms of group behavior may require substantial change before group learning benefits the whole range of students. Without such changes the group process may further advantage the few while contributing to the disillusionment of the many.

Does group learning foster cognitive knowledge construction?

Unguided and explicit group learning can each foster the construction of certain forms of knowledge in certain learners. Unguided group activity can lead to knowledge construction in highly motivating, open-ended situations such as described by Nix (Nix and Spiro 1990), and in voluntary collaborative situations such as joint authorship of a chapter or joint construction of a product (e.g., Pea, in press). Successful collaborations are characterized by (1) volunteer participants who presumably have the interpersonal skills necessary for success, (2) activities students are highly motivated to perform, such as creating a video, and (3) evaluation standards that are set by the participant rather than by authority. Explicit activities such as reciprocal teaching help groups of learners construct understanding of well-specified material as well as helping them construct productive comprehension strategies. This form of tutoring is more efficient as a vehicle for fostering knowledge construction than is collaboration. Skill in tutoring does not, however, generally result from unguided discovery. If tutored construction is desired, it is best to help those who will be tutoring to develop skills in fostering knowledge construction and to orchestrate the group activity such that it exercises the desired cognitive skills.

Explicit constructive activities may also fail if no group member has the desired knowledge (Brown, et al., in press; Linn and Songer, in press-b) or if group norms prevent consideration of certain views (Chi and Van Lehn, in press; Janis 1972, 1982a, 1982b, 1982c). Self-explanations suffer in the same way. Instruction should be designed to ensure that groups have necessary resources, can access appropriate feedback, can engage in healthy debate, and can monitor progress at regular intervals.

Does group learning foster democratic ideals?

Unguided group learning tends to recapitulate social roles and stereotypes so that those with high status within the group stand to benefit disproportionately. In some cases, group learning activities can be designed to teach students to respect each other and to overcome stereotypes. In other cases, the best solution is to design groups and activities to reduce the effects of status differences and to inhibit stereotyped behavior (see Cohen 1986). These influences on group interaction are not limited to classrooms; they are often even more powerful in informal group situations. For example, clubs still discriminate against certain racial and ethnic groups and gender groups. One approach to lessening status effects is to design activities that vary the expertise required for solution and, therefore, vary the status rankings of the participants.

Group learning will also be unproductive for certain learners who have dysfunctional views of group interaction. In many situations, the best response is to teach group interaction skills and discourse skills (Cohen 1986; Janis, 1972, 1982a, 1982b, 1982c; Slavin 1987a). In other situations, the best response is to give students the choice to work alone, rather than with the group.

Does group learning foster workplace skills?

Workplace skills involve the respect for others (characteristic of democratic ideals) as well as the ability to take the role of leader or follower depending on the situation. Another workplace skill is learning to work with difficult people. Finally, workplace success depends on seeking and getting feedback concerning the norms of behavior and expectations of the group.

These are important potential goals for group learning that require innovative programs in order to be accomplished; frequently, however, group learning activities fail to incorporate them.

In spite of these drawbacks, there is a growing need for groups to learn to work together, as in contemporary science where we find increasingly large and complex projects involving many participants. From software design to the human GENOME project or the space shuttle, in all branches of science, large collaborations are common. Although group learning skills are neglected in school, they are more and more necessary in the workplace.

Yet our discussion suggests that both classrooms and workplaces also feature many examples of unproductive group interaction. Unguided group activity in classrooms may reinforce unproductive workplace behaviors such as refusal to follow the designated leader, adherence to exclusionary norms, and reliance on stereotyped beliefs. In order for group learning to foster workplace skills, substantial intervention is needed.

Successful collaborations are more common among small groups when the goals are mutually beneficial and when group members contribute diverse skills. Thus, small numbers of investigators can successfully collaborate to produce projects. For example, there is a growing number of educational innovations such as the CLP project that have resulted from the collaboration of classroom teachers, natural scientists, technology experts, and educational researchers. These collaborations involve mutual respect, distributed knowledge, and common goals. Implementing similar collaborations in classrooms would be appropriate for some educational goals but not others.

Is group learning equally effective for all educational goals?

Advocates of group learning often assume that it is the best approach for learning cognitive skills, democratic ideals, and workplace practices. But why should we assume that the same teaching and learning activities necessarily serve all three? Claims that all these goals can be met at one and the same time are overly optimistic. More commonly, educational methods achieve some aims better than others and force us to choose among competing valued ends. Specifically, we have suggested, fostering cognitive and social benefits sometimes conflict, and fostering "authentic" workplace skills and dispositions often conflicts with both of these.

Group learning does seem uniquely suited to imparting democratic ideals. Constructing a view of one's self as respectful of others and able to work under different social arrangements is a democratic ideal that can be fostered by well-designed group learning opportunities. Such instruction should seek to ensure that individuals respect all group members. In well-planned groups, each participant could count on others in the group to help answer questions and monitor performance. When groups succeed in operating according to democratic ideals, they also have the opportunity to co-construct understanding. Groups can be especially effective at brainstorming and at helping each other solve problems. Even groups following democratic ideals may jointly construct unproductive ideas or compromise on a poorly integrated plan for a problem solution.

In this chapter, we have suggested that while group learning does offer mechanisms that can foster specific cognitive and social outcomes, it is not equally useful for all of them, or for all of them at the same time. Trade offs and priorities continually vie for attention. Moreover, we have suggested that where workplace skills have become the focal point, educationally significant cognitive and social outcomes of group learning have often been either neglected or pursued for the wrong reasons. There is a danger in framing social and cognitive goals primarily in terms of workplace skills when they are, in fact, *educational* goals that happen to have relevance to the workplace, as well as to many other areas of a full and effective human existence.

REFERENCES

Azmitia, M., and M. Perlmutter. "Social Influences on Children's Cognition: State of the Art and Future Directions." *Advances in Child Development and Behavior*, 22, (1989): 89–144.

Brown, A.L., and J.C. Campione. "Interactive Learning Environments and the Teaching of Science and Mathematics." In M. Gardner, et al. (Eds.), *Toward a Scientific Practice of Science Education*. Hillsdale, NJ: Erlbaum, 1990. 111–140.

Brown, A.L., J.C. Campione, R. Reeve, R.A. Ferrara, and A.C. Palincsar (in press). "Interactive Learning and Individual Understanding: The Case of Reading and Mathematics." In L.T. Landsmann (Ed.), *Culture, Schooling and Psychological Development*. Hillsdale, NJ: Erlbaum.

Brown, A.L., and A.S. Palincsar. "Guided, Cooperative Learning and Individual Knowledge Acquisition." In L.B. Resnick (Ed.), *Knowing, Learning, and Instruction: Essays in Honor of Robert Glaser*. Hillsdale, NJ: Erlbaum, 1989. 393–451.

Bruner, J.S. *Child's Talk*. New York: Norton, 1976.

Burbules, N.C., and M.C. Linn (in press). "Science Education and the Philosophy of Science: Congruence or Contradiction?" *International Journal of Science Education*.

Burbules, N.C., and P. Reese. *Teaching Logic to Children: An Exploratory Study of "Rocky's Boots"*. Berkeley, CA: University of California, Lawrence Hall of Science, Assessing the Cognitive Consequences of Computer Environments for Learning (ACCELL), 1984.

Chi, M.T.H., and K.A. Van Lehn (in press). "The Content of Physics Self-explanations." *Journal of the Learning Sciences*.

Child Development Project. *Evaluation of the Child Development Project: Summary of Findings to Date*. San Ramon, CA: Developmental Studies Center, 1991.

Clement, J., D. Brown, and A. Zietsman. "Not all Preconceptions are Misconceptions: Finding "Anchoring Conceptions" for Grounding Instruction on Students' Intuitions." *International Journal of Science Education*, 11, (1989): 554–565.

Cohen, E.G. *Designing Groupwork: Strategies for the Heterogeneous Classroom*. New York: Teacher's College Press, 1986.

Collins, A., J.S. Brown, and S.E. Newman. "Cognitive Apprenticeship: Teaching the Craft of Reading, Writing, and Mathematics." In L.B. Resnick (Ed.), *Cognition and Instruction: Issues and Agendas*. Hillsdale, NJ: Erlbaum, 1989. 453–494.

Damon, W., and E. Phelps. "Critical Distinctions Among Three Approaches to Peer Education." *International Journal of Educational Research*, 13(1), (1989): 9–19.

de Bono, E. *Lateral Thinking: Creativity Step by Step*. New York: Harper and Row, 1973.

DeVillar, R.A., and C.J. Faltis. *Computers and Cultural Diversity: Restructuring for School Success*. Albany: SUNY Press, 1991.

Dewey, J. *The School and Society* (rev. 1961, 2d ed.). Chicago: University of Chicago Press, 1915.

Dewey, J. *Democracy and Education: An Introduction to the Philosophy of Education*. New York: MacMillan Publishing Co., 1916.

Dewey, J. *The Child and the Curriculum*. Chicago: University of Chicago Press, 1920.

Dockterman, D. "The Truth about Cooperative Learning." *CUE NewsLetter*, 12(5), (1990): 1–8.

Doise, W., and G. Mugny. *The Social Development of the Intellect*. New York: Pergamon, 1984.

Elshout, J. (in press). "Formal Education Versus Everyday Learning." In E. De Corte, L. Verschaffel, M.C. Linn, and H. Mandl (Ed.), *Computer-based Learning Environments and Problem Solving*. Brussels, Belgium: Springer-Verlag.

Eylon, B., and M.C. Linn. "Learning and Instruction: An Examination of Four Research Perspectives in Science Education." *Review of Educational Research*, 58(3), (1988): 251–301.

Forman, E.A. (in press). "Discourse, Intersubjectivity and the Development of Peer Collaboration: A Vygotskian Approach." In L.T. Winegar and J. Valsiner (Ed.), *Children's Development Within Social Context*. Hillsdale, NJ: Erlbaum.

Fraser, B.J. "Research Syntheses on School and Instructional Effectiveness." *International Journal of Educational Research*, 13, (1989): 707–718.

Grice, H.P. "Further Notes on Logic and Conversation." In P. Cole (Ed.), *Syntax and Semantics 9: Pragmatics*. New York: Academic Press, 1978. 113–128.

Janis, I.L. *Victims of Groupthink: A Psychological Study of Foreign-policy Decisions and Fiascoes*. Boston: Houghton Mifflin Co., 1972.

Janis, I.L. "Counteracting the Adverse Effects of Concurrence-Seeking in Policy-Planning Groups: Theory and Research Perspectives." In *Stress, Attitudes, and Decisions*. New York: Praeger Publishers, 1982a. 291–308.

Janis, I.L. "Effective Interventions in Decision Counseling: Implications of the Findings from Twenty-three Field Experiments." In *Stress, Attitudes, and Decisions*. New York: Praeger Publishers, 1982b. 217–257.

Janis, I.L. "Group Identification Under Conditions of External Danger (1963)." In *Stress, Attitudes, and Decisions*. New York: Praeger Publishers, 1982c. 93–111.

Johnson, D.W., and R. T. Johnson. "Conflict in the Classroom: Controversy and Learning." *Review of Educational Research*, 49, (1979): 51–70.

Kidder, T. *The Soul of a New Machine*. New York: Avon Books, 1981.

Kulik, J.A., and C.C. Kulik. "Meta-analysis in Education. *International Journal of Educational Research*, 13, (1989): 221–340.

Lammers, S. *Programmers at Work: Interviews*. Redmond, WA: Microsoft Press, 1986.

Lave, J., and E. Wenger. *Situated Learning: Legitimate Peripheral Participation*. Cambridge: Cambridge University Press, 1991.

Leonard, M. (in press). "Learning the Structure of Recursive Programs in Boxer." *Journal of Mathematical Behavior*.

Levin, J.A., M. Boruta, and M. Vadconcello. "Microcomputer-based Environments for Writing: A Writer's Assistant." In A. Wilkinson (Ed.), *Classroom Computers and Cognitive Science*. New York: Academic Press, 1983.

Levinson, S.C. *Pragmatics*. New York: Cambridge University Press, 1983.

Linn, M.C. "The Computer as Lab Partner: Can Computer Tools Teach Science?" In L. Roberts, K. Sheingold, and S. Malcolm (Eds.), *This Year in School Science 1991*. Washington, DC: American Association for the Advancement of Science, 1991.

Linn, M.C., and M.J. Clancy (in press). "Can Experts' Explanations Help Students Develop Program Design Skills?" *International Journal of Man-Machine Studies*.

Linn, M.C., and N.B. Songer (in press-a). "Cognitive and Conceptual Change in Adolescence." *American Journal of Education* (Special Issue: Development and Education Across Adolescence).

Linn, M.C., and N.B. Songer (in press-b). "Teaching Thermodynamics to Middle School Students: What are Appropriate Cognitive Demands?" *Journal of Research in Science Teaching*.

Linn, M.C., N.B. Songer, E.L. Lewis, and J. Stern (in press-c). "Using Technology to Teach Thermodynamics: Achieving Integrated Understanding." In D.L. Ferguson (Ed.), *Advanced Technologies in the Teaching of Mathematics and Science*. Berlin: Springer-Verlag.

Mann, L. (Personal Communication 1991). University of California, Berkeley. Berkeley, CA.

Meadow, A., and S.J. Parnes. "Evaluation of Training in Creative Problem Solving." *Journal of Applied Psychology*, 43(3), (1959): 189–194.

Miller, G.A. *Spontaneous Apprentices: Children and Language*. New York: Seabury Press, 1977.

Minstrell, J. "Explaining the 'at rest' Condition of an Object." *The Physics Teacher*, 20, (1982): 1014.

National Commission on Excellence in Education. *A Nation at Risk: The Imperative for Educational Reform*. Washington, DC: U.S. Government Printing Office, 1983.

Newman, D. "Using Social Context for Science Teaching." In M. Gardner, et al. (Eds.), *Toward a Scientific Practice of Science Education*. Hillsdale, NJ: Erlbaum, 1990. 187–202.

Nix, D. "Should Computers Know What You Can Do With Them?" In D. Nix and R. Spiro (Eds.), *Cognition, Education, & Multimedia*. Hillsdale, NJ: Erlbaum, 1990. 143–162.

Nix, D., and R. Spiro. (Eds.). *Cognition, Education, and Multimedia*. Hillsdale, NJ: Erlbaum, 1990.

Pea, R. (in press). "Augmenting the Discourse of Learning with Computer-based Learning Environments." In E. De Corte, M.C. Linn, and H. Mandl (Eds.), *Computer-based Learning Environments and Problem Solving*. Berlin: Springer Verlag.

Phelps, E., and W. Damon. "Problem Solving with Equals." *Journal of Educational Psychology*, 81(4), (1989): 639–646.

Piaget, J. *Science of Education and the Psychology of the Child*. New York: Orion Press, 1970.

Robinette, W. *Rocky's Boots*. Fremont, CA: The Learning Company, 1982.

Sadker, M., D. Sadker, and S. Klein. "The Issue of Gender in Elementary and Secondary Education." In G. Grant (Ed.), *Review of Research in Educa-*

tion. Washington, DC: American Educational Research Association, 1991. 269–334.

Salomon, G., and T. Globerson. "When Teams do not Function the way They Ought to." *International Journal of Educational Research*, 13(1), (1989): 89–99.

Schoenfeld, A. H. "Ideas in the Air: Speculations on Small Group Learning, Environmental and Cultural Influences on Cognition, and Epistemology." *International Journal of Educational Research*, 13(1), (1989): 17–88.

Sharan, S., and Y. Sharan. *Small-group Teaching*. Englewood Cliffs, NJ: Educational Technology Publications, 1976.

Shneiderman, B. *Software Psychology: Human Factors in Computer and Information Systems*. Cambridge, MA: Winthrop Publishers, Inc., 1980.

Siegel, H. *Educating Reason: Rationality, Critical Thinking, and Education.* New York: Routledge, 1988.

Slavin, R.E. *Cooperative Learning*. New York: Longmann, 1983.

Slavin, R.E. "Cooperative Learning: Where Behavioral and Humanistic Approaches to Classroom Motivation Meet." *Elementary School Journal,* 88, (Special Issue: Students' Models and Epistemologies of Science). (1987a): 29–37.

Slavin, R.E. "Grouping for Instruction in Elementary School." *Educational Psychologist*, 22, (1987b): 109–122.

Songer, N.B. *Promoting Integration of Instructed and Natural World Knowledge in Thermodynamics*. Unpublished doctoral dissertation, University of California, Berkeley. Berkeley, CA, 1989.

Songer, N.B., and M.C. Linn (in press). "How do Students' Views of Science Influence Knowledge Integration?" *Journal of Research in Science Teaching* (Special Issue: Students' Models and Epistemologies of Science).

Stevenson, H.W., S.Y. Lee, C. Chen, J.W. Stigler, and S. Kitamura. "Contexts of Achievement: a Study of American, Chinese, and Japanese Children." *Monographs of the Society for Research in Child Development*, 55(1;2), 1990.

Treweek, S. *Beam Times and Lifetimes: The World of High Energy Physicists*. Cambridge, MA: Harvard University Press, 1988.

Von Fange, E.K. *Professional Creativity*. Englewood Cliffs, NJ: Prentice Hall, 1959.

Vygotsky, L.S. *Thought and Language*. Cambridge, MA: MIT Press, 1962. (Originally published in Russian in 1934).

Vygotsky, L.S. *The Collected Works of L. S. Vygotsky: Volume 1, Problems of General Psychology*. New York: Plenum, 1987.

Webb, N.M. "Peer Interaction and Learning in Small Groups." *International Journal of Educational Research*, 13(1), (1989): 21–39.

Wertsch, J.V. *Vygotsky and the Social Formation of Mind*. Cambridge, MA: Harvard University Press, 1985.

7

The Role of Negotiation in Mathematics Learning

Grayson H. Wheatley

This chapter considers the nature of mathematics learning with particular attention to the role of negotiation. This issue rarely arises in direct instruction (Confrey 1990) since the teacher is viewed as an authority who prescribes how mathematics is to be done. However, when instruction that is informed by constructivism (von Glasersfeld 1987) is considered, negotiation plays a prominent role.

NEGOTIATION

Negotiation takes place in a social and physical environment that greatly influences the beliefs, intentions, stances, and actions of the participants. Thus, we cannot easily speak of negotiation in general, but only analyze negotiations within particular settings. Political negotiation is characterized by compromises. In negotiating mathematical meaning there is little room for compromise. Unless there are different interpretations of the task, there is usually one answer or one set of answers. This chapter describes a negotiation between two ninth grade boys who had participated in a mathematics program called "Mathematics Achievement through Problem Solving" (MAPS) (Wheatley, et al. 1988) designed for noncollege-intending students. The central instructional strategy of MAPS is problem-centered learning (Wheatley 1991). A goal of problem-centered learning is the construction of mathematical knowledge by students. A construction (1) can be explained and justified, (2) has internal consistency, (3) can be reflected upon, and (4) is embedded in other knowledge.

PROBLEM-CENTERED LEARNING

Problem-centered learning (PCL) has three components: tasks, groups, and sharing. Primarily, in preparing for class, a teacher selects tasks that have a high probability of being problematical for students (tasks that may cause students to find a problem). A rich mathematical activity has the following characteristics:

- is accessible to everyone

- invites students to make decisions

- encourages "what if" questions

- encourages students to use their own methods

- promotes discussion and communication

- is replete with patterns

- has a mathematical conclusion

- has an element of surprise

- is enjoyable

- is extendable

Secondly, the students work on these tasks in small groups, often extending the task or setting a new problem to solve. Successful small group problem solving requires considerable negotiation of social norms, the meaning of mathematics, and what it means to do mathematics. The negotiation process continues, although it may take a different form as the group works together over time. As social norms are negotiated, more attention can be focused on negotiating mathematical meanings. During small group problem solving, the teacher is engaged in conceptualizing the individual and group activity. It is particularly at this time that the teacher-as-researcher metaphor applies. Finally, the class is convened as a whole for students to present their solutions to the class for discussion. Groups present their solutions to the class, not to the teacher. The role of the teacher in these discussions is that of facilitator, and every effort is made to be nonjudgmental and encouraging. Interventions are made, but care is exercised to promote student autonomy.

Problem-centered learning contrasts sharply with Direct Instruction (Confrey 1990). Direct Instruction is characterized by (1) expecting relatively short products, (2) teachers executing their plans and routines without constructing models of the students' reasoning, and (3) the teacher being the sole evaluator of students' knowledge. Problem-centered learning is characterized by (1) autonomy and commitment, (2) negotiation of social norms, (3) facilitation of reflection, (4) recognition that meaning must be negotiated (knowledge cannot be transmitted), (5) teacher-building models of students' mathematical knowledge, (6) recognition of the inherently interactive nature of learning, and (7) students judging the viability of their own constructions. Problem-centered learning attempts to promote sense making and help students develop their autotelic selves (Csikszentmihalyi 1990).

NEGOTIATION WITHIN A PROBLEM-CENTERED LEARNING ENVIRONMENT

As students attempt to communicate mathematical ideas in a problem-centered learning environment, negotiation assumes prominence. However, the process of negotiation is quite complex. In order to successfully negotiate, two persons must put themselves in the position of the other, or as Sless (1986) says, "construct the other person." Actually, one cannot put oneself in the place of another, one can only construct, using one's limited experiences with the other person and others, an imperfect conceptualization of the person's world (Goodman 1984). The more viable this construction is, the more useful it will be in the process of negotiation. A successful negotiation is reached when the two persons have no further reason to believe their positions are different. This sentence is stated with a double negative because one can never know that one's separate constructions "fit" (von Glasersfeld 1984). One may reach, however, a point in communication in which one comes to believe one's constructions do fit, even though they are likely to be different since two persons almost never make exactly the same construction, just as two sets of genes do not result in a person like any other.

Negotiation also requires the intention to negotiate. Frequently, a person will appear to assume a nonnegotiatory stance. Such a stance may be taken for a variety of reasons. This chapter will discuss this matter in some detail. An intention of teachers in problem-centered learning is to facilitate the negotiation of social norms that foster negotiation of mathematics meaning.

One of the major goals in problem-centered learning is establishing a learning environment in which students form the goal of making sense of mathematical ideas. Historically, making sense of mathematical ideas has not been an intention of most students in American schools. The teacher often presents a method or procedure and the student is expected to practice the procedure until she can carry it out flawlessly. A student who attempts to make sense of mathematical ideas may find herself in conflict with peers, the teacher, and the school system. Attempting to make sense of a mathematical statement may take the learner in directions that do not fit the narrow expectations of the teacher. If making sense of mathematical ideas is a goal of the learner, then the teacher must act in ways that make possible viable constructions rather than practiced routines.

In Lo, et al. (1991a), a classroom episode is examined in which Peter made a presentation to the class and negotiation of mathematical meaning was not accomplished. Several class members verbally criticized him with accusations of mumbling, repeating himself, not speaking clearly, and in general, failing to make himself understandable. An analysis of this episode revealed that Peter had explained a method that required constructions many students had not made. Peter was quite clear on his solution method and could explain it to a knowledgeable person. Some students may have tacitly concluded that Peter was taking a nonnegotiatory stance by failing to consider the knowledge of individual class members; he seemed to assume what was clear to him would be clear to others. He may have inadvertently communicated the message: I know my way is right; if you don't understand, then it is your problem.

On the other hand, several vocally dominant students took a nonnegotiatory stance in considering Peter's explanation. They seemed to have abandoned the intention of making sense of mathematical ideas because of their construal of Peter's intentions. By volunteering to present his solution and then, in their view, failing to give a clear explanation, he had violated the class norm of honesty. The episode ended with Peter returning to his seat in disgust, feeling he had been treated unfairly and the class feeling Peter had not given a satisfactory explanation. Both sides in this situation assumed nonnegotiatory stances; thus, Peter's mathematical explanation could not be considered because a consensual domain (Richards 1991) was not established. While Peter was attempting to make an explanation, members of the class were trying to get him to follow what they perceived to be the rules of discussion.

This impasse may have been reached, in part, because the teacher, in trying to facilitate effective presentations, attempted to get students to agree to a set of conditions for a "good" presentation. If the class had been allowed to negotiate the social norms of class discussion, this conflict may have been avoided. Through negotiation, the members of the class could have come to agree on acceptable actions. They would not have criticized Peter for "mumbling" if they could have understood what he meant. Consequently, Peter would not have been offended by what he considered to be trivial criticism.

MATHEMATICS ACHIEVEMENT THROUGH PROBLEM SOLVING

Mathematics Achievement Through Problem Solving (MAPS) is a ninth grade general mathematics course designed for noncollege-intending students. Although the students may not have algebra and geometry as goals, MAPS is designed to provide a foundation upon which previously unsuccessful mathematics students can build as they elect to study college preparatory courses. Presently, the program is used in more than 100 high schools in seven states. The primary instructional strategy of MAPS is problem-centered learning. Emphasis is placed on problem solving, both as a strand and as a means of promoting learning in all strands. Five two-week problem-solving units are distributed throughout the year. A collection of specially designed nonroutine mathematics problems serve as the instructional material for these units.

There is an emphasis on learning to visualize mathematical concepts and relationships. It is well documented that imagery plays a significant role in mathematical understanding (Brown and Wheatley 1989; Presmeg 1985; Wheatley and Reynolds 1991; Zimmerman and Cunningham 1990). As Hallet (1990, 121) states, "Students who are using few mental pictures are not really learning mathematics." Activities in MAPS are designed to facilitate and encourage the development and use of imagery in mathematics. Geometry is approached through imagery and communicating images in words. Of course, there is an emphasis on drawing diagrams in problem solving. Each of the strands reflects an emphasis on imagery in building mathematics relationships.

In previous years, MAPS students have, typically, studied rule-oriented mathematics in which they are expected to become proficient with many procedures and facts. Consequently, most MAPS students have not found mathematics meaningful, and they try to memorize what they are asked to learn rather than make sense and relate the central ideas of the discipline. The presence of these students in a remedial course shows that the way they were taught previously was not effective. The level of mathematical knowledge of ninth grade general mathematics students is frighteningly low. Our research has shown that many students have not given meaning to such fundamental concepts as one-third, even though they have for years been assigned exercises which require the use of computational procedures with mixed numbers. In

MAPS, there are opportunities for students to construct schemas for whole numbers and fractions. They are encouraged to develop their own procedures for computing rather than being required to use an algorithm dictated by the teacher. The strands of MAPS are listed below:

- mental arithmetic and estimation
- heuristic problem solving
- whole number concepts
- rational number concepts
- spatial visualization
- statistics
- probability
- applied problem solving
- measurement
- intuitive algebra
- computer problem solving

SAM AND BRETT'S NEGOTIATION

The nature of negotiation in mathematics learning can be seen in the analysis of an exchange between Sam and Brett, two ninth grade students who were in a MAPS class. In designing and refining additional activities for MAPS, these two boys were asked to work on written tasks while being video recorded. The exchange was primarily between the two students with the teacher intervening or speaking only when necessary. The tasks were all sequence tasks similar to the example shown in Table 1.

Table 1. Typical sequence task presented in the interview.

7, ___ , ___ , 8, ___ , ___ , ___ , ___ , ___ , 10
What is the pattern? _____

After the students had worked on these tasks for approximately 20 minutes, the teacher posed a task that proved to be quite problematic. The interviewer chose the task because he saw that the students had unresolved differences on this type of problem. The task he presented is shown in Table 2.

Table 2. The task that generated the episode to be analyzed.

15, ___ , ___ , ___ , 16, ___ , ___ , ___ , ___ , ___
What is the pattern? _____

Brett initially writes in 15-1/4, 15-1/2, 15-3/4 with confidence. Sam rejects the 15-3/4 and asks the interviewer, "Is that right, Mr. Lane?" When Mr. Lane responds noncommittally, he continues to insist Brett's numbers are wrong because, "three-fourths is lower than a half." Sam has in mind doubling the fractional part. He knows that if he doubles 3/4 he won't get a whole number, and thus Brett's method must be wrong.

S: That's wrong . . .

B: No,

S: Three-fourths is lower than a half.

B: You add one-fourth every time. 15 to 15 and 1/4, you add a fourth, 15-1/4 to 15-1/2, add a fourth. Know what I mean?

S: You know it's right?

B: Yeah.

S: I don't think it is. No, but that's wrong. (Points to 15-3/4)

B: No, look, you add one-fourth each time, see? 15-1/4 and 1/4 is 15-1/2 . . . Know what I mean?

S: (Shakes head no) Do you know you're right?

B: Yeah.

S: I don't think it is, cause you got three-fourths. That's lower than a half.

B: No it ain't, that's higher than a half.

In this episode, notice that on two occasions Sam asks, "Do you know it's right?" He is attempting to judge the strength of Brett's commitment to his method and whether to continue opposing it. He has confidence in the doubling method, which he can use, and is not ready to abandon it for a method he does not understand. Brett presents reasons for his position, stating, "You add a fourth every time."

S: Yes, but we ain't supposed to be adding. One-fourth and one-fourth equal a half, half and half equal one. Half and a half won't equal three-fourths, . . . will it?

Sam's use of "supposed to" suggests his orientation toward an external authority rather than determining what makes sense. He rejects Brett's method because it does not follow the rule they are supposed to be using. This move is consistent with his actions throughout the interview. It is interesting to note that it was Brett who initially used the doubling strategy but later began to question it as he had to write the pattern. Thus, Sam was using a rule developed by someone else. Brett was perturbed when "add a fourth" was written by Sam in filling in the line.

THE PATTERN IS _____ .

When using the doubling strategy, the same number is not added each time. Brett gradually shifted to the "add one-fourth" method because it has internal consistency. He was trying to make sense of the task. Sam, on the other hand, saw no conflict in writing "add a fourth" when part of a numeral (not number) is doubled.

In negotiating the solution to the sequence task, the two boys were actually operating from two incompatible orientations. Sam is rule-oriented, wanting to "get the problem right" as judged by some external authority. Brett insists on making sense of the task.

Up to this point, Sam has been rejecting Brett's method without formulating an alternative solution. Now he formulates a set of three numbers, which fit his doubling of the fractional part, to fill the blanks. He came to his solution by halving as the opposite of doubling; he thinks of taking half of one-fourth.

B: It's right, Sam. (Brett's method)

S: (Ignoring Brett's statement) Cause look, oh, oh. OK, so what is one-half of one-fourth?

B: (Cooperatively) One-eighth.

S: So 15-1/8, 15-1/4, 15-1/2, half plus half equals 16. (throws down his pencil triumphantly) Yeah, I know it's right, Brett.

B: OK, if you did this, 15-1/8 and then 15-1/4, 15-1/2, . . .

What are you doing, just doubling it each time?

(Brett makes sense of Sam's method.)

S: 15 . . . Then a half and a half is 16, a whole number. Wouldn't that work?

B: Yeah, I guess that's it. I guess I was wrong.

S: I told you so! (Ego orientation)

You would've got that one wrong, Brett!

(Now Brett erases his numbers and begins writing Sam's solution.)

B: So that's 15 and . . . what? (acting as a scribe) 15 and 15-1/8, no 1/4

S: (takes pencil and continues) A fourth and a fourth equals a half . . . 15, and a half and . . . 16 and a fourth . . .

B: (reflectively) I think that first one is right. (Brett's original solution)

S: Well, see, we got an eighth and a eighth equals a fourth, one-fourth equals a half . . .

B: That's if you double it.

S: Ain't we supposed to . . . (to Brett) Ain't we supposed to, Mr. Lane?

Mr. Lane refrains from answering Sam's questions. Once Brett has had an opportunity to reflect and compare the two methods, he rejects Sam's solution. Brett negotiates in good faith. When he thinks Sam is right, he says so, but he continues to test the viability of his construction. Sam, on the other hand, does not question his rule. Instead, he asks the teacher whether he should be using that rule.

B: You don't have to double it, though.

Brett, in effect, was saying that you don't have to use that rule if another method makes more sense. He is arguing for sense making over rule following. After more discussion and changing of the solution, the interviewer asks them to write each of their solutions and come to agreement on a single solution.

In an attempt to communicate with Sam, Brett makes up a new problem (using whole numbers 1 and 2 in the sequence instead of 15 and 16) and asks Sam to do it by adding a fourth each time. He is, in effect, trying to get Sam to reflect on his method. Unfortunately, Brett's attempt fails because Sam has no idea how to add 1/4 to 1/2. Their exchange is shown below.

1, __, __, __, 2, __, __, __, (the task Brett sets for Sam)

B: OK, if you add a fourth to everyone of these, it would be—what would it be? (hands the pencil to Sam)

S: It would be 1/8. (using his doubling strategy)

B: No, 1/4! Add a fourth, not a eighth!

S: It would be 1/4. (compliance)

B: 1 and 1/4. (Brett corrects Sam in a helpful manner. Sam is beginning to use fourths so Brett encourages him.) Then, what would be the next one?

S: One-half, then SEE . . .

B: OK, what is 2/4? 2/4. er . . . what is 1/4 + 1/2?

S: One-half. . .

B: All right, what is 1/2 + 1/4? (forcing Sam to think in terms of fourths)

S: 1-1/2.

(Brett smiles—knows Sam cannot follow his reasoning but doesn't know what to say. He doesn't want to embarrass Sam. He does not claim victory as Sam would have done.)

S: You're wrong. Just tell us who is right, Mr. Lane.

(Sam misinterprets Brett's silence as capitulation, but when Sam appeals to the teacher, Brett becomes more assertive.)

B: OK, 1/2 + 1/4, just give me the answer. (Lays down pencil emphatically as a challenge)

S: Two-sixths which is one-third. You're wrong (reasoning that 1/3 could not possibly be correct, so Brett's method fails). Look at it my way (shifts from Brett's constraints to his method).

B: OK. Go ahead, Sam (in a resigned manner).

After Sam argues his case one more time, Brett makes a third attempt to help Sam understand the add-a-fourth method.

B: OK, but listen, . . . say there's fourths over here, as many times as you want them. (writes) 1/4, 1/4, 1/4, 1/4 . . . that's all you can pick from.

S: That won't work out. If you have nutin' but one-fourths . . .

B: 1-1/4, there is as many as you need. Like 1/4, 1/4, 1/4, . . . Know what I mean?

S: No.

B: All these over here are one-fourths.

S: Write them out, don't just scribble.

B: (Writes in a cluster) 1/4, 1/4, 1/4, 1/4, 1/4

S: How many do you need?

B: I don't know. OK, then (ducking Sam's question). OK, take one of these away; it will be 1-1/4, and then if you take another away, and add it to that, it will be 1-1/2, and then if you take another (crosses off a fourth each time) 1-3/4, 2.

S: Yeah, but look at it my way (ignoring Brett's explanation).

B: Give me the collection, too (play by my rules).

S: All right. Say they give you 1-1/8, 1-1/8 (note the use of *they*).

B: (helpfully) No, just one-eighth.

S: If you add 1/8, it would be 1-1/4. If you add another eighth, it would be 1-1/2.

B: No, it wouldn't.

S: Uh huh!

B: No.

S: (pauses, looks at paper and says) Oh, you're right, ain't ya?

S: (announces to the teacher) We made a decision. Use the top one (Brett's).

B: He understands now.

In the end, Sam gives in and weakly supports Brett's solution, but he is far from convinced. He realizes that he is likely to be further embarrassed if he continues to pursue his method.

Throughout this episode, the two boys engaged in negotiation of mathematical meaning. They repeatedly attempted to refute the other's solution and argue for their own method. In most cases, they listened to the other person and responded to his statement or question. While the two held conflicting views about the solution, they attempted, over a considerable period of time, to come to a consensus. Sam finally said he agreed, but there was no evidence that he had made sense of Brett's method nor that he felt his solution was flawed. He may have begun to question the viability of his method.

Sam is typical of many, many mathematics students in our schools. Conducting clinical interviews with students of all ages has led me to believe that rule-following characterizes most students' stance, and it is a rare student who attempts to make sense of school mathematics. Conventional mathematics instruction demands rule-following and anyone who attempts to make sense will likely find disfavor in most classrooms. Yet Brett is evidence that sense making is not beyond most students. Brett is in general mathematics because he had not performed well enough in previous mathematics courses to be recommended for algebra. Yet given an opportunity to give meaning to a task in a way that makes sense to him, he reasons mathematically. Over the hour interview, Brett learned mathematics. Did Sam?

SAM'S INTENTIONS

Sam's goal was to actively participate without becoming too embarrassed by any "errors" he made. He wanted to get the task done, to get THE correct answer, and to do what was expected. His activity was rule-oriented. His construction of Brett lacked depth; he did not try to view the task from Brett's perspective. He does recognize that Brett has mathematical knowledge beyond his own. On several occasions, he attempted to determine how sure Brett was of his method. In Nicholl's (1983) terms, Sam is ego-oriented. That is, he is more concerned about not looking stupid, showing off, and appearing to be knowledgeable, whether his ideas are sound or not. He very much wants to be "right." He would like to be victorious over Brett in having his method judged correct and Brett's wrong. His stance is greatly influenced by his lack of mathematical knowledge. He senses that he does not know enough to justify his position. His attempts to cover up his lack of knowledge results in an ego orientation.

BRETT'S INTENTIONS

In Nicholl's terms, Brett is task-oriented. He attempts to make sense of the task. He is not concerned or even thinking about how he will be viewed by the stances he takes; he is thinking about the mathematics. He reflects on his actions and those of his partner. He revises his position in light of new experiences; his thinking evolves. He gives meaning to the method Sam is using. Once he has the opportunity to reflect on his method as Sam writes, he realizes an internal contradiction and forms the intention to make Sam aware of the difficulty in the method. He has constructed mathematical meaning for the method he has devised and can provide an explanation for it. He listens carefully and responds to what his partner says, even to the point of giving up his method. He does not want to be victorious, embarrass Sam, or in any way appear superior. He simply argues for what he believes makes sense. Brett very much wants Sam to understand his method. He constructs three strategies to help Sam see why his method is reasonable. Each of these fails because Sam has insufficient mathematical knowledge to participate in the discussion to its conclusion. Sam has not constructed 1/4 as a unit and thus cannot think about "one-fourth more than one-half," a necessary construction to give meaning to Brett's method.

SAM'S NEGOTIATORY STANCE

Considering the *stance* of a communicator seems to have considerable explanatory power in analyzing classroom discourse. Because of the situation, Sam accepts the obligation to participate and cooperate. He does maintain his position and argues it strenuously. He gives his partner opportunities to express his ideas fully. He listens to Brett's explanation but is limited in making sense by his mathematical knowledge. In working previous problems with Brett, he had accepted a rule for solving them; his thoughts about these tasks had become static. He does not reflect on the method, he simply uses it. There is no evidence that he has taken his thought as an object of reflection. At times, with some justification, he thinks Brett has agreed with him. Sam rarely gives a mathematical justification for his assertions. He will announce he is right without giving a rationale: "It is right because I did it." He sees the locus of control outside himself as evidenced by his frequent appeal to the teacher for adjudication. He wants the teacher to tell them whether or not what they have done is correct. Sam wants to be right, which means do it the way the teacher wants it done. On four occasions he appeals to authority. For example, "Is this right, Mr. Lane?" He does not dominate, nor does Brett. Sam wants to be right so that he will be seen as smart. Sam gloats over an apparent "victory," telling Brett, "You would've got that one wrong if it hadn't been for me." At one point, when he sees he cannot convince Brett to accept his method and that Brett seems firm about his, he tries to settle the matter by declaring them both right and asking the teacher to support him. In the end, he gives in to avoid embarrassment.

BRETT'S NEGOTIATORY STANCE

Throughout the episode, Brett strives for consensus but takes a strong stand for his beliefs. His assertions are supported by mathematical justifications. Brett gives his full attention to Sam's statements and attempts to give meaning to them. He constructs an explanation of Sam's method which Sam interprets as agreement. When he first understands Sam's method, it seems to be a plausible alternative to his method, so he agrees to let Sam change the answers. But as Sam is writing, he reflects on both methods and reasserts his belief in his method. He subsequently tries three approaches to help Sam give meaning to his method. Two of these are rather creative constructions. His action is driven solely by what makes sense. There is no appeal to authority or rule-following. As demonstrated earlier in the episode, he is willing to change his position based on new evidence. Prior to work on this problem, he had evolved a mathematically sound solution method. He does not want to embarrass his partner. At one point, he challenges Sam to give him the answer to $1/2 + 1/4$. When Sam is unable to correctly add the two fractions, he drops the matter rather than call further attention to Sam's mathematical limitations. However, he uses his knowledge of Sam's mathematical constructions in devising other ways to explain his method to Sam. His construction of Sam is quite elaborate.

TYPES OF NEGOTIATORY STANCES

While not exhaustive, shown below is a list of types of negotiation positions that can be taken in a mathematics discussion. The list is based on the Sam and Brett transcript. The first three facilitate negotiation while the others tend to block communication.

- Attempt to convince others by presenting a mathematical justification.

- Construct a new task to test each method.

- Devise a new argument based on your understanding of both methods.

- State agreement without being convinced.

- Judge how sure the other person is.

- Appeal to authority, e. g., the teacher, or take as correct the answer given by someone in the class known to be capable.

- Follow the rules without necessarily understanding what they mean.

- Hold a view without giving reasons.

- Consider both methods correct.

- Challenge another person on her word-use rather than the substance of the message (a diversionary tactic).

- Refuse to listen to the other person's argument.

SUMMARY

As seen from the analysis of an episode between two ninth grade students solving a mathematics task, negotiation is a complex phenomena. In direct instruction, as opposed to problem-centered learning, negotiation, if it occurs at all, is usually about how many problems are to be completed, how long is to be allowed by the teacher, or perhaps on what day the test is to be given, but it is rarely about methods to be used or mathematical meanings. A student may ask about a concept she does not understand, but the teacher's goal then is to explain it fully, and the student takes as her task to "understand" the teacher's explanation. Even in discovery learning, negotiation of mathematics rarely occurs since the responses students can make to the teacher's questions are quite constrained (Voigt, in press).

There are many potential learning opportunities in problem-centered learning. In particular, there are a variety of opportunities to negotiate social norms and mathematics meanings (Lo, et al. 1991b). Through negotiation, students can test the viability of their constructions, reflect on mathematical relationships, elaborate or refine their constructions, and move toward becoming autonomous learners.

In the exchange between two ninth grade general mathematics students described in this chapter, the complexities of the negotiation process become apparent. The participants' intentions are major factors in shaping the negotiation. Sam is generally ego-oriented while Brett is task-oriented (Nicholls 1983). Sam wants to "win" while Brett wants to make sense of the task. It is quite clear how task orientation positions the learner to profit from negotiation and achieve considerable satisfaction from the involvement. As teachers, we can do much to facilitate negotiation and thus provide rich learning opportunities. Actually, the facilitation itself is a negotiatory process. By encouraging the negotiation of social norms whereby students form the intention to make sense and feel comfortable to do so is an important component in establishing a productive learning environment. In this process there will be times when the focus of the discussion will be the social norms. The negotiation of social norms is necessary for the negotiation of mathematical meaning to occur.

In the episode discussed, the value of a task orientation, rather than an ego orientation, becomes clear. Brett is guided by his intention to make sense. Once this intention is formed and Brett is in an environment where others will negotiate, the potential for him to become a powerful mathematics student exists. As long as Sam maintains his ego orientation, his learning opportunities are severely limited. By being in a classroom where social norms are negotiated rather than rules being imposed by the teacher, a student like Sam could shift from his ego orientation to a more productive stance.

A thesis of this chapter is that mathematics learning (all learning?) is greatly facilitated by negotiation. Analysis of a discussion between two students allows us to see the complexity and power of negotiation. By striving to establish the conditions for negotiation, teachers can increase the probability that meaningful learning will occur.

REFERENCES

Brown, D., and G. Wheatley. *Spatial Visualization and Arithmetic Knowledge.* Proceedings of the Eleventh Annual Meeting of the North American Chapter—International Group for the Psychology of Mathematics Education. New Brunswick, NJ, 1989.

Confrey, L. "What Constructivism Implies for Teaching." *Journal for Research in Mathematics.* Monograph No. 4 (1990): 107–122.

Csikszentmihalyi, M. *Flow: The Psychology of Optimal Experience.* New York: Harper and Row, 1990.

Goodman, N. *Of Mind and Other Matters.* Cambridge, MA: Harvard University Press, 1984.

Hallet, D.H. "Visualization and Calculus Reform." In W. Zimmerman, and S. Cunningham (Eds.), *Visualization in Teaching and Learning Mathematics.* Washington, DC: Mathematics Association of America, 1990.

Lo, J., G. Wheatley and A. Smith. "Learning to Talk Mathematics." Paper, annual meeting of the American Educational Research Association, Chicago, 1991a.

Lo, J., G. Wheatley and A. Smith. "Negotiation of Social Norms in Mathematics Learning." Paper, thirteenth meeting of the Psychology of Mathematics Education—North American Chapter, Blacksburg, VA, 1991b.

Nicholls, J. "Conceptions of Ability and Achievement Motivation: A Theory and its Implications for Education." In S.G. Paris, G.M. Olson, and H.W. Stevenson (Eds.), *Learning and Motivation in the Classroom.* Hillsdale, NJ: Erlbaum, 1983.

Presmeg, N.C. *The Role of Visually Mediated Processes in High School Mathematics: A Classroom Investigation.* Unpublished doctoral dissertation, University of Cambridge, Wolfson College, 1985.

Richards, J. "Mathematical Discussions." In E. von Glasersfeld (Ed.), *Constructivism in Mathematics Education.* Holland: Klewer, 1991.

Sless, D. *In Search of Semiotics.* London: Croom Helm, 1986.

Voigt, J. (in press). "Social Functions of Routines and Consequences for Subject Matter Learning." *International Journal of Educational Research.*

von Glasersfeld, E. "An Introduction to Radical Constructivism." In P. Watzlawick (Ed.), *The Invented Reality.* New York: Norton, 1984. 17–40.

von Glasersfeld, E. *The Construction of Knowledge: Contributions to Conceptual Semantics.* Seaside, CA: Intersystems Publications, 1987.

Wheatley, G. "Constructivist Perspectives on Mathematics and Science Learning." *Science Education,* 75(1), (1991): 9–12.

Wheatley, G., E. Yackel, J. Brice, and K. Lucas (Eds.). *Mathematics Achievement through Problem Solving.* West Lafayette, IN: Purdue University, 1988.

Wheatley, G., and A. Reynolds. "The Potential for Mathematical Activity in Tiling: Constructing Abstract Units." Paper, fifteenth annual conference of the International Group for the Psychology of Mathematics Education, Assisi, Italy, 1991.

Zimmerman, W., and S. Cunningham (Eds.). *Visualization in Teaching and Learning Mathematics.* Washington, DC: Mathematics Association of America, 1990.

8

Constructing Potential Learning Opportunities in Middle Grades Mathematics

Elizabeth Jakubowski

With a renewed emphasis on mathematics curriculum (e.g., National Council of Teachers of Mathematics (NCTM) 1989) in schools causing educators to reexamine what is occurring in classrooms, the multitude of people involved with education are being challenged to consider questions such as, What is mathematics? What does it mean to do mathematics? and What are appropriate experiences for teaching and learning mathematics? In addition, the images individuals hold of mathematics and teaching and learning mathematics are being challenged by research efforts that continue to focus on mathematical learning and how students construct various mathematical concepts (e.g., Cobb et al. 1991; Hiebert and Behr 1988). In particular, research efforts in mathematics specific to middle grades are receiving increased attention (Hiebert and Behr 1988). The purpose of this chapter is to provide an opportunity for the reader to reflect on what does occur in mathematical learning environments and what could occur if students are provided experiences that are conducive to promoting potential learning opportunities. Potential learning opportunities occur when a person attempts to make sense of an experience and to communicate this experience.

In describing mathematical activities, words such as *experiences*, *activities*, or *tasks* are used instead of *classwork* or *homework*. This is done to deemphasize the work notion that appears to accompany such words and to emphasize a learning atmosphere that should characterize classrooms (Marshall 1988).

As students move into middle level classrooms, certain characteristics of the emerging adolescent have implications for teaching and learning middle grades mathematics. Since these students are at various stages in their mathematical development, the mathematics they are engaged in should continue to be conceptually oriented, with logical and/or abstract reasoning a goal of a mathematics program.

Physically, socially, emotionally, and intellectually, middle level children are going through tremendous upheavals in their lives. These changes have implications for teaching and learning in the middle school. For example, because there are fluctuations in basal metabolism, middle level students may be extremely restless at times and listless at others. Thus, activities that encourage hands-on and minds-on, and promote active, dynamic learning rather than passive work should be included in daily mathematical experiences. Intellectually, emerging adolescents display a wide range of skills and abilities unique to their developmental patterns. Therefore, a variety of approaches and materials in teaching and learning mathematics should be utilized. Because they tend to prefer active learning experiences involving peer interactions over passive activities, a mathematics program should be exciting and meaningful with active involvement of the students.

In grounding my perspective in constructivism, I have followed the consensual description of constructivism provided by Noddings (1990, 10). The four tenets identified by her as being agreed upon by constructivists are

- All knowledge is constructed. Mathematical knowledge is constructed, at least in part, through a process of reflective abstraction.

- There exist cognitive structures that are activated in the processes of construction. These structures account for the construction; that is, they explain the result of cognitive activity in roughly the way a computer program accounts for the output of a computer.

- Cognitive structures are under continual development. Purposive activity induces transformation of existing structures. The environment presses the organism to adapt.

- Acknowledgement of constructivism as a cognitive position leads to the adoption of methodological constructivism.

- Methodological constructivism in research develops methods of study consonant with the assumption of cognitive constructivism.

- Pedagogical constructivism suggests methods of teaching consonant with cognitive constructivism.

The implications of radical constructivism for learning and teaching mathematics are subtle and multifarious. Since no person has a God's eye view of reality, it seems almost axiomatic that teachers and students ought be more considerate of the views, observations, and explanations of others. Rather than representing mathematics as a search for truth, it is incumbent on teachers to portray it as the negotiation of meaning and development of consensuses among members of the mathematics club. What should students know about mathematics? And what should teachers do if students have a different explanation or understanding than members of the mathematics club? Is it appropriate to characterize alternative explanations as wrong? What knowledge should comprise the mathematics curricula? Questions such as these spring to mind when radical constructivism is considered in relation to the school curriculum. Of course there are no definitive answers to questions such as these. However, learning mathematics as a process of knowledge generation seems to have considerable merit, and experiencing mathematics as problem solving seems to be a viable way of implementing a curriculum.

Since potential learning opportunities occur within learning environments, it is necessary to examine the status of these as well as the potential for changing them. Included in an examination of learning environments are constructs which influence the promotion of opportunities for learning. In order to provide an environment where an atmosphere of trust has been established, thereby allowing participants to feel free to participate in classroom discussions, many current situations would have to change.

LEARNING ENVIRONMENTS

The milieu in which learning and teaching occur is multifaceted. An examination of the construction of learning environments is really an examination of the curriculum. In order to understand and describe what occurs with curriculum, one has to look at the culture of the school, that is, the social context of the school. At this level, we find that the composition and organization of the school culture influences educational practices. Delving deeper, the culture impacts the curriculum, its meanings, and practices engaged in by members of the community. In education, the members of a community, that is, teachers, students, administrators, and parents, define and shape the culture. Helping to define the curriculum are customs, epistemologies, practices, interactions, metaphors, beliefs, values, and images held by the members. As a result of these factors, learning environments are defined and shaped.

My premise is that to understand and describe learning environments the factors affecting them have to be understood. Grundy (1987, 7) stated that

> Educational practices, and curriculum is one set of these, do not exist apart from beliefs about people and the way in which they do and ought to interact in the world. If we scratch the surface of educational practice, and that implies organizational as well as teaching and learning practices, we find, not universal natural laws, but beliefs and values.

Habermas' (1972) theory of "knowledge-constitutive interests" has been used as a framework in order to make meaning of curriculum practices (Grundy 1987). (Habermas posits a theory about the fundamental human interests that influence how knowledge is "constituted" or constructed.) It is used here to help make meaning of the culture, curriculum, and practices of teachers and students involved with the construction of learning environments. Habermas (1972) identifies three basic cognitive interests: technical, practical, and emancipatory.

Individuals guided by technical interests have a basic orientation towards controlling and managing the environment. Control is achieved through rule-following actions based upon empirically grounded laws. A primary concern is with the product or outcome of the curriculum. In the school context, technicists identify these laws as state or county mandates, curriculum guides, and/or standardized tests. Control exerted by community members and their actions can be justified as a result of these external factors.

Contrasted with the technical interests of control are practical interests that have a basic orientation towards understanding (Habermas, 310). Practicists have an interest in understanding the environment so that one is able to interact with it. Interaction is based upon a consensual interpretation of meaning. Negotiation and consensus making are important notions within practical interests. A practical curriculum would be one where the major interest lies in the processes involved in achieving meaningful learning.

The third cognitive interest is that of emancipatory. When found in school cultures, interests lie in empowerment which allow individuals or groups to engage in autonomous actions. These actions arise out of reflection on the social construction of human society, and in this case a construction of mathematical learning environments. People, that is, teachers and students, are empowered to take control of their own lives in autonomous and responsible ways.

The manner in which the curriculum is visualized and actualized is important in the study of learning environments. In many studies the context in which teaching and learning occur is what is ignored as researchers explore close relationships between teaching and learning variables. In other studies context is considered, but is recognized as being "out there," having an external reality that influences teachers and learners.

Some research groups (e.g., Fraser 1986) have spent more than a decade investigating psychosocial environments and developing instruments to measure them. When learning environments are considered as a construction of individuals, constructed from external cues attributable to interactions that occur between the physical environment, teachers, and students, there are additional factors to consider. The context matters, thus, each person holds her own image of the context in which particular meanings are constructed, an image shaped by the social setting of the classroom, but nonetheless unique to each individual. Factors such as epistemologies, knowledge, beliefs, values, and experiences of participants in a learning environment determine the nature of the context and must be identified and understood in any study that endeavors to take account of the context of learning.

Past approaches to changing and improving learning environments have been oversimplified. Based on behaviorist models of learning, it was assumed that making sure that teachers knew what the learning environment should be like and then providing feedback on what it was like ought to be sufficient for change to occur. Given a discrepancy between an ideal learning environment and an actual learning environment, it was assumed that teachers could and would change their practices, which in turn would result in improvements in the learning environment experienced by students. These assumptions do not take into account the reasons underlying what teachers do when they implement a curriculum. Why do teachers do what they do? And is it reasonable to expect them to change their practices to accommodate discrepancies between an ideal and measured learning environment? In addition to the obvious necessity for teachers to change their classroom practices, the advocated changes need to make sense to them.

ROLES IN LEARNING ENVIRONMENTS

Role of Students

The student's role in a classroom built on constructivist principles is to accept responsibility for her own learning. This requires students to make sense out of what they are learning by reflecting and testing their ideas with peers and the teacher. As students explain what they understand by a concept or procedure, students should ask questions and explain differences in their own understandings. Above all, a focus should be on making sense of the explanations of others. Disagreement is to be expected, and arguments viewed as a productive means of reaching consensuses. Manipulation of objects and materials often will be desirable as students endeavor to solve the puzzles of their minds. Provision of materials ought to be undertaken by teachers, however, the hands-on aspects of learning should not be prescribed for everyone. For some it might not be necessary, or even relevant, for the problems they are endeavoring to solve. However, if materials are needed by individuals, they ought to be accessible. Perhaps the greatest responsibility that students can take is to monitor their own learning. How do students learn? And can they reflect on what they are thinking? A priority for students in mathematics classes is to learn about learning.

Role of Teacher

Most traditional teaching roles are not consonant with constructivist principles. To begin with, the teacher would be a facilitator of student learning with understanding rather than a transmitter of knowledge. The facilitating role implies that teachers provide an environment in which students can do what is necessary to construct knowledge with understanding. Depending on the circumstances, teachers will provide tasks, ask questions, provide explanations, make suggestions, evaluate student ideas, and generally be an interested co-learner in the classroom. Teaching and learning both should be purposeful, not random and ad hoc. Acceptance of radical constructivism as a guiding epistemology does not carry a penalty of abrogating a role in structuring the learning environment. Learning does not have to lack purposeful and deliberative actions. Radical constructivism is *not* a synonym for discovery learning.

Role of Instruction

When teachers adapt their teaching in order to have a learning environment whose characteristics reflect emancipatory interests, then potential learning opportunities may abound. Specific prescriptive instructional strategies are not a characteristic of such learning environments. While recent research agendas have not prescribed instructional strategies, they have delineated some parameters for effective instruction. However, dependent upon your philosophical perspective, these may be interpreted in various ways. The three features of instruction for middle grades mathematics are first, the broadening of the scope of mathematics to "include the more fundamental conceptual bases of the quantitative notions and to include increased attention to the semantics of the symbol systems" (Hiebert and Behr 1988, 12). Second, because of the complexity of the mathematics content, middle level students should not be expected to develop this knowledge spontaneously. However, the third feature does acknowledge that the content cannot be prepackaged and delivered in a finished form to students. Students must be provided with appropriate experiences in order to construct their own knowledge. Thus, choosing appropriate learning experiences for middle level students becomes important in establishing learning environments.

PROBLEM-CENTERED LEARNING

One instructional strategy that provides potential learning opportunities is problem-centered learning. Problem centered-learning is a method of teaching and learning that focuses on students' ability to construct their own meanings for concepts. Problem-centered learning involves students in activities that have the potential to stimulate students to think about and make sense of mathematics concepts in their own way, while at the same time developing concepts in agreement with the consensual body of knowledge present in mathematics. Through problem-centered activities children

- learn to view mathematics as a meaningful activity
- learn to appreciate mathematics as a dynamic and active subject
- can see the reason for studying mathematics
- are intrinsically motivated to learn
- come to view mathematics as a human endeavor in which they can participate, rather than as a set of unrelated facts determined only by experts in the field
- learn that mathematics content can be applicable in a variety of life situations

Problem-centered learning is designed to facilitate active engagement of students in the learning process by encouraging them

- to invent their own ways of attacking and working out problems
- to exchange points of view rather than reinforce correct answers and correct wrong ones
- to think rather than to compute with paper and pencil (Kamii and Lewis 1990)

Wheatley (1991) describes problem-centered learning as having three components: tasks, groups, and sharing. He notes:

> In preparing for class a teacher selects tasks which have a high probability of being problematical for students - tasks which may cause students to find a problem. Secondly, the students work on these tasks in small groups. During this time the teacher

attempts to convey collaborative work as a goal. Finally, the class is convened as a whole for a time of sharing. Groups present their solutions to the class, not to the teacher, for discussion. A goal of the discussion is consensus. The role of the teacher in these discussions is that of facilitator and every effort is made to be non-judgmental and encouraging. (15-16)

A teaching model described in *Elementary School Science for the 90s* (Louks-Horsley et al. 1990) presents a four-stage format that reflects the nature of problem-centered learning as described by Wheatley. These stages include

- engaging the learner in an invitation to learn. This process is sparked by a child's spontaneous question or by a teacher-generated question about science or technology. (The task)

- engaging the learner in exploration, discovery, and creation of ideas. In this stage children try to answer their questions using materials they have collected or those provided by the teacher. (Group work)

- learners proposing explanations and solutions based on their own explorations. Students persuade themselves, as well as their peers, that conceptions they have generated are based on convincing data from their activities. (Sharing)

- children taking action on what they have learned. Students demonstrate that they have integrated the new information and proposed explanations into their existing conceptual frameworks.

A model for elementary mathematics teaching, based on problem-centered learning, is described by Yackel et al. (1990). This model is just as applicable to middle grades mathematics teaching.

- The children working in pairs engage in activities that, given what is known about their level of development, are believed to be problematical for them. This phase generally lasts about 25 minutes.

- The children present their solution methods to the class. The teacher encourages all students to explain their ideas and elaborate on them. Children are encouraged to agree and disagree with one another.

- The children reach consensus as to a correct solution or solutions, but learn to recognize that there are multiple ways for reaching a solution.

Although inquiry and problem solving are at the heart of problem-centered learning, there are certain qualitative features of problem-centered learning that differentiate it from typical inquiry-based or problem-solving activities as they are generally implemented in schools.

- Problem-centered learning focuses on problems that *interest students* and that students *want* to solve.

- Problem-centered learning focuses on the importance of *communication* in learning because all activities are carried out by students working in cooperative or collaborative groups.

- Problem-centered learning focuses on the *process* of inquiry and *reasoning* in problem solving rather than on getting the correct results of an experiment or the right answer to a problem question.

- Problem-centered learning focuses on facilitating students' development of confidence in using mathematics and applying it as they make sense of everyday life situations.

POTENTIAL LEARNING OPPORTUNITIES

Potential learning opportunities occur when a person attempts to make sense of a mathematical experience and to communicate this experience. Each part of problem-centered learning affords students potential learning opportunities. Lo (1991) identified various opportunities that are likely to occur during elementary mathematical activities, thereby indicating the existence of potential learning opportunities for students.

When participating collaboratively in seeking a solution to a mathematical task, potential learning opportunities exist for the collaborative group. These could occur during the time in which a method of solution is jointly constructed. A student would have to attempt to make sense of her understanding of the task and communicate this to her peers. In communicating her thinking, opportunities exist for the student to develop more understanding of the task. In elaborating or justifying a method, a student is also provided with a potential learning opportunity. As a member of the group is sharing her way of interpreting and making sense of the task, the other members have to attempt to make sense of this method or interpretation.

During the whole class discussion (or sharing portion) conducted after small group activities, many potential learning opportunities exist. A primary goal of whole class discussion is to set up the opportunity for students "to present their solution methods to their peers and to compare/contrast different mathematical ideas" (Lo et al. 1991). When a student assumes the role of presenter, the potential learning opportunities for her occur as she prepares to make a presentation. First, the presenter will need to reflect on the mathematical task just completed and organize her thoughts for presentation. In the process of reflection, a person's awareness of her thought processes are increased, thus creating a potential for revision and elaboration (Bruner 1986; Duckworth 1987; Lo et al. 1991). A second opportunity occurs when the presenter is preparing a communicable explanation and has to "create a reader" (Sless 1986). Lo, Wheatley, and Smith (1991) explain this as meaning that the presenter "has to consider peers' expectations." When two or more students have to negotiate what counts as an explanation, the opportunities for learning abound. Problematic situations arise in the process of negotiation whereby potential learning opportunities are created.

The other participants in the whole class discussion, the nonpresenters, also have potential learning opportunities. In problem-centered learning, all students have either solved or attempted to solve the task. "Thus it is natural for them to compare and contrast their ideas with the presenter's explanation, or more accurately, their construction of the presenter's explanation (Lo et al. 1991). Through the reflection on their explanation and the presenter's, opportunities for learning occur.

CONCLUSION

As educators move toward the twenty-first century, we are challenged by calls for change. What is advocated in this chapter is the construction of appropriate learning environments where problem-centered learning promotes the existence of potential learning opportunities. When teaching and learning middle school mathematics are approached as activities that allow middle school students to make sense of mathematical tasks, and construction of meaning is an individual activity that relies on social interactions, then learning environments are in contrast to many current practices.

The middle level child, a child in the midst of transition, is experiencing changes in her development. With this in mind, the teacher has to provide a learning environment where students "feel free to explore mathematical ideas, ask questions, discuss their ideas, and make mistakes" (NCTM 1989). Through problem-centered learning such an environment can be maintained. Teachers can listen to students' ideas while encouraging the students to listen to each other's ideas. The learning environment's atmosphere can be created where many answers are accepted and explored. Moreover, students can be encouraged to seek a verification of their own answers and not search for a verification of a right or wrong solution from an outside authority. Through small group experiences and whole class discussions, students are provided the possibilities for potential learning opportunities. By being challenged in their thinking, students are afforded the opportunity to pull prior experiences and mathematical ideas together and to learn from these opportunities.

REFERENCES

Bruner, J. *Actual Minds, Possible Worlds*. Cambridge, MA: Harvard University, 1986.

Cobb, P., T. Wood, E. Yackel, J. Nicholls, G. Wheatley, B. Trigatti, and M. Perlwitz. "Assessment of a Problem-centered Second-grade Mathematics Project." *Journal for Research in Mathematics Education*, 22(1), (1991): 3–29.

Duckworth, E. *The Having of Wonderful Ideas*. New York: Teacher's College Press, 1987.

Fraser, B.J. *Classroom Learning Environments*. London: Croom Helm, 1986.

Grundy, S. *Curriculum: Product or Praxis*. London: Falmer Press, 1987.

Habermas, Y. *Knowledge and Human Interests* (2d ed.). London: Heinemann, 1972.

Hiebert, J., and M. Behr (Eds.). *Number Concepts and Operations in the Middle Grades*. Reston, VA: National Council of Teachers of Mathematics, 1988.

Kamii, C., and B.A. Lewis. "Research into Practice: Constructivism and First Grade Arithmetic." *Arithmetic Teacher*, 38(1), (1990): 36–37.

Lo, J. *Student Participation and Potential Learning Opportunities in Grade Three Mathematics Class Discussion.* Unpublished doctoral dissertation, Florida State University, Tallahassee, FL, 1991.

Lo, J., G. Wheatley, and A. Smith. *Learning to Talk Mathematics.* Paper, meeting of the American Educational Research Association, Chicago, 1991.

Louks-Horsley, S., R. Kapitan, M.A. Carlson, P.J. Kuerbis, R.C. Clark, G.M. Melle, T.P. Sachse, and E. Walton. *Elementary School Science for the '90s.* Washington, DC: Association for Supervision and Curriculum Development, 1990.

Marshall, H. "Work or Learning: Implications of Classroom Metaphors." *Educational Leadership*, 17(9), (1988): 9–16.

NCTM. *Curriculum and Evaluation Standards for School Mathematics.* Reston, VA: National Council of Teachers of Mathematics, 1989.

Noddings, N. "Constructivism in Mathematics Education." In R.B. Davis, C.A. Maher, and N. Noddings (Eds.), *Constructivist Views on the Teaching and Learning of Mathematics.* Reston, VA: National Council of Teachers of Mathematics, 1990.

Sless, D. *In Search of Semiotics.* Beckenham, UK: Croom Helm, Ltd., 1986.

Wheatley, G.H. "Constructivist Perspectives on Science and Mathematics Learning." *Science Education*, 75(1), (1991): 9–21.

Yackel, E., P. Cobb, T. Wood, G. Wheatley, and G. Merkel. "The Importance of Social Interaction in Children's Construction of Mathematical Knowledge." In T.J. Cooney and C.R. Hirsch (Eds.), *Teaching and Learning Mathematics in the 1990s.* Reston, VA: National Council of Teachers of Mathematics, 1990, p. 1221.

9

Construction Sites: Science Labs and Classrooms

Wolff-Michael Roth

VIGNETTE

When I began teaching more than a decade ago, I had just completed a masters degree in physics, but I did not have any background in educational psychology or methodology. Pondering how I should approach the teaching of science, it occurred to me that I wanted my students to learn in a setting like the one in which I had experienced the excitement of scientific discovery, and I recalled my days in graduate school: I remembered those long days and nights when I sat over my results, eagerly trying to understand the meaning of all those graphs and charts; I remembered the long discussions I had with my peers who worked on similar problems for their theses to come to understand our own and each other's work; I remembered the discussions with post-doctoral students and professors through whose critical questioning I came to reflect on my own understanding; and I remembered those weekly seminars where we (professors, post-doctoral, doctoral and masters students) critically examined, in turn, each of the research projects in the department of atomic physics. I was determined to create classroom environments in which my students could experience all the excitement of real science and the excitement of finding out for themselves, and with others, in a setting that nurtured abilities and where science was conducted authentically. Since then, I have taught all my courses in laboratory, nonlecture settings that emphasize science as a process of meaning-making, and knowledge as individual and negotiated construction.

INTRODUCTION

A traditional view of the teaching-learning process regards knowledge as a body that is transferred from a teacher to a student. This view of teaching is so deeply rooted in our culture that learning is mostly described in terms of the conduit metaphor such as in "I can't get it," "I was cramming," or "it didn't sink in." More recently, constructivism began to be accepted as a more appropriate conception of knowledge in the fields of education, epistemology, history and philosophy of science, cognitive and social psychology, philosophy, or sociology of science (Bruner 1986; Gergen 1985; Goodman 1978; Knorr-Cetina 1981; Toulmin 1982; von Glasersfeld 1987). Constructivists recognize that, rather than being transferred from one individual to another, knowledge has to be constructed by each individual through his or her active engagement with the physical and/or social environment. The dialectic between the world and its representations in a human mind was emphasized by Piaget who held that "intelligence organizes the world by organizing itself," (in von Glasersfeld 1984, 24) and by Goodman (1978, 22) who stated that "comprehension and creation [of the world] go together." These idiosyncratic constructions can then be negotiated in shared discourse which may, in turn, lead to an internal reconstruction when there is a clash between the individual constructions of the participants (Piaget 1970). This process has been likened to the construction of stories by individuals that can later be discussed, negotiated, and checked for fit (Davis and Mason 1989). From a Piagetian viewpoint, knowledge is always constructed first intrapsychologically before it can be negotiated. The meanings which emerge during the interactions between individuals are assumed to be compatible from a cognitive perspective (von Glasersfeld 1987), but are shared from a social psychological view (Gergen 1989).

Past research has shown that this view of the negotiation of knowledge has limitations that account for situations where individuals display problem-solving abilities in collaborative groups that are over and above their individual abilities (Kroll 1989; Newman et al. 1989; Rogoff 1990). Our research along with that of others indicates that discourse in collaborative situations goes beyond mere negotiation. Discourse is used as a resource to collaboratively construct (co-construct) knowledge interpsychologically which can then be appropriated by the individual, that is, constructed intrapsychologically by the individual partners in the discourse situation. The processes of co-construction and individual construction of knowledge occur simultaneously as "new, more powerful structures may be constructed interpsychologically and the new structure can interact with the child's intrapsychological structures to result in individual cognitive changes" (Newman et al. 1989, 68). For many situations of collaboration, this social constructivist perspective (see Wertsch 1985 and accompanying note) seems to be more appropriate than the perspective that developed out of Piaget's work (Cobb 1989). Albeit this is particularly true for situations where individuals interact with more able individuals such as parents, teachers and peers. The model can also be used for analyzing the work of collaborative peer groups (Newman et al. 1989; Rogoff 1990).

The purpose of this chapter is to describe the inter- and intrapsychological constructive work of students in physics classes when they work on their own, in peer groups, in small groups interacting with a teacher, and in whole class discussions. Both the cognitive constructivist and the social constructivist perspectives will be used where appropriate to account for the constructive processes in the laboratory and classroom setting.

TEACHING-LEARNING SETTING

It has been shown that the beliefs and metaphors of teachers are determinants of the interactions and transactions in their classrooms (Tobin

1990, 1991). Accordingly, knowledge of my basic beliefs, epistemological commitments, and central metaphors will help the reader construct a better understanding of the classroom strategies I use, and they will help the reader better appreciate the teaching-learning environments in these classrooms.

Beliefs and Metaphors

Knowledge about and meaning of our world are constructed individually and negotiated in transactions with others, that is, all meaning is constructed by specific in-context human practices (Hawkins and Pea 1987; Lemke 1990). In the classroom, teachers and students are involved in building recognized patterns of activities and use language to build the special meaning relations of specific subjects. Thus, the students are initiated into the special ways of talking, writing, and doing within a subject, that is, they are initiated to its specialized forms of social discourse (Lemke 1990). Built on this basic belief are two metaphors that effectively determine the type of classroom interactions between students and me. The two metaphors characterizing my teaching are *cognitive apprenticeship* and *enculturation*.

Learning is a process of cognitive apprenticeship. I try to set up learning environments that will enculturate students into the practices of scientists. Within the context of these activities and social interactions, I take a position that can be likened to a master in craft apprenticeship, or that of an adviser of graduate physics students, who, through a close working relationship, allows students to enter the culture of physics practice. The metaphor of cognitive apprenticeship also suggests the practices of situated modeling, coaching, and fading (Brown et al. 1989) whereby teachers first model their strategies in the context and/or make their tacit knowledge explicit. Then, the teacher supports the students' attempts at implementing the strategies (coaching). And finally, they leave more and more room for the student to work independently (fading). As the research by Schoenfeld (1985) and Lampert (1986) has shown, students will gain more self-confidence, become more autonomous in collaborative situations, and participate consciously in the culture of the subject. Through this active participation in the transactions of a culture, students will not only develop the language and belief systems (Brown et al. 1989) but, in return, will shape these very systems (Lemke 1990).

Learning is a process of enculturation similar to growing up in a particular society and learning its sign systems such as language, behaviors, and other culturally determined patterns of communication. Learning the language of physicists in the classroom is similar to the learning of a language by a child. The original, very limited understanding of a concept and its meanings are developed and extended through negotiations with teachers and fellow students. The concepts, thus, receive more and more texture as they are applied in an ever larger number of settings. From these beliefs and metaphors follow some necessary implications. One key implication of learning as cognitive apprenticeship and enculturation in a social setting is the necessity for collaboration. The other main implication is that the implementation of learning environments compatible with the above beliefs and metaphors will have to operate with time lines different from those of traditional curricula which are usually formulated in terms of low-level, short-term goals.

Because knowledge is a social phenomenon, negotiated and constructed through transactions, collaboration is of primary necessity to the science classroom. It has been shown that such collaboration has positive effects on students' learning, motivation, and attitudes (Sharan 1980; Slavin 1980; Schoenfeld 1985; Kroll 1989). Provided with appropriate environments and activities that allow for interactive exchanges, the participants will exceed that which they might achieve as individuals, thus amplifying the learning of each individual. In collaborative environments, students can make their understanding public to be reflected upon and critically evaluated. Subsequently, students can make necessary refinements and extensions in this understanding. Continuous feedback from peers and teachers is necessary to delimit permissible and effective diction (action). As with the acquisition of natural language, one should not expect a novice to use scientifically acceptable language overnight. Given the opportunity, the skills and practices of apprentices, whether they are high school students, crafts persons or Ph.D. students of physics, will not only become increasingly complex, but also better adapted to solve the problems and questions facing them within each of their respective cultures. Seen myopically, this takes more time than school officials are traditionally willing to allow for "measurable outcomes." However, the apparent success of my past and present teaching seems to indicate that over a longer term, students in a constructivist classroom develop viable conceptual frameworks.

The Classroom

The following examples are all taken from my current school, a private college preparatory school, where I teach five classes of junior- and senior-year physics. My experience of teaching science in various public schools and colleges has shown that the events reported here are not specific to a private school. The enrollment per class ranges from 14 to 19 boys, between 16 and 18 years of age. The scheduled classroom time consisted of nine 40-minute periods per two-week cycle.

At the beginning of each unit, the students received a written outline of compulsory and suggested activities. There were four types of activities: experiments, reading and concept mapping, textbook problems, and essay assignments. Experiments made up the core of the activities around which the rest of the course revolved. For each unit, the students read the relevant chapters from the textbook and at least one other source and constructed a concept map that contained the key concept of the unit. Each week, the students also did from six to eight "Think and Explain" and/or word problems from at least two sources. The selection of the specific problems was most often left to the students. About once a month the students prepared an essay on special topics in physics not covered in their textbook. The topics ranged from the biography of scientists to essays on "Knowing and Learning Physics" or "Objectivity in Science." We used one or two periods per week for whole class discussion, review, the sharing of experimental results, and concept mapping activities. Most of the other classes were spent on the investigation of experimental questions. Although the scheduled classroom time consisted of nine periods per two weeks, the students used the lab and its facilities during their free periods, after school, in the evening, or during the weekend. Attending a residential school, many students made use of this availability of the lab during "off-class" hours. The physics laboratory was an open classroom in the sense that students used any other facility on campus if the activity so required. Conversely, whenever the circumstances permitted, any student from another section used the laboratory facilities, which sometimes led to interesting discussions across course sections. On occasion, other teachers used the facilities, such as computers, when classes were going on.

Students spent about six of their nine bi-weekly periods planning, designing, and executing experiments, followed by analysis and interpretation of the results. To facilitate the students' entry into a new unit, it was introduced with a demonstration of some key concepts. This demonstration also included measurement instruments which the students had not encountered previously, or a new piece of software. For the most part, however, new instruments and software were introduced as the need in students' investigations arose. Pupils usually tinkered with the materials during and after the demonstration, which helped them in thinking about the investigations that they later designed. Although I sometimes suggested an investigation, particularly at the beginning of a unit, most of the experiments were designed and developed by students, which often led to new and innovative activities (Roth 1991). As the students investigated, I served as a facilitator, trouble-shooter for broken equipment, and as a sounding board for student ideas.

The classroom-laboratory described above (for this purpose there is no distinction) is replete with situations that illustrate the constructive activities in which students engage, thus the title for this chapter. Throughout the year, I have brought both video- camera and tape-recorder (which were operated by an assistant) to the classroom to record students during their activities. Sometimes these recordings were made in class. At other times, recordings were made at quieter moments in the evenings to study specific tasks in different, out-of-class settings.

ON THE CONSTRUCTION SITE

Among other definitions of construction in *Webster's University Dictionary* are "explanation or interpretation given a particular statement" and "arrangement of words to form a meaningful phrase, clause, or sentence." In both senses, science classrooms offer many opportunities to become sites for constructive activity. Implicit in both definitions is the active nature of the learner; she has to do the explaining, interpreting, and arranging to make meaning. In the following sections, I will describe students as they are constructing knowledge in four different social settings: as individuals, in peer groups of equal status, in a dyad of unequal status (teacher and student), and in a whole class discussion.

An Individual Makes Sense of Graphs

Michael was a student who had always struggled, both in mathematics and science to develop an understanding of science. During his two years in junior- and senior-year physics, he had always tried to understand rather than resort to other strategies for coping, such as memorizing or relying on partners to do the work. In the event discussed here, Michael and his partner had done an experiment during which they collected distance-time, velocity-time, and acceleration-time graphs of a cart which oscillated back and forth between two springs on a frictionless airtrack. Because he had difficulties following his partner's interpretations, Michael returned to the lab one afternoon after school, bringing with him his data tables and graphs. He set up the experiment, and repeatedly shifted between moving the cart slowly from one side to the other, running the experiment in real time, and going back and forth from his graphs to the experiment. Observing him from afar, it became clear that he was struggling to come to an understanding. A few days later he submitted a laboratory report that contained distance-time, velocity-time, and acceleration-time graphs to which he had added (1) colored points labeled with their coordinates, (2) hand drawn lines, and (3) text. The following is a section from his report:

> This graph has been plotted with the results acquired from the experiment. All important points I will be referring to have been labelled on the graph and have received asterisks beside them in the "Table of Calculated Values." I will now describe the situation that allowed me to achieve the data which formed the first complete cycle [of oscillation]. Please understand that I have treated the airtrack like a number line. Motion to the right I will refer to as positive and motion to the left I will refer to as negative. I begin my description when the cart is closer to the left end of the airtrack. At data point (0.606[s],0.106[m/s]) the cart's spring on the left was contracted and the spring on the right was expanded. As the expanded spring's potential energy transforms into kinetic energy, the cart begins to accelerate in a forward positive direction, this is represented on the velocity-time graph by the first positive slope. At the same time this is happening the contracted spring on the left is beginning to expand.
>
> The maximum point (0.934[s], 0.484[m/s]) on the velocity-time graph and any other maximum points on any other half cycle represents the cart's maximum velocity during the cycle's period. This point (crest) also represents the instant where both horizontal forces exerted by the springs acting on the cart are balanced. From this point on the cart's resistance to change its state of motion (inertia) propels it forward. After this point the spring on the left has expanded so much that it begins to force the cart to accelerate in the opposite direction; this is represented on the graph by the first negative slope.

This excerpt is typical for the reports received from my students in that it shows how a student "talks" or "writes" his way to understanding. In the process of writing, new connections may emerge. For two reasons I encourage students to use their own voice in writing. First, I believe that these personal reports encourage students to see the knowledge constructed throughout these reports as products of their own processes; that is, the students develop a feeling of ownership of this knowledge. Second, they can always learn in college to extricate themselves from their products if they need to learn to write formal reports.

Michael's effort in making sense is apparent from the cited paragraphs. First, he readied the readers to construct their own understanding by indicating the use of number pairs and color to denote those data points that he wanted to discuss (important points), and by indicating that the airtrack was like a number line with positive and negative directions. This strategy not only helped the reader, but also it helped him in constructing the understanding at which he arrived in the process. These preparations also show how Michael used prior knowledge, even if it reached as far back as the early years of schooling when he used the number line to learn addition and subtraction. Second, as Michael described the motion of the cart, he made connections between several levels of conceptual abstraction. These levels included (1) the actual concrete events and objects with which he was dealing (L1); (2) the description of motion in the form of data tables and graphs (L2); (3) the mathematical symbolic framework (L3); and (4) the explanatory, conceptual framework (L4).

Michael observed the cart in real time and in slow motion (L1). As he followed the movement of the cart, he matched its speed (L1) with the speed-time graph (L2), indicating that "at data point $(0.606[s], 0.106[m/s])$ the cart's spring on the left was contracted and the spring on the right was expanded." In the next sentence he immediately referred to combining the changes in speed (L2) with its explanation, the change of the spring's potential energy into the kinetic energy of the cart (L4). He also connected this relationship with the mathematical symbolic framework by referring to acceleration as the slope of the velocity-time graph (L3), by indicating that "as the expanded spring's potential energy transforms into kinetic energy, the cart begins to accelerate in a forward positive direction, this is represented on the velocity-time graph by the first positive slope." Subsequently, he correctly observed that the described motion led to a concurrent change in the length of the spring (L1). In the second paragraph, Michael made a connection between the "important point" of maximum velocity (L1, L2) with the state of balanced forces from the two springs that accelerated the cart (L4). Finally, he invoked inertia, the "resistance to change its state of motion" (L4), to explain the continued movement of the cart to the right (L1); he connected this to an acceleration "in the opposite direction" (L4); and he linked acceleration to the negative slope of the velocity graph (L3).

In this example a student constructs and coordinates multiple representations of the same phenomenon. The physical system (frictionless cart suspended between two springs), the graphs and tables, and the mathematical equations describing the motion (not shown here) not only evoked cognitions, but also served as objects for the active interpretive process (Kaput 1988). Although Michael was working here on his own, it seems as if the fact of "talking himself through the problem" facilitated his construction of knowledge from this experience. When he used the physical system, Michael presumably built and elaborated his associated mental representations which he could express in natural language. Language appears between the material and the subjective worlds (Kaput 1988). It is that which is shared with others. These others are designated members or representatives of a discourse community, such as teachers, or potential future members, such as peers (Pea et al. 1990; Roschelle 1990). Constructive processes such as those in which Michael engaged are facilitated when more than one person is present. Then dialogue almost inevitably occurs, which means that natural language is invoked and produced. The advantage and pedagogical value of concrete systems such as the one featured here

> lie in the richness and robustness of the cognitive structures that students typically have available to work with to build [a cognitive representation of the physical system], the extent to which these are shared among students, and most importantly for this part of the story,

the specific kinds of cognitive processes that they evoke, support, and constrain. (Kaput 1988, 15).

This quote particularly fits our situation because Michael had available computer-generated tables and graphs, which were not available in the case that Kaput discussed. These graphs and tables, as well as the equations, together with the concrete objects and events of the experiment provided for multiple feedback systems in the construction of knowledge in Michael's physics course.

Construction of Meaning in Small Groups

Student-centered classrooms provide for discourse situations where knowledge can be constructed as small numbers of individuals interact to solve problems. These interactions can be analyzed according to the subjects' relative status in the small group related to their contextually specific knowledge, and according to the degree of collaboration (Granott 1991). For the present purposes I will focus on two settings. The first setting is constituted by a high degree of collaboration of two individuals with unequal status, a student and a teacher. Under the heading of guided participation, I will discuss two excerpts from a mentor relationship. In the second setting, peers of equal status share in a highly collaborative activity to produce a common product. Under the heading of peer interaction, I will discuss a situation in which three senior physics students are engaged in constructing a concept map.

Guided Participation

During the summer of 1990, I was the mentor of Rod, an outstanding grade 10 student, who wanted to get his grade 11 equivalent in physics, chemistry, and mathematics. Some of our sessions revolved around word problems from a grade 12 textbook. Although the following transcript was recorded outside of the regular classroom, it is not too different from many of the interactions I have had with students in the classroom on similar problems. Prior to the episode featured below, Rod had worked on a textbook problem with torques. His answer did not match that from the textbook because he had applied the ratio rule in reverse, that is, he had associated the larger mass with the larger distance.

Teacher: (Reads problem from book. There is a drawing of a person making push-ups with distances from toes to hands and from toes to center of gravity indicated.) Calculate the force at the hands and toes of a 58-kilogram athlete holding a push-up position.

Rod: Hands and toes, OK. Ah (begins some calculations)

Teacher: Why don't you make a line with the arrow, the arrows of force. What are the forces acting?

Rod: OK, at the toes, force, ahm, here is the center of gravity.

Teacher: Why is the person not, sinking into the ground? There must be some force holding that person up (pointing to the drawing in the textbook)?

Rod: OK, there is, ahm, that is the center of gravity right there, and this should be, center of gravity, times y.

Teacher: Just tell me about the forces, where are the forces?

Rod: Force? There, there, there (points to toe, center of mass, and hands)

Teacher: So why don't you make a drawing. (As Rod is drawing:) So this is the person, so there is a force up here (pointing), at the feet, there is a force at the hand. Now the question is, where would you choose the pivot to be?

Without a scaffold, Rod immediately began to calculate. I made a first attempt to get him to draw an abstract sketch of his understanding of the problem situation. This attempt was driven by three considerations. First, from a conceptual point of view, problems such as the one in the example above can be solved by considering that in equilibrium, the sum of all forces and the sum of all torques are both zero. Therefore, all forces and torques have to be identified, which can be done in stick drawings. Second, I know from my own experience that such sketches help in reconstructing the physical situation which the word problem describes in that they allow me to construct an understanding of the essential parts of the problem (Hake 1987; Heller and Reif 1984; Reif 1981, 1986). Third, such drawings help in the construction of a discourse space shared between the participants, and serve as tools to mediate the construction of meaning. For the tutor, this means that she comes to a better understanding of the student's thinking, and the sketch becomes a checking device to determine whether the student constructed the problem from a physicist's perspective. For the student, it means that something concrete becomes the object of conversation with the tutor. Rather than making the drawing, Rod focused on two of the forces (at the toes and at center of gravity) and returned to attempt to calculate an answer without constructing an understanding of the problem. However, I didn't give up and helped Rod to identify all the forces, that is, to construct the problem from a perspective of canonical physics. By asking him to identify all forces, I structured the activity for Rod such as to encourage him to generate the sketch with all forces indicated.

The excerpt gives a good indication of the types of processes involved in the middle of a student's development of a complex skill. Initially I modeled the solution of such a task by supplying extensive support or providing the reasoning myself. Then students do such tasks on their own, but with me providing support and direction (scaffolding the students' problem solving). The above transcript gives an example of the support a student typically might receive. Finally, as students become more self-sufficient, I withdraw more and more support (fading). This last stage sometimes leads into joint inquiry when both teacher and student engage in a problem that the teacher had not seen before. An example of such an interaction is provided below. A description similar to the present was provided for McArthur et al. (1990) for the apprenticeship model of tutoring. This model, which uses a strategy of "modeling-scaffolding-fading," is used in many cultures to introduce novices to its authentic practices through activity and social interaction (Brown et al. 1989; Rogoff 1990).

In the present setting one might question the authenticity of doing word problems from a textbook. However, when I ask students to do such problems, it is not to keep them busy or merely to help them develop skills in solving such problems. Rather, doing word problems together with peers or a teacher affords opportunities to engage in physics discourse, to reflect on the use of general physical principles, and to engage in meta-discourse about solving problems. The latter includes general strategies for constructing an understanding of the problem as intended by its creator or strategies on negotiating the conventions of interpretations. One example of the use of general principle can be seen in my question, "Why is the person not . . . sinking into the ground?" Implicit in this question is my understanding of the situation that static equilibrium is only maintained, in accordance with Newton's Second Law, when the sum of the forces and torques, respectively, are both zero. Later on in this conversation I engaged Rod in a discussion about the relevance of Newton's Second Law to equilibrium problems.

During the session reported in the second transcript, Rod and I were talking about a problem involving forces and friction. The transcript picks up toward the beginning of the session, after we had constructed a diagram of the situation.

Rod: Because what I had done is set that, OK, I see, I had forgotten to put the cosine thirty-seven times force, OK so that works out as.

Teacher: Let's see what sort of things we have, we have the frictional force, Ffriction, this way you have to pull this way.

Rod: But isn't Ffriction [friction force] opposite to that force [pull]?

Teacher: No, it's opposite to, it's only opposite to,

Rod: OK, the movement is that way.

Teacher: You have to, this force which is mg, force of gravity

Rod: Right

Teacher: and they sort of have to balance.

Rod: Right, and that Fnormal [normal force] as well because of the ground and when you go against this force. You increase this force when it decreases this?

Teacher: All you have to consider here is, Ah, you don't have to consider, let me see, the normal force.

Rod: Because it's part of the static friction component.

Teacher: Ahm.

Rod: Something has to oppose gravity as well.

Teacher: All we have to consider is, Ahm, the normal force, this normal force is that

Rod: Right, but this is that, that force is this.

Teacher: Yeah, that's right and then Fnormal equals to

Rod: mg [weight] minus

Teacher: minus that vertical component

Rod: Right.

In this situation, again, the problem provided for a joint problem space. It is obvious here that as the student had developed, the discourse was more evenly distributed. There was coordinated and collaborative inquiry, during which the student contributed to the structuring of the task, and there was less guidance than during the scaffolding phase.

As the specific problem was novel to me, and while Rod had already tried to solve it, both of us contributed to the co-construction of the problem space, the problem and its solution. In the first half of the dialogue, one can see how I was constructing an understanding of the situation myself, how I engaged in an inquiry, "Let's see what sort of things we have," "You have to, this force is mg," and "and they sort of have to balance." At the same time, Rod was not listening passively but was engaging in this inquiry, thereby constructing his own understanding. Initially, he thought that the friction force was opposite to the pull, but he seemed to modify his construction after realizing that friction was opposing the motion ("OK, the movement is this way"). About halfway through the excerpt, Rod took over the inquiry, trying to separate the various forces acting horizontally ("because it's part of the static friction component") and vertically ("something has to oppose gravity as well"). In this inquiry, his arguments ("but this is that, that force is this") were as valid as my own and I acknowledged his move with "Yeah, that's right." During the exchange, "and then Fnormal equals to," "mg minus," "minus that vertical component," we completed each other's sentences. Such interlocking of discourse is evidence for the co-construction of a taken-to-be-shared understanding (Granott 1991). Here, our understanding was constructed interpsychologically at the same time as we constructed our individual versions. This example highlights the difficulty of using only an internalist approach to the construction of knowledge that neglects the social aspect of all knowledge construction (Cobb 1989; Lo et al. 1991; Newman et al. 1990). For Newman et al., situations as the one described have an *interpsychological* character, arising from the interaction between people, which produces cognitive changes. These changes can be understood to arise out of the interpretive processes in which the participants are engaged. This interpretive work can be of both intrapsychological as well as of interpsychological nature.

Peer Interaction

In our student-centered classroom, collaboration within peer groups is the most common form of interaction. Although the interaction with a more skilled partner may be the most effective learning situation, the collaboration with individuals of equal or even less advanced skills also yields cognitive benefits (Rogoff 1990). To see how peer groups negotiate meaning in small groups, I studied peer interactions during concept mapping activities. In the following episode, Peter, Eldon, and Michael were constructing a map including, among other things, the concepts wave, interference, constructive interference, destructive interference, crest, trough, phase, node, and antinode. At that time, the students had completed three experiments related to the wave character of light and had done some cursory reading of the relevant chapters in their textbook. I had asked the students to construct the maps without any other resources such as textbooks. The results of their effort in this episode can be seen in Figure 1.

> Eldon: See, what we should actually do is put something like a, because we were talking about this earlier, put crest and trough like here, when the crest and trough miss there, is constructive, right?

> Peter: They can, they can. They have to be totally in line to be. Well it depends, they don't have to be exactly canceling each other out and they can still be constructive or destructive. Like you have a wave like this (Paul is drawing overlapping waves). Well, not exactly like that. I don't know.

> Michael: Like the interference still exists what you probably have, any crest you have is gonna be constructive.

> Peter: Yeah

> Michael: It still holds.

Eldon suggested linking the concepts of crest and trough to describe constructive interference. In his opinion, crest and trough have to "miss" to

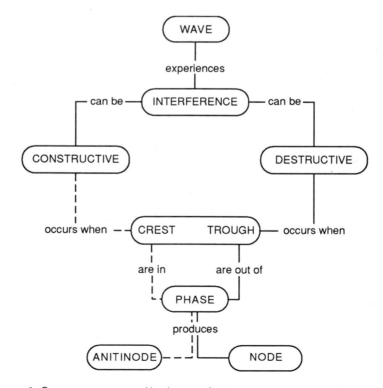

Figure 1. Concept map prepared by three students

make constructive interference. Peter was not satisfied with such a connection. He remembered that unless a crest and trough are exactly in phase, the waves do not completely cancel to form a node. Michael agreed with Peter, and Peter's "Yeah" confirmed that both thought they had the same understanding, or at least they acted as if their understanding was the same. According to Newman et al. (1989), such vagueness should not trouble us, and they ascribed to it the key element for the process of appropriation through which interpsychologically constructed meaning is reconstructed by the individual. As the following continuation of the dialogue shows, Eldon realized that he did not construct the same understanding as Michael and Peter.

Eldon: How did we want to do it, how did we want to do it here?

Peter: See, we can only link that over there.

Eldon: Sorry?

Peter: We can always link like, which is caused when two, when two ...

Eldon: When two crests of light (looks at Peter for confirmation) constructive or nodes, you have to like that (orders concepts on table), then to constructive, when the two crests align, when they are in phase, right?

Peter: Yeah

Eldon: Right, so you can get antinode, and then when the crest and trough are out of phase, and they cancel each other, you get nodes. I guess this makes sense.

Although Peter's first attempt to help him back into the joint effort failed, Eldon was able to complete Peter's second sentence. Peter confirmed both verbally ("Yeah") and with a nod of his head that Eldon did construct the same meaning. From the fact that Eldon completed Peter's sentence, we could have inferred the co-construction of meaning (Granott 1991). Peter's double confirmation solidifies this inference. Not only did Eldon align his understanding with that of the others, but he also elaborated his own prior understanding. His earlier claim that crest and trough have to miss for constructive interference to occur developed into the claim that crests have to align or have to be "in phase." He then continued to propose connections to the terms "node" and "antinode." Thus, Eldon developed from using a simplistic expression to one that would be acceptable by the community of canonical science. Such developments in the ability of a student to use the language of science are the reasons for demands to increase student participation in classroom discourse (Lemke 1990).

In the next segment, Michael, who had not contributed very much in the earlier excerpts, made a significant move which would significantly affect the look of the concept map. (Underlining marks show overlapping speech.)

Michael: Constructive occurs when, maybe we just bring, put together crest and trough

Eldon: Yeah

Michael: Yeah, in phase produces <u>you</u>

Peter: <u>You align</u>, how do you want to do it? Constructive when two

Michael: Constructive occurs when crest and trough, like they shouldn't be together.

Eldon: Make crest and trough one concept

Michael: Yeah

Peter: You want it to be one?

Eldon: Yeah

Michael: Yeah, so that you can say like you can say something like this, interference can be destructive, destructive occurs when crest and trough in phase produces node, and then you can say, when constructive occurs, when crest and trough are out of phase, Ah

Peter: Constructive produces an antinode.

As Figure 1 shows, the group ultimately accepted Michael's proposal to combine the concepts of crest and trough into one concept circle. The ease with which the proposal was accepted indicates that the participants understood that their meanings were the same. This excerpt also shows Michael going through a transformation in his expression similar to that through which Eldon went on earlier. His "like they shouldn't be together" changed to "in phase," and "produces a node." In essence, Michael formulated the key propositions that constituted this part of the concept map. The following exchanges confirmed and stabilized this formulation:

Eldon: Yeah, put, put crest and crest and trough across right here and two arrows coming in from constructive and an arrow coming in from destructive.

Peter: Why don't we do it just here, because it's still part of

Eldon: <u>Yeah, go ahead</u>, put just (Peter writes something).

Yeah, it's OK.

Peter: OK, so we want it there, is everyone saying?

Michael: OK, ah, constructive occurs when crest and trough are in phase, put like in phase there, are in

Peter: Constructive does? (Looks at Michael) Because I've seen, what I've done, when you say in phase that means that lines are (word inaudible) against each other. (To Michael) Is that what you wanted? Constructive and crest and trough are out

Michael: And like interference, like produces node, produces antinode, respectively.

These exchanges also served as feedback to ascertain that all members of the group constructed the same understanding. Peter again expressed his reservations about the fact that even if the phases of two waves are shifted, it is not necessarily a case of destructive interference. But ultimately he accepted the formulation which they had arrived at because the dialogue shifted to other areas after Michael's last remark. The excerpt does not indicate to us whether Peter agreed because of conviction, or to avoid conflict and get on with the task.

The group's acceptance of Michael's formulation is remarkable because both Peter and Eldon showed much higher achievement in all their courses. Peter's achievement in physics had always been one or two grades above Michael's, while Eldon had demonstrated superior language competence at his grade level. This incidence was only one of many in which I have observed that higher ability students do not necessarily monopolize the discourse within a group. In fact, weaker students (in terms of more traditional tests of individual achievement) often contribute as much to the discourse and thus to the co-construction of knowledge in a group as do the high-ability students. These contributions are not always the ones that will be accepted, but they help these students develop their own facilities in using scientific language. The above excerpts also illustrate how the work in small groups permits students to make the transition from colloquial to scientific language. Such transitions are usually continuous and have to be facilitated by the teacher (Lemke 1990). In a sense, these transitions correspond to a reconstruction of students' cognitive frameworks from prescientific to scientific ones.

A CLASS NEGOTIATES THE MEANING OF DISCREPANT EXPERIMENTAL RESULTS

In our physics courses, students do not receive ready-made procedures ("cookbook recipes") for an experiment. Thus, discussions about the meaning of the focus question and the necessary experimental design begin immediately in each student group. Consequently, different groups will decide on varying experimental designs, which lead to the "same" or, more interestingly, to "different" results and claims. For example, while investigating the question "What is the relationship between the mass and the acceleration of falling bodies?" the students designed three types of experiments, all leading to apparently different results.

Before the experiment, the students in each group discussed possible designs for their experiment, then went about setting up for the investigation. The students in the various groups soon realized that other groups had different experimental designs. Because this experiment was conducted about four weeks into the course, the students fell back into their traditional mode of schooling and asked the teacher for "the right set-up," or the set-up that would give them the correct answer. I did not, however, provide any answer and encouraged students to develop their own experiments. To find the relationship between mass and acceleration, the students developed three designs (Fig. 2).

For all three experimental designs, the students had chosen a "picket fence" to trigger a photo gate run in conjunction with a computer resident timer and software package in order to determine the time needed for the object to travel the distance from one bar to the next. In Design 1, the students changed the mass attached to the picket fence and dropped it through the photo gate. The designers of Experiment 2 used a frictionless airtrack and accelerated the picket fence mounted on a cart by using varying masses on a string connected to the cart via a low friction pulley. In Design 3, the picket fence was mounted on a frictionless cart to which varying masses could be attached. The cart (and its cargo) moved on an inclined airtrack. Using the data they had collected, the students generated—with the aid of the software package—velocity-time and acceleration-time data tables and graphs. The students recorded the acceleration for each mass and then analyzed the data by using another software package for graphical analysis. Once all reports had been submitted, I produced an acetate sheet for overhead projection (Fig. 2) that included both the experimental designs and the various results to serve as a basis for a class discussion.

The ensuing discussion was enlightening for all of us. Virtually all of the students participated in the exchange, which reminded me of the discussions sometimes found in professional journals. Recognizing that the experiments had yielded different relationships, the students immediately focused on the concept of "free-falling." Those who performed the experiment according to Design 1 argued that theirs was the only one in which the whole system was falling freely. The defenders of Design 3, however, argued that in their experiment the only force, and thus acceleration, was that due to gravity. Consequently, their results also represented a free-fall experiment. They interpreted the differences as stemming from the incline which deflects acceleration. I pointed out that Design 3 was of historical relevance because it was essentially the one used by Galileo, one of the fathers of modern science. Design 2 was the most difficult to be integrated. However, the students concluded, supported by careful questioning, that part of the accelerated system, the cart, did not contribute to the accelerating force. For large weights,

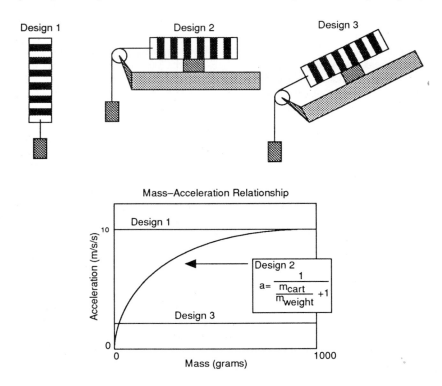

Figure 2. What is the relationship between mass and acceleration?

the mass of the cart is negligible and the results approach those of the free-falling picket in Design 1. For smaller and smaller masses that provide the force for acceleration of the system, the acceleration goes to zero.

At this point, I introduced the argument based on Newton's Second Law that the whole system in Design 2 (cart and mass) was accelerated by the weight (force) of the hanging mass. This leads to a relationship between mass and acceleration that varies with the hanging mass as indicated in Figure 2. By using a graphing utility, the students plotted this function, permitting them to recognize the pattern they had achieved in their data. With questions such as "What would happen to the acceleration if the mass of the cart was zero?" "What would happen to the acceleration if the mass of the weight was zero?" and "What would happen to the acceleration if the mass of the weight was large compared to the mass of the cart?" I provided the starting point for an exploration that permitted students to construct an understanding of the equivalence of the representations at hand—both the equation and the graph being generalizations from the data they had collected. Then, for a meaningful integration of the three experiments, we focused on questions such as "What would happen if the Galileo groups had raised their air tracks until they reached the vertical?" "What would happen if the Design 2 groups had increased the mass 10, 50, or 100-fold?" and "How can the differences between the designs and results be integrated into one explanatory scheme?"

The discussion was very lively, sometimes to the point of breaking apart into small group interactions. It was obvious that all the students wanted to contribute and that the format of the whole class discussion, which demands turn-taking, did not allow them to contribute to the extent that they intended. Some groups found it difficult to defend their experiments and results given the severity of the attacks of other groups. At other times, silence followed a particularly well-formed argument to which others did not have an equally strong counter argument.

In the end, the students in all three classes had arrived at the following consensus. The accelerating force is due to gravity. If one takes account of the angle of the airtrack (Design 3), or of the mass of the horizontally moving cart (Design 2), then all three experiments will yield an acceleration due to gravity of about 10 m/s^2. However, only the groups using Design 1 measured acceleration of a free-fall directly, while the other groups did not take account of the additional factors. This discussion then led us directly into the next experiment in which we wanted to find out how the acceleration on an inclined airtrack depends on the angle of incline.

DISCUSSION

The purpose of this chapter was to illustrate the constructive efforts students engage in during their work in a physics laboratory and classroom. The perspectives I am providing here are those of a practicing classroom teacher who is also actively involved in conducting research. From a methodological point of view, my perspectives are those of a participant researcher, with all its advantages and shortcomings. Although I am not consciously thinking of research while I am teaching, and vice versa, the effect of each perspective on the other cannot be neglected. This experience of both perspectives affecting my work is most noticeable when I am planning the experiences for my students and when I am thinking and writing about the classroom implications of my own work and that of others. In this dialectical tension between the two ways of looking at teaching and research, I am developing further; developing also is the teaching-learning environment that I can make available to my students. Thus, as the environment shapes me and helps me learn more about teaching-learning contexts, I begin to affect and change this very environment. In my mind, the real impact that my research has is not that it enables me to talk about the findings in various public forums, but that it brings about changes in the classroom environment of which I am a part.

The examples I have shown clearly illustrate a science classroom/laboratory as a construction site—a site where students individually, in peer groups, with the teacher in small groups, and as a whole class are involved in the active process of constructing knowledge. Although individuals construct knowledge on their own without direct social interaction (such as Michael), a more common situation in our classroom is the collaborative construction of knowledge. Even when students learn in a group, there still exist problems. For although many opportunities for the construction of relationships in collaborative settings exist, students do not spontaneously exploit many of the logical features that are possible (Pea et al. 1987; Roschelle 1991). Initially in my teaching career, I naively thought that students could learn the ways of science through pure and unguided discovery, without any assistance. I am now much less optimistic, although I know from personal experience that one can learn a new subject by tinkering and interacting with books. But here again the construction of knowledge is a social event, owing to the social nature of language and other sign systems (such as mathematical formalisms or graphs). Working with others is especially important when the research on students' naive conceptions or alternative frameworks is considered. When left to her own spontaneous discoveries, "the child as intuitive scientist [all too often] arrives at theories of how the physical or mathematical world works that are at odds with appropriate formal theories" (Pea 1987, 135). The first example illustrated how an individual, Michael, constructed new knowledge by interacting with his surroundings, that is, the physical arrangement of the apparatus and the printouts of distance-, velocity-, and acceleration-time graphs. This construction was characterized by the effort and time which Michael was willing to invest in order to achieve the understanding he desired. Most often, however, our students are not willing to invest as much as Michael did. Fortunately, those situations where students learn by interacting with others are much more common and much more efficient for classroom learning. The work in collaborative groups both facilitates and accelerates the construction of new conceptual and procedural knowledge.

Ideally, students find themselves in one-on-one learning situations with a teacher. For practical reasons, however, ordinary classrooms cannot provide such situations on a regular basis because the teacher cannot be available to all students at all times. Here, two other schemes work. Either a more advanced peer collaborates with a less advanced student, or peers of equal status work together. In both cases, the construction of knowledge and negotiation of meaning may still be supported as skills and conceptual knowledge are assembled in a social situation and are interpreted by the individual group members in their own terms. In the context of this book, both the notion of scaffolding and that of appropriation of socially constructed knowledge by individuals ask for clarification. To do this, it is necessary to look at the notion of the "zone of proximal development."

Vygotsky created the concept of the zone of proximal development to account for the difference between a child's "actual developmental level as determined by independent problem solving" and the level of "potential development as determined through problem solving under adult guidance or in collaboration with more capable peers" (Vygotsky 1978, 86). To Newman et al. (1989), this zone becomes the "construction zone," where individuals construct new knowledge by interacting with others. As the participants in a teaching-learning situation co-construct new conceptual knowledge or skills publicly, they appropriate knowledge and skills, that is, they reconstruct their personal knowledge. Whatever it is that the members of a group co-construct in public, each participant has to construct his own version of it. Both the internal and external constructive processes occur simultaneously.

When an adult, an expert, or a more capable peer facilitates the performance of a student so that achievement is at a higher level than was achieved individually, the process is known as "scaffolding." Recently, the concept of "cognitive apprenticeship" has emerged to describe the learning through expert-guided experience on cognitive and metacognitive skills (Brown et al. 1989; Collins et al. 1989; Rogoff 1990). Cognitive apprenticeship emphasizes authentic practice. "Conceptual and factual knowledge thus are learned in the terms of their uses in a variety of contexts, encouraging both a deeper understanding of the meaning of the concepts and facts themselves and a rich web of memorable associations between them and the problem-solving contexts" (Collins et al. 457). The point to emphasize is that the more skilled partner also gains understanding in this process of coaching the less skilled learner in the zone of proximal development. Research in peer coaching has shown that more skilled partners elaborate their own understanding as they attempt to facilitate. Skilled partners, whether these are mothers interacting with their children, teachers interacting with students in their classrooms, or architects in a design studio, gain an understanding not only of the topic, but also of the communication and the needs of the student with whom they interact (Rogoff 1990; Schön 1987). The metaphor of apprenticeship also emphasizes the active role a student has to take as she engages in learning the ways of a (sub)culture by being involved in collaboration, shared activity, and problem solving.

I had conceptualized my mentorship of Rod in part as an apprenticeship. During the eight weeks of interaction, we spent about 100 hours in informal conversations about mathematics, physics, and chemistry. In addition, we met formally for a total of 56 hours during which we discussed textbook topics, worked on textbook-related problems, and constructed a series of experiments to resolve questions to which the textbooks did not provide answers. During this relationship, my goal was for Rod to increasingly use symbols, to think, to argue, to act, and to interact with me according to the rules of canonical science, all basic goals of apprenticeship (Brown et al. 1989; Lave 1988; Rogoff 1990). When we worked on problems in a new topical area or with new problems that Rod brought to the session, I often modeled the problem solving by thinking aloud, thus making available for reflection my construction of the problem. Subsequently, we reviewed the problem, highlighted general principles, and invoked general heuristics that could help in the solution of similar problems. In this way, Rod participated as observer in expert behavior at two levels: first, he observed my approach to the problem; second, he experienced metacognitive activities through shared reflections on the process of searching a solution. We subsequently used the same two-tiered approach for his attempts in solving problems. First, Rod thought through a problem, alone or with my support (coaching); second, I assisted him to reflect on his own problem-solving process. Thereafter, we discussed alternate problem-solving strategies, multiple forms of representation, or alternate conceptualization of the same problem. For example, in high school problems involving torques, the systems are usually at equilibrium. When Newton's Second Law is invoked, both the net force and the net torque, that is, the sum of all forces/torques, equal zero. Thus, I offered for discussion different situations in which torques or forces did not add to zero. Our reviews also addressed the issue of multiple representations, or multiple routes to a solution of a problem. Finally, the transcripts revealed that I had modeled general heuristics such as classifying problems by general principles, qualitative assessment of answers, and order of magnitude assessment of answers.

The evidence provided here illustrates the power of the "guided participation" and "cognitive apprenticeship" metaphors: they encourage teaching strategies that help students connect the new to what they already know (Rogoff 1990). Because both expert and novice have to strive for mutual understanding, the expert learns more about the student's current framework. This allows for adjustments in the strategies that the teacher employs toward achieving a shared understanding, that is, toward achieving agreement that the participants' meanings are shared or fit. In respect to this situation, Tolstoy had said that "A teacher must . . . by regarding every imperfection in the pupil's comprehension, not as a defect of the pupil but as a defect in his own instruction, endeavor to develop in himself the ability of discovering new methods" (in Schön 1987, 105). Throughout the mentorship, I learned not only about the process of communication, but also more about the subject matter, and about strategies of teaching this subject matter. My experience of learning in this situation is not singular. Rogoff (1990) pointed out that experts also still develop breadth and depth of skill and understanding by getting involved in joint activities with the learner.

Enculturation

Closely linked to the metaphor of cognitive apprenticeship is the concept of enculturation. Historically, the primary means of bringing a newcomer into a community of knowers is through shared participation in the authentic practices and discourse patterns of the culture. Through shared participation, we learn our mother tongue; traditional apprenticeship training, medical, and law practice, and the training of research scientists in post-doctoral programs introduce novices into their work. "When someone learns in practice, he is initiated into the traditions of a community of practitioners and the practice world they inhabit. He learns their conversations, constraints, languages, and appreciative systems, their repertoires of exemplars, systematic knowledge, and patterns of knowing-in-action" (Schön 1987, 36). In the context of my classroom, I represent the community of physicists. As such, all interaction between students and me serves to scaffold their fluency and flexibility of using the language accepted by canonical science. Both the sessions with Rod and the classroom discussion to negotiate the discrepant results of an experiment offered a glimpse into my practices of engaging students in conversations. To achieve such purpose, the activities in which students engage are never an end in themselves, but are ways of getting students to talk the language of the subject. Although the discourse may be lopsided initially, such as during Rod's first attempts, the conversations become much more balanced as students develop greater expertise in the conversational and work practices of the field. As the second excerpt with Rod shows, once a student has acquired some skill proficiency, both teacher and student can get involved in joint inquiry.

The apprenticeship metaphor can be extended beyond the situation of a single expert and a single novice. Apprenticeship systems "often involve groups of novices (peers) who serve as resources for one another in exploring the new domain and aiding and challenging each other" (Rogoff 1990, 39). Although collaborative peer groups lack the scaffolding guidance of a more skilled expert, they can, nevertheless, achieve high levels of expertise. Levin (cited in Collins et al. 1989) found that in contemporary computer clubs, nonexperts can bootstrap their knowledge about computers to high levels of expertise by pooling their individual fragments of knowledge. In their effort of co-constructing new knowledge, the members in such clubs serve each other as scaffolds for carrying out new tasks. In our concept mapping example, Michael proposed an arrangement of concepts that was ultimately accepted by the group. It was not certain, however, whether he arrived at this construction on his own, apart from the agency of the group. As I have shown, Michael, Eldon, and Peter, through verbal and nonverbal cues, confirmed each other's intersubjectivity, that is, they confirmed the fit between their individual constructions and the publicly available statements. In such cases, it is difficult to say who is the originator of an idea, although it may seem that one person contributes more to the discourse than another. But in the concept mapping example, Michael who contributed least to the discourse initially and who was also the weakest of the three in overall academic achievement makes a decisive move toward constructing the map. When people collaborate, they increase the social resources available for solving problems, thereby making it hard to say who did what (Newman et al. 1989). More so, Newman et al. hold that the change in social organization may change the source of the problem and thus the nature of the task. The benefits of collaborating peer groups, even in the absence of a more capable partner, may arise from the distribution of the task to various members of the group. While some members may perform part of the task, others may be observing, guiding, or offering critique. Such was the case when Peter questioned Eldon's statement concerning constructive interference and when he offered a more refined interpretation of this concept. Although learning through interaction with an expert or a more skilled peer seems to be superior to the interaction with peers of equal status, Forman and Cazden (1985) indicate some advantages in the latter learning situation. They argued that for optimal learning to occur, learners have to reverse the interactional roles of the teaching-learning situation. "The only context in which children can reverse the interactional roles with the same intellectual content, giving directions as well as following them, and asking questions as well as answering them, is with peers" (344). This contention is substantiated by research findings on reciprocal teaching (Brown and Palincsar 1989). The question that poses itself in the present situation is, how much of this jointly constructed knowledge can or should be appropriated by the individual. Pea (1987) suggested that we should not expect all cognitive tasks to be constructed intrapsychologically for solo performance. Rather,

> solo performances are not realistic in terms of the ways in which intelligent activities are organized and accomplished in the real world. They are often collaborative, depend on resources beyond an individual's memory, and require the use of information handling tools. If we took away from practicing thinkers and practitioners what we take away from children to assess their cognitive functioning, scientists could not do science, mathematicians could not do math, historians could not do history, and policy makers could not make policy. (136–137)

Large group and whole class activities also present opportunities for co-constructing meaning in a social setting. Although the opportunities for individuals to contribute decrease with the increase in class size, the diversity of inputs from individuals seems to increase with class size. Such a tension is probably similar to that which scientists experience themselves. When they negotiate the products of their work within the laboratory group, individual contributions to the discourse are much more frequent than at the level of journal publication or conference contributions. The important aspect with large group discussion is, however, that we have an opportunity for a class to act as a community, a community that has to negotiate not only the meaning of cognitive objects but also the social norms under which these negotiations take place. Cobb et al. (1991) showed that students who negotiated social norms like "explaining and justifying solutions, trying to make sense of explanations given by others, indicating agreement and disagreement, and questioning alternatives in situations where a conflict between interpretations or solutions has become apparent" (7), tend to relate success in mathematics to their individual responsibility to construct meaning and skills. From teacher and student point of views, there are advantages and disadvantages to using large group discussions. One advantage for students is that they have to consider a larger variety of ideas which may challenge their own understandings of the task at hand and at the same time encourage them to generate more ideas for further investigations. The major disadvantage is—and I am thinking here of when I attend conferences—that large groups afford fewer opportunities to test one's own understanding. Rather, other forms of interacting within the classroom community arise, such as private discussions, talking out of turn, and expressing aggressive attitudes. Others have termed these behaviors "problems" (Lo et al. 1991) when these behaviors really seem to express failed opportunities for students to elaborate on their own understanding. From my own observations in many classrooms, small groups give rise to fewer problems of the type mentioned. However, suppression of these problems may neither be necessary nor beneficial, because a certain degree of chaos seems to be essential for establishing shared norms (Lo et al. 1991). For a teacher, the large group allows her to submit the views of canonical science, which can then be discussed by the classroom community. Small groups also afford these opportunities, but the teacher has to submit the canonical view many times. On the other hand, small groups allow for the deconstruction and reconstruction of these views with special attention to the needs of individuals. Thus, small group discussions between teacher and students may lead to very different interactions and to very different results in students' constructions (Newman et al. 1989).

Tools for Negotiating Meaning

In all examples of knowledge construction within a group, we used diagrams to facilitate the discourse between the participants. These diagrams (drawings, concept map, graphs) can be regarded as both *social* and *cognitive* objects (Pea et al. 1990). Rod and I used diagrams to substitute perceptual actions on these diagrams for logical actions on text propositions. Of equal importance to our conversations was the diagrams' function to facilitate the co-construction of the problem space, that is, the diagrams facilitated our efforts in making meaning of each other's utterances. By being able to refer to a diagram in both verbal and nonverbal ways, the participants in the discourse can reduce the cognitive load of the task. Similarly, the concept cards and the final concept map facilitated the discourse between Eldon, Michael, and Peter. This facilitation was promoted by the existence of the diagram as something that could be pointed to and talked about, which served to align and repair meaning between the participants. In this way, diagrams and concept maps had become "conversational artifacts" (Pea et al. 1990) or "mediational tools" (Roschelle 1990) for the construction of new meaning in a social setting. When such mediational tools or other material objects are not available, research should be able to show problems in the construction of meaning. Such a lack in mediation means may have contributed to the problems of constructing meaning that students in the Eichinger et al. (1991) study experienced. Lack of mediational means can devastate the efforts of even the best teacher (Roschelle 1990); and by extension, such a lack can devastate the efforts of many collaborative groups in their communicative efforts, and thus in their effort to construct meaningful knowledge.

CONCLUSION

At this point, I have come full circle. I began in the teaching profession as a naive scientist who wanted school children to experience science as I did. Through a phase of conceptualization in the wake of recent movements in education, epistemology, philosophy and history of science, cognitive and social psychology, and the sociology of knowledge, I have returned to the classroom to teach the way I always did, modified by the findings of my classroom research: a scientist modeling research skills, coaching students, and scaffolding their efforts to higher levels. In this process, I function as a resource and as a facilitator in the students' quest for new knowledge. As we have seen, this classroom offers opportunities for the construction of knowledge in various social settings, contexts that are similar to the ones I had encountered in graduate school: students working individually, in peer groups, individually or in groups with a teacher-facilitator, or in large group discussions. The students have come to function more independently, choose their own paths of inquiry, determine their own research agendas, and make their own discoveries in the laboratory, which I see as a construction site. Both small and large group interactions permit students to engage in, and develop, the language of science. By constraining the tasks, to some extent, and through our presence as facilitators and mentors, we can provide settings that are constrained and have minimal complexity so that students can construct conceptual and procedural knowledge with low risks of failure. We can provide settings in which students may experience the frustrations and the successes of those building new knowledge. We can provide settings in which our students think critically and generate new knowledge as a matter of course.

REFERENCES AND NOTE

Brown, A.L., and A.S. Palincsar. "Guided Cooperative Learning and Individual Knowledge Acquisition." In L.B. Resnick (Ed.), *Knowing Learning, and Instruction: Essays in Honor of Robert Glaser*. Hillsdale, NJ: Erlbaum, 1989. 393–451.

Brown, J.S., A. Collins, and P. Duguid. "Situated Cognition and the Culture of Learning." *Educational Researcher*, 18(1), (1989): 32–42.

Cobb, P. "Experiential, Cognitive, and Anthropological Perspectives in Mathematics Education." *For the Learning of Mathematics*, 9(2), (1989): 32–42.

Cobb, P., T. Wood, E. Yackel, J. Nicholls, G. Wheatley, B. Trigatti, and M. Perlwitz. "Problem-centered Mathematics Projects." *Journal of Research in Mathematics Education*, 22(1), (1991): 329.

Collins, A., J.S. Brown, and S. Newman. "Cognitive Apprenticeship: Teaching the Craft of Reading, Writing, and Mathematics." In L. Resnick (Ed.), *Cognition and Instruction: Issues and Agendas*. Hillsdale, NJ: Erlbaum, 1989. 453–494.

Davis, P.J., and J.H. Mason. "Notes on a Radical Constructivist Epistemethodology Applied to Didactic Situations." *Journal of Structural Learning*, 10(2), (1989): 157–176.

Eichinger, D.C., C.W. Anderson, A.S. Palincsar, and Y.M. David. "An Illustration of the Roles of Content Knowledge, Scientific Argument, and Social Norms in Collaborative Problemsolving." Paper, annual meeting of the American Educational Research Association, Chicago, 1991.

Forman, E.A., and C.B. Cazden. "Exploring Vygotskian Perspectives in Education." In J.V. Wertsch (Ed.), *Culture, Communication, and Cognition: Vygotskian Perspectives*. Cambridge: Cambridge University Press, 1985. 323–347.

Gergen, K.J. "Social Psychology and the Wrong Revolution." *European Journal of Social Psychology*, 19, (1989): 463–484.

Goodman, N. *Ways of Worldmaking*. Indianapolis, IN: Hackett Publishing Co., 1978.

Granott, N. "Play, Puzzles, and a Dilemma: Patterns of Interaction in the Co-construction of Knowledge." Paper, annual meeting of the American Educational Research Association, Chicago, 1991.

Hake, R.R. "Promoting Student Crossover to the Newtonian World." *American Journal of Physics*, 55(9), (1987): 878–884.

Hawkins, J., and R. Pea. "Tools for Bridging the Culture of Everyday and Scientific Thinking." *Journal of Research in Science Teaching*, 24(4), (1987): 291–307.

Heller, J.I., and F. Reif. "Prescribing Effective Human Problem-solving Processes: Problem Description in Physics." *Cognition and Instruction*, 1(2), (1984): 177–216.

Kaput, J.J. "Truth and Meaning in Representation Situations: Comments on the Greeno Contribution." Paper, annual meeting of the North American Chapter of the International Group for Psychology of Mathematics Education, DeKalb, IL, 1988.

Knorr-Cetina, K.D. *The Manufacture of Knowledge: An Essay on the Constructivist and Contextual Nature of Science*. Oxford: Pergamon Press, 1981.

Kroll, D.L. "Cooperative Problem-solving and Metacognition: A Case Study of Three Pairs of Women." *DAI,* (49), (1988): 2958A.

Lampert, M. "Knowing, Doing, and Teaching Multiplication." *Cognition and Instruction*, 3(4), (1986): 305–342.

Lave, J. *Cognition in Practice: Mind, Mathematics and Culture in Everyday Life*. Cambridge: Cambridge University Press, 1988.

Lemke, J.L. *Talking Science: Language, Learning and Values*. Norwood, NJ: Ablex Publishing, 1990.

Lo, J.J., G.H. Wheatley, and A.C. Smith. "Learning to Talk Mathematics." Paper, annual meeting of the American Educational Research Association, Chicago, 1991.

McArthur, D., C. Stasz, and M. Zmuidzinas. "Tutoring Techniques in Algebra." *Cognition and Instruction*, 7(3), (1990): 197–244.

Newman, D., P. Griffin, and M. Cole. *The Construction Zone: Working for Cognitive Change in School*. Cambridge: Cambridge University Press, 1989.

Pea, R.D. "Integrating Human and Computer Intelligence." In R.D. Pea and K. Sheingold (Eds.), *Mirrors of Minds: Patterns of Experience in Educational Computing*. Norwood, NJ: Ablex Publishing, 1987. 128–146.

Pea, R.D., M.D. Kurland, and J. Hawkins. "Logo and the Development of Thinking Skills." In R.D. Pea and K. Sheingold (Eds.), *Mirrors of Minds: Patterns of Experience in Educational Computing*. Norwood, NJ: Ablex Publishing, 1987. 178–197.

Pea, R.D., M. Sipusic, and S. Allen. "Seeing the Light on Optics: Classroom-based Research and Development of a Learning Environment for Conceptual Change." Seventh Annual Tel Aviv Workshop on Human Development: Development and Learning Environments, Tel Aviv, 1990.

Piaget, J. *Genetic Epistemology*. New York: Norton, 1970.

Reif, F. "Teaching Problem-solving: A Scientific Approach." *The Physics Teacher*, (May 1981): 310–316.

Reif, F. "Scientific Approaches to Science Education." *Physics Today*, (November 1986): 48–54.

Rogoff, B. *Apprenticeship in Thinking: Cognitive Development in Social Context*. New York: Oxford University Press, 1990.

Roschelle, J. "Designing for Conversations." Paper, AAAI Symposium on Knowledge-Based Environments for Learning and Teaching, Stanford, CA, March 1990, and at AERA Symposium on Dynamic Diagrams for Model-Based Science Learning, Boston, 1990.

Roschelle, J. "Microanalysis of Qualitative Physics: Opening the Black Box." Paper, annual meeting of the American Educational Research Association, Chicago, 1991.

Roth, W.M. "Open-ended Inquiry: How to Beat the Cookbook Blahs." *The Science Teacher*, 58(4), (1991): 40–47.

Schoenfeld, A. *Mathematical Problem-solving*. Orlando, FL: Academic Press, 1985.

Schön, D.A. *Educating the Reflective Practitioner*. San Francisco: Jossey-Bass, 1987.

Sharan, S. "Cooperative Learning in Small Groups: Recent Methods and Effects on Achievement, Attitudes, and Ethnic Relations." *Review of Educational Research*, 50, (1980): 241–271.

Slavin, R.E. *Cooperative Learning*. New York: Longman, 1983.

Tobin, K. "Changing Metaphors and Beliefs: A Master Switch for Teaching." *Theory into Practice*, 29(2), (1990): 122–127.

Tobin, K. "Constructivist Perspectives on Teacher Learning." Paper, annual meeting of the American Association for the Advancement of Science, Washington, DC, 1991.

Toulmin, S. "The Construal of Reality: Criticism in Modern and Post-modern Science." *Critical Inquiry*, 9, (1982): 93–111.

von Glasersfeld, E. "An Introduction to Radical Constructivism." In P. Watzlawick (Ed.), *The Invented Reality*. New York: Norton, 1984. 17–40.

von Glasersfeld, E. "Learning as a Constructive Activity." In C. Janvier (Ed.), *Problems of Representation in the Teaching and Learning of Mathematics*. Hillsdale, NJ: Erlbaum, 1987.

Vygotsky, L.S. *Mind in Society: The Development of Higher Psychological Processes*. Cambridge: Harvard University Press, 1978.

Wertsch, J.V. (Ed.). *Culture, Communication, and Cognition: Vygotskian Perspectives*. Cambridge: Cambridge University Press, 1985. The social constructivist perspective developed out of the work of the Russian psychologist Vygotsky, which is extensively reviewed in Wertsch, and is being successfully applied to collaborative learning situations in mathematics (Cobb 1989; Newman et al. 1989; Smith and Confrey 1991) and in reading (Brown and Palincsar 1989).

10

Learning Science in Multicultural Environments

Alejandro José Gallard

I feel, in many ways, inadequate to write this chapter. My inadequacy does not come from a lack of experience as a Hispanic and minority as I have been both since 1946. My discomfort comes more from the idea that I can only speak for myself, and not for other Hispanics or minorities. The notion that each of us has a set of individual experiences to draw upon daily tempers me from making broad-sweeping assumptions about Hispanics or minorities. Each student (minority and nonminority) has a unique set of experiences from which the science teacher can draw in order to facilitate learning, an important idea to consider. This is especially true in an educational system that tends to lump together students of diverse backgrounds under the heading of minority and to provide them with labels such as disadvantaged, at-risk and underrepresented in the sciences. Such labels thereby group minority and nonminority students together and further reduce an emphasis on the individual. It is important to consider the blurring of individuals' needs, because our efforts to improve the learning of science for individuals is focused on finding universal classroom solutions that are applicable to all students.

The ideas to be discussed in this chapter focus on the individual learner who has a need for a very specific set of *personalized* learning experiences. Personalized learning experiences are those that utilize and build upon a student's cultural tools. These cultural tools include language, cognitive referents such as myths, personal beliefs and metaphors, and images. I am suggesting that the focus for teaching science needs to be on the learner and learning, and that the emphasis needs to occur at a much deeper level than what current practice dictates. In my view, if the science teacher focuses on learners and learning with the attitude that what they know is not valid and thus must be replaced with appropriate knowledge, then science teaching is meaningless. It is meaningless because "whatever a student says in answer to a question (or problem) is what makes sense to the student at that moment" (see von Glasersfeld, Chapter 2). In essence, it is the student's way of understanding the world or her knowledge base. On the other hand, if the science teacher focuses on the learner and learning with the attitude that what a student knows is not only valid, but an integral part of teaching, then the teacher has reached the much deeper level in the schooling process alluded to previously. This suggests that a science teacher would have to closely define how she views learning, and by implication, knowledge.

All students come to school with cultural tools such as language, mental images, and personal beliefs that they use to make sense of the world in which they live. However, all students do not come to school with the same types of referents for making sense about school and society. Some students come to school with language, mental images, and personal beliefs that are more in line with what is valued and considered appropriate in American society. Their cultural tools are more closely represented in textbooks and in formal as well as informal assessment efforts. Other students come to school with languages, mental images, and personal beliefs that may, in some cases, be reflected and tolerated in the classroom, but never accepted. For example, some students may attend school with a rich description of science phenomena provided by their parents and based on cultural experiences, beliefs, and historical antecedents. A case in point are the daughters and sons of migrant workers. Their parents and grandparents have accumulated vast amounts of experiences (knowledge) about plants (botany) and pesticides (chemistry and ecology) that they have passed on to their children. Though the science teacher may be sympathetic and even tolerant of the student's understanding of science phenomena and the words used to express understanding, it is not considered real science. Thus, science teachers, by discounting this knowledge, i.e., not using it to further learning, de-legitimize a student's way of making sense.

Learning is a social process of making sense of experiences in terms of extant knowledge (Tobin 1991). Hence, learning and knowledge construction are synonymous: one learns what one knows, and one knows what one has learned. In other words, students use extant knowledge, based on culturally diverse experiences, to make sense of new experiences. Accordingly, existing knowledge comes from students interacting directly with the world around them. Consequently, one can separate neither the learner from knowledge, nor knowing from knowledge. Yet, this is precisely what we attempt to do when we inhibit people from making sense on the basis of what they know by discouraging inappropriate language and behavior that we perceive as unsuitable in the school context. For example, why do we insist that limited English-proficient students start communicating in the English language as soon as possible in school? Why not provide students with opportunities and time to develop essential conceptual understandings as opposed to having them learn science vocabulary words that are detached from real contexts?

The following scene is one that I witnessed while teaching science at a middle school. Though I never questioned her about what I saw, I can still vividly picture the scenario and hear her words.

> A sixth grade teacher had a Vietnamese student backed up against the wall, her face inches away from his, assuming much the same posture as a drill sergeant in the military. With her index finger wagging in this student's face, she admonished him in a voice loud enough for everyone in the hallway to hear because he dared speak Vietnamese in her class. "How dare you speak another language in front of me? We speak English in this school, and you had better stop embarrassing me and being disrespectful [by], speaking Vietnamese."

Another type of behavior that can be appropriate or inappropriate is the noise level in the classroom when students are working. Quite frankly, if you put me in a room with three of my friends, we will make enough noise for ten people. Yet, I have learned how to keep my voice low and check my enthusiasm while talking or interacting with Anglos.

In order for students to be able to use their cultural tools, science teachers must plan to facilitate this happening in their classrooms. Facilitating learning is a difficult task at best. However, it becomes even more problematic when ethnic or language minorities are present in the classroom. The following vignette offers an insight as to how a science teacher may use what students know in order to facilitate the learning of science.

The setting was a grade seven classroom in a rural combination middle and high school in the northern part of Florida. The classroom teacher was Mrs. Jones (pseudonym), a second year teacher who had majored in social studies. In the class period that I was observing, all of the students used Spanish as their first language, and they were the daughters and sons of Mexican or Central American migrant workers. Many of these students were already working in the fields. They had many years of exposure to botany through agriculture: they had direct experience through working in the fields and having listened to hours of their parents' and friends' conversations.

Mrs. Jones spoke Spanish with a fair degree of fluency. I was observing in her middle school classroom as part of an effort to help her teach science. Mrs. Jones was concerned because she was not certified to teach science, and she felt that her science content was deficient. Her questions and requests for help indicated that she was interested in whatever tips, tricks, or techniques I could share with her about teaching science.

In one of our interaction group sessions, I asked Mrs. Jones if she would consider doing a botany unit. Her response was that she did not know enough about botany to teach it. However, if I were to help and call in university teaching experts in botany, she would give it a try. At this time I suggested to her that her students were experts, and perhaps they needed an opportunity to share experiences they had accumulated about agriculture.

Mrs. Jones consulted with her students, and within a two-week period, they were designing experiments in the school's greenhouse. Furthermore, Mrs. Jones decided they could do the design and implementation of the experiments in either English or Spanish.

It was during a classroom observation period that I became most impressed with how she carried forth the planning for this science unit. Students had formed their own working groups and were discussing the plants they were growing—measuring, drawing, and recording their observations.

While students were working in their groups, Mrs. Jones would go about the room and participate in the various groups. The exchange between teacher and students was interesting in several ways. First, she did not insist that students use any descriptive vocabulary about plants other than what the students were familiar with. Second, whenever students engaged in off-task conversations, she participated.

Only when an outsider intervened and recommended that Mrs. Jones consider using her students as content experts did she realize that students must be allowed to draw on their own experiences if they are to learn. For her, content experts are those who write textbooks and work as scientists, not students. Thus, knowledge for her was not what students knew, but what the experts had published. This is a critical differentiation because as long as she viewed knowledge in this way, she could not facilitate learning that focused on what people knew as they negotiated and made sense of specific experiences.

The consequences of her beliefs about knowledge also affected her professionally. Mrs. Jones had a fear of teaching science, primarily because of her content inadequacy, and secondarily due to her lack of teaching experience. Her solution to the problem was for someone else to devise neat and colorful ideas on teaching science and be surrounded by content experts or people who had legitimate knowledge. Her beliefs about knowledge were like those of prospective students in science methods classes who accumulate reams of science activities to use on people they have never met so that they can be more successful teachers. I also do not believe that Mrs. Jones' beliefs about her students, experts, or knowledge, were very different from that of many science teachers who conceptualize their roles as dispensers of knowledge.

Notwithstanding, when Mrs. Jones did respond to the idea of students as experts, she provided an atmosphere that facilitated learning. She facilitated learning by not only encouraging students to demonstrate what they knew, but also by communicating this knowledge in the language with which they were most comfortable. Her actions were also sensitive because they took into consideration the complexity of the whole situation, which included encouraging the use of a language the student understood best. This is a critical point to consider because "in general, language is learned in the course of interaction with other speakers, because speaking is a form of interaction, and we modify our use of words and utterances when they do not yield the expected results. Insofar as a classroom is interactive it contributes to the students' linguistic development and also provides them with opportunities to witness the use of words in the context of the experiences to which they refer" (von Glasersfeld, this edition). Mrs. Jones' actions were also very difficult because she was sharing her authority as a teacher with the students and making students responsible for their learning.

Mrs. Jones' actions were an attempt to break away from traditional ways of teaching science; such a break meant questioning and acting on her new beliefs about the ways learning is managed in a classroom. In traditional teaching, we disenfranchise students when we insist they learn the formal language of science but ignore their initial conceptual understanding and disregard the rich experiences they bring to the classroom. Walsh (1991) voices questions that must be considered:

As teachers, how frequently do we explore the diverse and multiple realities of our students? That is, do we endeavor to understand the histories they bring with them and all the tensions these histories may imply? And, what about the sociocultural perspectives, semantic understandings, and intentional actions that are partially fashioned from history and from present lived relations? Do we interpret them only in light of our own? In the rush of daily school life, such considerations are probably also uncommon, despite the fact that they could help us better comprehend our students, guide instruction, and offer insight into how it is that students come to know. (1)

In the end, Mrs. Jones thought about teaching and learning in ways much different from those to which she was accustomed. While Mrs. Jones was grappling with her beliefs about knowledge, she also was empowering her students by recognizing the validity of their voice. Student voice is an important point to consider because "language is one of the principal ways people define themselves; through language we establish alliances with a community, undertake interactions with others, and communicate and receive information" (Walsh 1991, 1).

VOICE AND LANGUAGE IN THE SCIENCE CLASSROOM

A critical component of any classroom is language. In the United States, the use of a language other than English for instructional purposes has been very controversial. Researchers such as Cummins (1981, 1986), Cuevas (1984), Hakuta (1986), Ramirez (in press), Walsh (1991), Wong Fillmore (1991), and Wong Fillmore and Valadez (1986) have demonstrated that students' use of their primary language in the classroom adds to their ability to learn and excel in the English language. In the science classroom, the issue of language becomes a bit more complicated. Not only is there the issue of whether to use the English language but also whether to use the language of science and of the teacher (Carroll and Gallard 1991).

Communication is contextually constrained (Goldman 1989; Taylor 1987; Walsh 1991). Classroom communication is a matter of whether a science teacher stresses learning (as learning has been defined previously), science vocabulary accumulation, or English vocabulary (Secada 1991). For example, in the middle school science classroom that Carroll and Gallard (1991) studied, they found that science seems to be treated as either inexplicable magic or as a body of facts that can be understood simply through knowledge of key vocabulary words. Communication in the particular classroom studied is an example of the type of communication environment which Cummins (1981) refers to as context-reduced communication.

Cummins suggests that a learning environment can place an emphasis on context-reduced or context-embedded communication:

> Context-embedded communication derives from interpersonal involvement in a shared reality that reduces the need for explicit linguistic elaboration of the message. Context-reduced communication, on the other hand, derives from the fact that this shared reality cannot be assumed and thus linguistic messages must be elaborated precisely and explicitly so that the risk of misinterpretation is minimized. (11)

Reducing the risk of misinterpretation (context-reduced communication) seems to be another way of increasing one's appropriate vocabulary and making a person scientifically literate. Scribner (1984) (as cited in Goldman 1989) describes three metaphors for literacy: "literacy as adaptation, literacy as power, and literacy as a state of grace" (2).

Literacy as adaptation seems to parallel the notion of context-reduced communication. "Critical issues that arise within this metaphor do not concern definitions of literacy per se; rather, functional competencies and cultural and linguistic variation in language proficiencies necessary for successful functioning are the foci" (Goldman 1989, 2). I think this is an important idea to consider because behind all of the rhetoric of developing citizens who will participate in the decision-making process, being sensitive to minority students, and developing multicultural environments, we are really using language for function's sake as the framework for learning.

STUDENT DIFFERENCES ARE COMPLEX

An individual lives and grows within a milieu of very complex life patterns. For example, when I first arrived in this country and started schooling, I was a language minority. This is to say that Spanish was the language in which I communicated and was my basis for attaching meaning to the world. However, English was the language of instruction in school. As a consequence, every time instruction took place, I was linguistically disenfranchised—disenfranchised because I had no idea what was going on in the classroom. I was at a loss because my common-sense knowledge was in Spanish, yet what I heard, spoke, or was expected to do was in English. As a consequence, it was extremely difficult for me to make connections between ideas in Spanish while people used only English and insisted that I do the same. The situation was difficult for me because my teachers never provided opportunities for me to make sense in personally meaningful ways (i.e., the use of Spanish, or the opportunity to collaborate with others who spoke Spanish).

It would have been a simple solution if the school had hired a Spanish-speaking teacher. In the short run, this would have helped; however, my experiences were such that ultimately it probably would not have mattered. I came from a home where people were well educated, literate in English and Spanish, secure, spoke Spanish proudly, and maintained the old-country ways. Also, when we left Central America, it was not because we were a part of those who were socially and economically oppressed. On the contrary, we were one of the oppressors of the poor. Even though we were not wealthy from the standpoint of American standards, my family's legacy was so strong that economics was a matter of inconvenience. I was always reminded by my family that I was still part of the elite in Central America.

Thus, my experiences were not like other Hispanics in the United States. First, I am Central American and not from one of the Caribbean countries, Mexico, or South America. This means that my historical, social, and cultural past, including language, is different from those individuals from other Spanish-speaking parts of the world. My experiences were also different from those who have had to suffer a legacy of discrimination based on where they come from and who they are. For example, unlike large numbers of first- and second-generation Mexican-Americans with low socio-economic status, I have not had to suffer social and economic oppression and discrimination from either my native country or the United States.

It was not until I moved to California to live, study, and work that I became cognizant of issues such as the large number of Mexican-Americans in special education, disproportionate retention rates of Mexican-American students, and the bloody political battles fought against the notion of bilingual education. Monolinguals sadly lost their grandparents' and parents' languages; in order to fit in America, they chose to speak only English, and many were involved in the bilingual education resistance movement.

In the beginning, I did not understand the issues surrounding Chicanos because I had not experienced them. It was not until I was exposed to discrimination that is based solely on who you are, what you look like, how you speak, and where you come from that the notion of being a Chicano made sense. By becoming resistant to the Anglo view of an English-speaking Caucasian society as the only possibility for the United States, I developed pride for who I am. Words like *ese, trucha, chota* and *bato* became part of my vocabulary. In later years, these same words also happened to be part of my students' vocabulary when I taught in the public school system. In essence, I had become a part of a new American culture, one that did not fit with an Anglo- or Euro-centric view of life. This culture was also appearing in my classroom as represented by students with low economic and social status, or who were non-English speaking, African-Americans, recent immigrants, and first- or second-generation American students.

The reason that I had become a part of this new American culture and developed ways of speaking and acting that were resistant was because it became abundantly clear to me that students were being penalized for how they talked and where they came from. Specifically, if students spoke Mexican-Spanish, or were from Mexico, they were denied access to higher educational opportunities and were thus disenfranchised. They received treatment that was more directly oppressive than did students from Central America (at this time, the huge waves of Central Americans had not arrived), South America, or Spain. For example, while working as a bilingual elementary teacher, I observed many instances where non-Mexican students and parents would treat those from Mexico with disdain and contempt. I have absolutely no idea how many times I had people say to me, "I am not worried about you—you speak Spanish but not Mexican-Spanish, and you do not look like one."

Actually, these comments (beliefs) served to reinforce my attempts at resisting the Anglo-centric view. I distinctly remember being upset that I was not of a darker color and would deliberately speak to Anglos using *barrio* vocabulary while defying those who spoke *proper* Spanish to correct me. However, my wishes and behavior of resistance did not take place in the classroom, especially not at the university level. By this time, I had learned what schooling expected of me and that the culture of teaching would tolerate only one way of looking at the world. This was especially true of science classes where, for example, I would spend hours laboring over microscope drawings because the teacher kept insisting that what I was depicting was not what I was seeing. One day I realized that to continue looking through the microscope was foolish because what the teacher wanted were the drawings from the textbooks, so I gave them to her.

I had become a part of another culture that defines the role of a student in an American school, or any school in the world. Upon finishing my fifth year and receiving teaching credentials, I now became part of still another culture—teaching. As a consequence of both of these experiences (successful student and a new teacher), I had gone through evolutions such that the cultural sensitivity I brought into the classroom became a teaching mechanism for survival. In other words, because I could talk like my students talked, and I ate the same foods, I was able to build a rapport with them. Yet, my successful schooling, along with my *training* as a teacher, were experiences that served to distance me even further from the culture of my students.

I was not the only one who had gone through change. My Chicano colleagues did also. Specifically, we were no longer minority students struggling to make it; in many ways, we already had made it. At least we proved to those in control that we had acquired enough of their customs and habits to be acceptable. As survivors, and as humans struggling to fit professionally in this world, we learned what is not acceptable to those who set the standards. However, we had to partially turn our backs on cultural values that do not belong in a white elitist world. For example, we would make snickering remarks about *ese* in the Guayabera shirt, or how *ese bato* talked. Mind you, we were ridiculing others who talked and dressed as we did while sitting in our favorite Mexicano bar, slipping into our slang dialect, and wearing Guayabera shirts. Yet, instead of trying to make sense of why we now believed and acted as we did, our time was spent trying to understand why our students failed their spelling tests and were not more enthusiastic about mathematics and science. It had never occurred to us that what we were teaching did not make sense to our students because, linguistically and culturally, it was not relevant. What we failed to realize was that even though we had identified with our students through language and cultural sensitivity, it was not their language nor culture that counted. What we were not cognizant of was that although we shared language, as teachers we were perpetuating the thrusts of the mainstream culture that discriminated against our students. In other words, we never planned for meaningful personal experiences of the type that drew upon the language and culture of our students in ways that legitimized their knowledge.

We had learned that in order to survive we had to behave in an appropriate way, at the appropriate time. And we had learned that we had no voice in defining what is appropriate. Too many of our minority students do not learn soon enough about appropriate behavior as they go about trying to make sense in a unicultural educational setting. Unfortunately, as time goes on, students who are different become even more aware of how out of place they are. Sadly, their behavior becomes similar to a drowning swimmer (i.e., thrashing about until they become so exhausted, they drown).

The notion of appropriateness has far-reaching consequences to our students. I believe that the notion of appropriateness is a complex issue that affects the whole of one's past, present, and future. This is especially true in the world of science where terms are so exact, and the notion of reality is so absolute. In the simplest of views, a student receives an F because she does not provide the correct answer. Yet, what a student provides is what she knows, and what she knows is based on a lifetime of experiences. This lifetime could be 5 or 18 years; age does not matter. What does matter is that what a person is sure she knows has been declared null and void by someone who swears by what experts really know. It is the direct experiences of students and their everyday language that should be used to provide experiences that lead to conceptual understanding of science phenomena. The language of science is thus connected through everyday language and to the student's direct experiences. This involves creating and maintaining science classrooms that are rich in opportunities for students to use their everyday language as they attempt to make sense of the world.

CONCLUSION

I cannot speak for all people who have had to develop into appropriate beings at the sacrifice of a way of being that they already considered to be appropriate. However, I can try to relate what it has meant to me. For example, my legal name is Alejandro José Gallard Martinez. However, I have gone through a social and cultural metamorphosis, and now my name is Alexander Joseph Gallard. I have undergone a change that is controlled by, and has taken place in, a white elitist world. As a result, many of the values that a white elitist world embraces have been impressed upon me, perhaps in ways so subtle that I am not yet fully aware of their consequences. I am not arguing that I have become a white elitist, or even that it is wrong to be a white elitist; however, just like my Chicano friends, survival has meant establishing two sets of values: one as a home boy and the other as an educator in a world where the white elite have all the power cards and decide what is sacred and what is not.

In the June 19, 1991, issue of the *Chronicle of Higher Education*, there is an article about the first Ph.D. graduate of the Temple University program in Black Studies. Temple University has the only doctoral program of Black studies in the United States. What was interesting to me about this article was not that Dr. Coker was the program's first graduate, but that there was national debate on whether the notion of Afrocentricity was a discipline in itself or an interdisciplinary part of social studies. Actually, the debate is so resounding that the Ford Foundation did not give the program any money because it is new, does not have a long history of academic work, and is controversial (*Chronicle of Higher Education*, June 19, 1991).

What this story suggests to me is that what is studied is Eurocentric and approved by those who are in control. What we are asking our minority students to study also has the same Eurocentric focus. The only difference is that what our students study is sprinkled with pepper in September (*Dia de la Independencia*), February (Black History Month) and May (*Cinco de Mayo*). Is it any wonder that many of our minority students become so frustrated that they drop out of a world that they have little understanding of and use for—a world that makes it obvious that what counts as knowledge is not what the student knows, but what is represented in a textbook, or what the teacher considers correct science, or what is represented in a national science education framework.

Students who do not come from the Anglo world are at a disadvantage—or at a loss—in science classrooms because teachers have been trained to recognize the validity of a certain type of knowledge and ways to express this knowledge. Once students are deprived of opportunities for drawing on their own experiences in order to negotiate and construct meaning, science, or any other field of study, becomes meaningless. Science teachers must reexamine philosophies about knowledge and learning, and must reconsider their own roles in the classroom. Authority (scientific and political or social) must be shared as must the responsibility of learning.

It is easy to say that we understand our students' needs and plan to meet them; but even situations that give us tangential connections to students (e.g., a language or a skin color) do not guarantee that we will communicate with them—where they are—because each of us has had to sacrifice our own cultural identities in order to be accepted into the community and culture of American science educators.

REFERENCES

Carroll, P.S., and A.J. Gallard. "What Does Discourse Mean in a Science Classroom?" Paper, National Association of Research In Science Teaching, Lake Geneva, WI, 1991.

Cuevas, G. "Mathematics Learning in English as a Second Language." *Journal for Research In Mathematics Education*, 15, (1984): 134–144.

Cummins, J. "The Role of Primary Language Development in Promoting Educational Success for Language Minority Children." In California Department of Education (Ed.), *Schooling and Language Minority Students*. Los Angeles: Evaluation, Assessment, and Dissemination Center, 1981.

Cummins, J. "Empowering Minority Students: A Framework for Intervention." *Harvard Educational Review*, 56, (1986): 18-36.

Goldman, S.R. Introduction: "Contextual Issues in the Study of Second Language Literacy." In S.R. Goldman and H.T. Trueba (Eds.), *Becoming Literate in English as a Second Language*. Norwood NJ: Ablex Publishing, 1989. 1–8.

Hakuta, K. *Mirror of Language: The Debate on Bilingualism*. New York: Basic Books, Inc., 1986.

Ramirez, J.D. *Final Report: Longitudinal Study of Structured English Immersion Strategy, Early Exit and Late Exit Transitional Bilingual Education Programs for Language Minority Children*. Washington, DC: National Clearinghouse for Bilingual Education, 1991.

Secada, W.G. "Teaching Mathematics and Science to Limited English Proficient Students." *NABE NEWS*, 14(7), (1991): 15–16.

Taylor, O.L. *Cross-cultural Communication: An Essential Dimension of Effective Education*. Washington, DC: Mid-Atlantic Equity Center, 1987.

Tobin, K. "Constructivist Perspectives on Teacher Learning." Paper presented at the 11th Biennial Conference on Chemical Education, Atlanta, GA, 1991.

Walsh, C.E. *Issues of Language, Power, and Schooling for Puerto Ricans*. New York: Bergin and Garvey, 1991.

Wong Fillmore, L. "A Question for Early-childhood Programs: English First or Families First?" *NABE NEWS*, 14(7), (1991): 13–24, 19.

Wong Fillmore, L., and C. Valadez. "Teaching Bilingual Learners." In M. C. Wittrock (Ed.), *Handbook of Research on Teaching* (3d ed.). New York: Macmillan Publishing Co., 1986. 648–685.

11

Secondary Science Teachers and Constructivist Practice

James J. Gallagher

Recently, I saw the following inscription on a bumper sticker: SUBVERT THE DOMINANT PARADIGM. I asked myself if that bit of contemporary popular philosophy had any relevance to our work in science teaching. To examine the question, I asked, "What is the dominant paradigm in secondary science teaching?" Since that question shaped much of my research over the past 10 years, I gave the following answer:

In the dominant paradigm in secondary (and tertiary) science

- Teaching is equated with transmitting information to students.

- Learning is equated with acquiring that information, quite frequently by memorization.

- Assessment of learning is summative, to determine which students have been successful in acquiring the information.

This paradigm is so commonly practiced in secondary school and university science classes that other paradigms are of only minuscule influence (Gallagher 1989). For most secondary science teachers in the United States, the behaviorist-positivist tradition, which underlies this paradigm, has been deeply ingrained in their own education, both in science and in teaching. This tradition has been such an integral part of the scene in education in the sciences from primary school through college as to render alternatives such as constructivism, strange and often unwelcome.

In the behaviorist-positivist tradition, knowledge is viewed as a commodity to be transmitted to students whose responsibility is to learn it in a way that is faithful. Learning is often viewed as receiving and storing knowledge, and little thought typically is given to the processes by which "acquisition" occurs. The task of the teachers is usually viewed as transmitting scientific knowledge to students, whose responsibility is to learn it (Gallagher 1989).

Given the pervasiveness of this tradition in science education in the United States and many other parts of the world, it should be no surprise that it is difficult for teachers to make the transformation from behaviorist-positivist teaching to practice that is consonant with constructivism. In our work with a group of 40 middle school science and mathematics teachers, the change process was slow and arduous. On the positive side, many members of this group of teachers have been able to make the change, but it required substantial support, both internally from colleagues in their four schools and externally from university researchers (Gallagher 1991).

By some standards our task was easy because we began with the approval of the teachers' union and the administration in the district. Our program was a joint project of these two groups and our university. The project was designed to bring about improvement in teaching and learning of science and mathematics in four of eight public junior high schools in Toledo, Ohio. The need for improvement was evident to most professional staff in the district. As a result of the changing attitudes and demography, the traditional methods of teaching were seen as less successful than they were previously. However, there was a tendency for some teachers to place much of the responsibility for decreased success in science and mathematics on the students and the general conditions in the "world outside" (Vernile and Monteiro 1991). When we began the program, little of the blame for low achievement and interest in science and mathematics was assumed by the teachers.

Eight of the 40 teachers were assigned a new role as a result of the cooperative project between the union, the district, and the university. This role was embodied in a new position within the district called the *support teacher*. Four science support teachers and four mathematics support teachers were given specific training over a period of eight months prior to their assumption of their new role. This training focused on the application of findings, from research on teaching and learning in science and mathematics to the improvement of teachers' practice. Following initiation of the new role, training for support teachers continued, and university researchers documented changes in practices, departmental climate, interactions among colleagues, and students' learning.

Support teachers were given a half-time teaching load, with the remaining half of their load assigned to helping their peers in their department improve their teaching. It should also be noted that support teachers worked in pairs (one in mathematics and one in science in each of four junior high schools in the district).

Much of the support teachers' training for their new role focused on constructivist ideals. The science support teachers were engaged in thinking about their students' understanding of the subject matter as taught by teachers. This activity involved reading selections from the alternative frameworks research literature and then replicating these studies with their students. For the science support teachers, these replications provided a rich opportunity to examine new ideas. First, they found that their students did not understand the subject matter as well as expected, based on the data acquired earlier from the tests that had been given as part of their regular school work. They found that their students still had misconceptions, even after they had taught them the subject matter and after the students had passed the tests that they had given them. Second, this opened the door to examination of different approaches to teaching and learning that were not part of their usual repertoire. Third, constructivism became a useful theoretical framework in the teachers' effort at trying to understand their students' work as learners, their own work in classrooms with students, and their work with their peers.

A CONCEPTUAL MODEL TO UNDERSTAND TEACHER CHANGE

In our work with the teachers in Toledo, it became apparent that a new paradigm of teaching and learning was needed to guide changes in the teachers' practice. The new paradigm we constructed was designed to fill that immediate need. Later, we expanded this paradigm to help us understand the sequence of transformations that teachers were making as they moved from the dominant paradigm to the new paradigm. What was clear from our studies was that teachers made these changes slowly, with difficulty, and in small steps. We also found that teachers based their changes on a "new vision" of teaching which they constructed out of their experiences. Further, we observed that their practices changed more slowly than their vision. Lastly, and most importantly, we have noted that the vision also continues to change, sometimes by small steps and sometimes by leaps (Madsen et al. 1991).

A specific model of teaching and learning was used in helping teachers make the change from a behaviorist-positivist tradition to constructivist tradition. This model, which represents the first transition of teachers from the dominant paradigm to the new paradigm, is shown in Figure 1. The model, based on the work of Driver (1989), was useful because it helped teachers identify the actions and approaches that were part of their familiar routine in classrooms and laboratories. Moreover, the way the model was presented, their present, familiar actions were accepted as necessary (although not sufficient) to aid students in attaining understanding of the subject matter being taught. Thus, the model became a point of discourse in which experienced teachers could identify what they were doing that was acceptable and important, while at the same time they could identify the areas of shortfall in their traditional practice.

As is evident in Figure 1, the model contains three elements, all of which focus on students' learning: acquisition of scientific ideas, integration, and application.

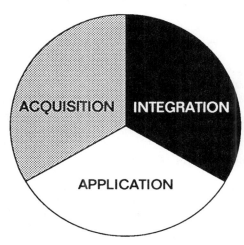

Acquisition is the point of identification for science teachers because they see it as descriptive of the work they did as students, and the work they require of their students. Further, they easily identify with the need for integration, as they recognize that students typically do not understand the subject matter taught in science classes. Lastly, science teachers also see the need for assisting students in applying or using scientific knowledge in their daily lives.

Figure 1. Transition Stage 1

With elaboration, the model can guide teachers to think about new practices, such as group work and writing to learn, and theories such as constructivism. This elaboration will follow, but first, it is useful to see how teachers can use the model to examine their own teaching and their students' learning.

The most common image of science teaching, found when observing science teachers at work, is a teacher presenting scientific ideas to a whole classroom of students in an oral mode. The lecture is a key method used in university science classes. Lecture-discussion, with occasional demonstrations, is most common in secondary science classes (Gallagher et al. 1991). In both of these cases, the teacher does most of the talking while students are passive recipients. Movies, the chalkboard, and overhead projectors add a visual dimension. These whole class methods of presentation typically are coupled with individual written work by students. In spite of much talk about "hands-on" approaches to science, they occur rather infrequently in most secondary science classes. Further, hands-on activities usually are designed to help students with acquisition, rather than to assist with integration or application of knowledge.

However, presentation is only one part of teaching science. Unfortunately, most science teachers operating in the dominant paradigm stop there. They help students acquire scientific information and then test them to see if they have acquired It. They then proceed to the next topic. While acquisition of scientific information is an important step in learning, students need to take at least two other steps if they are to understand and be able to use scientific knowledge. These steps are quite obvious, as presented in the model. Students need to *integrate* the new information with what they already know. They also need to learn how to *apply* their new knowledge beyond the classroom and connect their knowledge with their world outside school.

Most science teachers expect students to make sense of science and figure out its applications on their own through individual study and reflection. Both research and experience show that students are not very effective at either of these activities when left to do them individually, outside of class. Unguided, individual sense-making typically results in poor understanding and inability to apply knowledge. It is a source of many misconceptions that students develop. Also, it is a source of frustration to students because they are not able to understand and use scientific knowledge effectively. Using the model in Figure 1 as a guide, teachers can see that while acquisition is a necessary step, it is not sufficient. Moreover, they are guided to the need to require that students engage in activities that foster integration and application.

WHAT IS MEANT BY *INTEGRATION* AND *APPLICATION*?

Perhaps an elaboration of what is meant by *integration* and *application* would be helpful in determining actions needed by teachers to help their students in achieving these two goals. Following this elaboration, a more action-oriented approach will be considered.

Integration has two readily identifiable components: making sense and making connections. Making sense is the process by which students grasp relationships between what they already know and the new information that they are encountering as they study science. For most students, this involves elaborating naive concepts and renegotiating the meaning of many of their existing concepts as they study concepts that are currently scientifically acceptable. The knowledge that students bring with them to science classes is a powerful force in shaping their understanding of new information. Therefore, teachers must honor students' need to reconcile their existing ideas with the new information that is being taught.

Making connections is a second form of integration that students must also accomplish if they are to understand the subject matter of science. The connections referred to here are the connections that are a part of that subject matter. An important component of the power of scientific knowledge is its connectedness. As students learn science, comprehending these connections is an essential element in developing their understanding. Many of these connections are not obvious unless students develop the "habit of mind" to look for them. Others are less obvious and need to be arrived at through careful teacher guidance.

Application also has several components; one application of scientific knowledge is to learn more scientific knowledge. This application is a consequence of the connectedness of scientific knowledge—students can apply scientific knowledge that they possess to help them develop an understanding of additional scientific knowledge. Second, academic knowledge of science can be used to understand events in students' experiential world outside of school. Third, scientific knowledge, including the processes and values of science, can be used to solve problems that are part of students' experiential world. Unfortunately, many science teachers and most students have little experience in applying scientific knowledge in the manner sketched above. To be effective, teachers and teacher educators need to address this shortcoming in the knowledge base of most secondary science teachers.

HOW CAN TEACHERS HELP THEIR STUDENTS UNDERSTAND AND APPLY SCIENCE?

If students are to understand and apply scientific knowledge, including the processes and values of science, teachers will need to alter their approaches to teaching and testing. This does not mean science teachers' current work is incorrect; however, it is incomplete. Science teachers pride themselves on their ability at presenting scientific ideas to students. This can, and should, continue, as the methods that teachers use in presenting scientific ideas to students are very important in helping students acquire scientific ideas. It is at this stage of helping students acquire scientific information (or engaging students with scientific ideas) that all of the familiar teaching methods come into play. This is where hands-on activities play a very important role in helping students gain first-hand experiences with scientific phenomena and the ideas that help us to understand them. Lectures, demonstrations, labs, videos, films, etc. are all techniques that we use to help students to acquire scientific information and become engaged in thinking deeply about scientific ideas. But other approaches are needed to help students integrate and apply their newly forming scientific ideas.

What other approaches should science teachers use? A variety of strategies can be used to help students make sense of scientific ideas and to "test" the new scientific ideas against what they already know. *Group work* is a very important strategy when it is used to help students talk about their ideas with other students. Students will learn that their ideas frequently differ from ideas held by other students. As students work together to answer questions posed by the teacher, their understanding of scientific ideas is enriched by trying to make sense of their own ideas, those of other students with whom they are working, and the scientific ideas presented in class. The teacher's role in this type of group work is one of asking questions and causing students to think deeply about the match, or mismatch, between their own ideas and those presented by the teacher. Group work becomes even more powerful in helping students with sense-making when it is coupled with *group writing* (when students in groups are required to prepare one written answer to questions that challenge their understanding of an idea).

Concept mapping is another group task that helps students see connections and relationships in science. Most students, regardless of age, can learn how to construct concept maps to represent relationships among scientific ideas in graphic form. These visual images are very important in helping students to see relationships among scientific concepts. Even more important, having students construct their own concept maps engages students in deep, extended thought about the relationships, which is essential to building understanding.

Individual writing tasks need to be coupled with group tasks to cause students to think through and synthesize ideas independently. Both individual and group writing can be important aids in helping students develop understanding, and they are invaluable assessment tools to give teachers information about students' understanding.

These strategies also can be used to help students apply science group tasks and group writing; individual writing can help students learn how to apply scientific knowledge by simply giving students tasks that require application. Teachers can also increase students' capability in applying scientific knowledge by asking "the right question" in response to many of the questions students ask. For example, when a student asks a question about an illness, or an event in the news that has a scientific basis, the right question would be (after writing the student's question on the chalkboard to give it clear focus) "One of your classmates just asked this question. What have we learned that can help us in answering it?"

Applications of scientific knowledge can be increased if students search for applications of a particular concept or principle in newspapers, other media, or in the "real world" of their experiences. Teachers' questions can also generate applications of knowledge. For example, when studying evaporation, students could be asked why dishes rinsed in hot water are easier to dry than dishes rinsed in cold water.

If students are to integrate and apply scientific knowledge, assessment must also change (in both form and function). The form of assessment will need to change, with the familiar, easy-scoring, multiple-choice, completion, true-false, and matching tests being enhanced by assessment strategies that require students to construct written answers that examine their understanding and application of scientific knowledge. Other forms of assessment will also be needed, including oral presentations, assessment of students at work in groups, and longitudinal assessment of students using portfolios of their work.

The function of assessment will need to change in ways that allow teachers to learn more about students' learning, not just as a tool for giving grades, but also as a tool for gaining insight into students' understanding of, and capability to use, their scientific knowledge. Thus, assessment will take on a more formative role as an aid to students' learning, in addition to the traditional role of assigning grades.

The strategies described in the paragraphs above help students understand and use scientific knowledge. But these are only two of the benefits that accrue from this approach. Students will also learn higher-order thinking skills, skills of cooperative work in groups, and skills in discussing ideas and presenting them in oral and written form (all of which are important in the adult work place and to citizens living in a democratic society). Moreover, students who are engaged in this way of learning science will develop a more positive image of science and improved self-esteem as a result of their engagement in discourse with peers and teachers.

HELPING TEACHERS MAKE THE TRANSITION TO CONSTRUCTIVIST PRACTICE

Our experience with secondary science teachers strongly suggests that there are several stages involved in the transition from the behaviorist-positivist tradition of practice to constructivist practice. Three stages will be described below according to our present understanding of them. As we continue our work with other teachers in other settings, these ideas may be subject to revision.

Transition Stage 1 was described by four of the science teachers in our study as "a new vision of teaching and learning." It was their personal discovery, several months into our project, that teaching and learning required the three stages that comprise Figure 1. After nearly 10 months of cooperative work with us, they came to the realization that learning required acquisition, integration, and application if students were to understand and to be able to use scientific knowledge. Moreover, they recognized that teaching must establish the conditions whereby students cannot avoid extended engagement in activities fostering integration and application.

The consonant change in teaching caused the teachers considerable frustration. They recognized that students needed different kinds of classroom, laboratory, and homework experiences in order to integrate and apply the scientific subject matter that was being taught, and they also recognized that they had very limited skills in designing, planning, and delivering these experiences. The process of developing skills for successful inclusion of group work, group writing, individual writing, concept mapping, and appropriate questioning strategies as techniques for fostering students' integration and application of knowledge took even the most willing teachers more than a year. Teachers who were reluctant or more timid took longer. In addition to developing skills in these techniques, teachers also had to learn how to orchestrate them to form lessons that had continuity and made sense to students. (It must be remembered that the students of these teachers were also deeply entrenched in the behaviorist-positivist tradition since they were in grades seven and eight.)

This process was of long duration. We made the assumption that self-discovery is important to much of the process, as we did not think it was appropriate to lecture teachers on the use of constructivism in improving their teaching. We did supply teachers with readings and engaged them in seminars. We also encouraged and supported them as they explored new methods in their classes. We helped them to establish a new climate in their schools that fostered cooperative work among teachers. Our motto for this feature of the program was "Professionals Meeting to Resolve Problems." For most of the teachers in this project, the process of change is still continuing, even though we have been working together for four years now. Many of the 20 science teachers in the project are still working to make the transformation to Stage 1, and they are at various places in that transformation. At least two still appear to be resisting all the changes we are proposing.

The transition required in Stage 1 is a very substantial one for teachers, and it has a very significant effect on students. For teachers, the sense of rejuvenation that their "new vision of teaching and learning" brings is professionally and personally enriching. When students make sense of scientific knowledge, see connections among scientific ideas, and experience the use of scientific knowledge to understand and deal with the world around them, their excitement about science, learning, and themselves also leaps ahead. Attendance improves, achievement increases, classroom management problems are diminished, and the overall climate of the science classroom becomes more positive. Quite literally, classrooms of teachers who made the transformation into Stage 1 look and sound very differently from a year or two earlier.

Transition Stage 2 is represented graphically in Figure 2. It is characterized by two changes, and it is not clear if they are related to each other. One change is in the role of assessment, described earlier in this paper and summarized by the phrase, "assessment in the service of teaching and learning" (Kulm and Malcolm 1991). These changes in assessment are symbolized in Figure 1 by incorporating assessment with all three components of learning. Thus, assessment becomes formative as well as summative. Moreover, the nature of assessment, of necessity, will become broader to include more examples of students' construction of meaning and demonstrations of under-

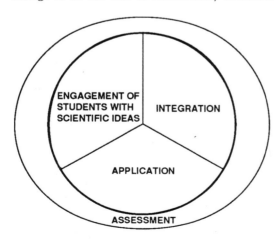

Figure 2. Transition Stage 2

standing, instead of mere recall and applications of knowledge. Also, it should have a cumulative character to it to measure growth over time. Thus, assessments will be more closely integrated throughout instruction, and they will be designed to give teachers and students feedback throughout the process of learning, instead of just at the end.

The other change that is included in this stage is represented by the replacement of the term *acquisition* by the phrase *engagement of students with scientific ideas. Acquisition* is a term that is closely aligned with the behaviorist tradition. It implies that knowledge is a commodity that can be acquired. In spite of its inappropriateness in a model of teaching that is based on constructivist premises, the term *acquisition* was included in our work with teachers after careful consideration. I could have used the phrase *engaging students with scientific ideas* instead of "acquisition" in nearly all places in this chapter, and the intention would have been expressed in terms that would have been more faithful to constructivism. However, that change would have resulted in a lessening of communication with the teachers for whom this model was intended, because the longer phrase does not form an easily comprehensible bridge for teachers between the two traditions (behaviorism-positivism and constructivism). In contrast, "acquisition" is immediately identifiable by teachers as a point of contact between their current work and the model.

That being said, there is a point, as teachers make their transformation from one tradition to the other, where their thinking alters from viewing learning as beginning with acquisition of scientific knowledge to viewing learning of science as engagement with scientific ideas. These ideas are then integrated by learners with what they already know, followed by application of their new knowledge to learning other new ideas and to their "real world" of experience. This part of the transformation is more than just semantic. It involves a transformation in teachers' views of the nature of knowledge and in their understanding of teaching and learning. While it is more subtle than Stage 1 in terms of demanding teachers' use of new techniques, it is very profound in terms of the nature of the interactions that will occur between students and teachers as they engage in discourse and other interactions about subject matter.

Transition Stage 3 occurs when teachers view learning as having starting points other than acquisition or engagement. For example, problem-focused learning often begins with application, as students learn about a "real problem" such as an environmental problem and then use it as a springboard for understanding its scientific basis. For many teachers who have learned and worked in the familiar paradigm, this stage represents another big step that requires new understandings of their scientific knowledge, new skills, and different teaching techniques. The difficulty of this transformation should be recognized by those who propose that teachers use problem-based teaching with their students. To do it effectively appears to require that the teachers also have successfully made Transformations 1 and 2.

CONCLUSION

Is all of this just "pie-in-the-sky?" No! It can work! It has been shown to work by the junior high school science and mathematics teachers in Toledo,

Ohio. For four years my colleagues and I have worked with and studied the science and mathematics teachers. in four junior high schools as they have adopted many of the strategies described above. They have changed the climate of their schools and their own classrooms. They have developed a new vision of teaching and learning and put it into practice in their classrooms. The effect has been remarkable. It has renewed their work as teachers. As one teacher put it, "This program has been the most important thing that has happened to me in 26 years of teaching." Even more important, it has resulted in substantial improvements in students' understanding of, and interest in, science. Students' self-confidence has also improved as teachers have employed a wider range of teaching and assessment strategies in their science classes.

Has it been easy? No. Nor has it been quick to occur. For the support teachers, the time needed for change to occur is measured in years of study and experimentation with new methods in their classes. First, they had to develop a new vision about teaching and learning based on their new knowledge base from research on teaching and learning, and then, they had to work to change their practices to achieve their vision. For their peers, who did not have the benefit of the extensive background in research-based knowledge learned through seminars and replication of studies, the process has been slower. It must be remembered that we have been working together for four years, and we have not fully reached our goals. For the support teachers and their peers, much of that time was filled with frustration, as they realized that they did not possess needed knowledge of the interconnections and applications of science that was central to the understanding that was desired for students. Moreover, even greater frustration occurred because they found they were deficient in skills and techniques for, among other things, effectively managing small group work and for using writing in science teaching, which are necessary parts of assisting students as they learn to integrate and apply scientific knowledge. However, they were buoyed throughout this by the belief that scientific understanding becomes a possibility for students whose teachers expand their vision about teaching and learning, and then develop new skills needed to attain that new vision. To do this is not easy, but for those who prevail, the rewards are great—for them, for their students, and for the society in which we live.

REFERENCES

Driver, R. "Changing Conceptions." In Philip Adey (Ed.), *Adolescent Development and School Science*. London: Falmer Press, 1989.

Gallagher, J.J. "Research on Secondary School Science Teachers Practices, Knowledge, and Beliefs: A Basis for Restructuring." In Matyas, M., K. Tobin, and B. Fraser (Eds.), *Looking into Windows: Qualitative Research in Science Education*. Washington, DC: American Association for the Advancement of Science, 1989.

Gallagher, J.J. "Framing the Science Support Teacher Program: Context, History, Assumptions, Training and Technical Assistance." Paper, annual meeting of the National Association for Research in Science Teaching, Fontana, WI, 1991.

Gallagher, J.J., L. Cavazos, Z.Y. Deng, D. Holland, and R. Yerrick. "Initial Report of a Survey of High School Science Teaching Practices." Unpublished working paper. East Lansing, MI: Michigan State University, 1991.

Kulm, G., and S. Malcolm (Eds.). *Science Assessment in the Service of Reform.* Washington, DC: American Association for the Advancement of Science, 1991.

Madsen, A., J.J. Gallagher, and P.E. Lanier. "A New Professional Role for Junior High School Science and Mathematics Teachers." Paper, annual meeting of the American Educational Research Association, Chicago, 1991.

Vernile, L., and R. Monteiro. "Another Pair of Eyes: On Reflectiveness of Ethnographic Observation of a Science Teacher." Paper, annual meeting of the National Association for Research in Science Teaching, Fontana, WI, 1991.

12

A Pragmatic View of Instructional Technology

Lloyd P. Rieber

Technology can be defined as the application of one or more knowledge bases for a useful purpose. Technology is represented in every practical object and activity around us. For example, the technology of "mobility" is illustrated in the doors we open, the stairs we climb, the elevators we ride, the cars we drive, the roads we drive on, and the maps we follow. Technologists consider what is known (i.e., basic science) and needed (i.e., real-life problems), and then they make decisions for action. Technology implies action and reflectivity as well as risk taking and inventiveness. Since knowledge bases and needs constantly change, revision and renewal are central to any technology. Consider how the design of many sidewalks, parking lots, and doors over the last decade have provided (at long last) at least minimal access for people with physical disabilities. Of course, change for the better is the intent, though not necessarily the result. Most early proponents of the automobile foresaw personal freedom and mobility, but few anticipated pollution and traffic.

Instructional technology can be defined as the creative application of what is known about learning and instruction (Romiszowski 1981; Knirk and Gustafson 1986). The term *technology* is best viewed in this context as a process tool for solving instructional problems. In this sense, instructional technology is a very practical and pragmatic field that is driven by clear goals. Interestingly, people outside of the field often equate instructional technology with one of many instructional media such as video or computers, or with models of instructional design such as those based on the instructional systems development (ISD) approach. However, these are mere instances of instructional technology that do not cover its breadth of scope and purpose any more than "addition" or "subtraction" defines mathematics or engineering. By definition, instructional technology is an *evolving* and *interactive process*. It is constantly reshaped by advances in the understanding of human learning and instructional practice.

Learning theory is arguably the most important foundation of instructional technology (Gagné and Glaser 1987). Advances in learning theory have had considerable influence on instructional technology. The roots of modern instructional technology are in behavioral learning through audio-visual devices (Saettler 1990; Reiser 1987). Most applications of computer-based learning have largely been extensions of the programmed instruction paradigm as set forth by Skinner (1968). The pioneering work of Patrick Suppes in the development of computer-assisted instruction (CAI) in arithmetic is a prime example (Suppes and Morningstar 1972). Instructional technology has proceeded from a behavioral to a neo-behavioral philosophy (Case and Bereiter 1984) that considers the learner to be a more active agent in the learning process. Cognitive psychology has been a major influence on instructional technology over the last 15 years, though probably more conceptually in the writings of the field than in actual practice (Gagné and Dick 1983; Hannafin and Rieber 1989; Shuell 1986).

Cognitive interpretations of instructional technology have usually taken either an "instructivist" or constructivist course. The term *instructivist* is defined here as approaches that involve refinements of existing ISD models to include advances from cognitive psychology while retaining the skill- or procedural-oriented flavor of the model. The identification of instructional goals is still the dominant influence on the resulting instructional designs, despite the attention placed on learner abilities and needs (Merrill et al. 1990). Decisions related to the learning goals and paths in achieving those goals are largely made for the learner. *Meaningful learning* is usually defined as a progression of stages from novice to expert or from lower- to higher-level learning outcomes. Instructional design is usually modeled on the selection and organization of a content area or domain, along with its continual integration with prior knowledge (Mayer 1983, 1984). Although the opportunity for a learner to become an active part of this process is never completely negated, the reality of instructivist models usually limits the role of the learner to surface-level decisions.

The second cognitive branch of instructional technology has been patterned after a constructivist philosophy closely aligned with the theories of Jean Piaget (Forman and Pufall 1988a; Fosnot 1989; Butts and Brown 1989; Goodman 1984; Watzlawick 1984). This approach is characterized by the creation of a rich variety of "cognitive tools," usually computer based, with which learners can experience and control a wide range of ideas. LOGO is probably the most well known computer-based application of constructivism (Papert 1980, 1987, 1988). LOGO was the result of a collaborative effort between the Massachusetts Institute of Technology (MIT) and Bolt, Beranek, and Newman (BBN), with initial funding by the National Science Foundation. Though LOGO involved the contributions of many people such as Wally Feurzeig, Daniel Bobrow, Hal Abelson, and Andy diSessa, Seymour Papert is usually credited as LOGO's chief developer. LOGO is a primary example of an application of constructivism based on "microworlds." As the name suggests, a *microworld* is a small, but complete, subset of reality, "a miniature universe" in which one can discover and explore a specific domain or area (Dede 1987; Papert 1981).

The purpose of this chapter is to suggest that instructional technology can be based on a confluence of instructivist and constructivist approaches, rather than necessitating a shift toward either paradigm. Microworlds represent an immediate application of this confluence by offering *guided-discovery* orientations to instructional design. Microworlds offer an appealing compromise between the potential excesses of instructivism (i.e., heavy top-down structure) and constructivism (i.e., instructional chaos). Learning environments based on microworlds are founded on personal discovery and experience, but yet are goal-oriented, because a learner's experience is naturally constrained by the microworld's parameters. These parameters, when carefully constructed, naturally direct the learning process toward predetermined goals.

In this chapter, the constructivist position on learning based on several milestones of Piaget's theories will be summarized and applied to computer applications. The concept of microworlds will be reviewed as well as two important areas that have a direct bearing on the design of microworlds for specific goals: intrinsic motivation and mental models. Intrinsic motivation is one of the important assumptions of the constructivist approach, but yet it is rarely described in detail by its advocates. Mental models comprise an important area of research in cognitive psychology. Mental model research is helpful in understanding human knowledge about the world and subsequent human interaction within it. Most of the microworlds created thus far, like most of the mental model research conducted to date, have been in technical areas such as physics and mathematics. Insights gained by mental model research closely parallels the goals of microworld design.

This chapter takes an eclectic view of instructional technology and is a beginning position statement based on my own personal experiences. I began my education career as an elementary classroom teacher at about the same time that the microcomputer "revolution" took place. My education as a teacher was heavily grounded in Piagetian learning theory; however, I found my attempt to translate those learning principles in the real world of the public school classroom difficult at times. Rather than concede to frustration when inconsistencies or confrontations arose between theoretical ideals and actual circumstances, I began a pattern of compromise and conciliation between philosophies and ideas in order to increase my understanding of learning and effectiveness as a teacher. Similarly, I encountered both the instructivist and constructivist interpretations of instructional technology almost simultaneously. Rather than claiming one interpretation and discounting the other, I merged the two as I designed and developed courseware for my classroom. So far, I have found the concept of computer-based microworlds, though a constructivist invention, to be an open channel for practicing this point of view.

Based on the philosophy that technologists are doers, not talkers, I will use a recently developed computer project, called *Space Shuttle Commander* (SSC) (Rieber 1991), throughout this chapter to present a set of unified examples of the many principles discussed. SSC is a "minicourse" that introduces students to Newton's laws of motion and is based on a merger of instructivism and constructivism. SSC is an adaptation of the "dynaturtle" microworld (diSessa and White 1982), but it is set in a simulated context of space exploration. SSC encourages students to imagine that they are the commander of the space shuttle and then takes them on a series of missions, as shown in Figure 1, during which students acquire a beginning understanding of the laws of motion. The course map of SSC is shown in Figure 2. At first glance, this resembles a flow chart typical of direct instruction. However, structure and sequence are only suggested, not imposed. Flight Lessons and

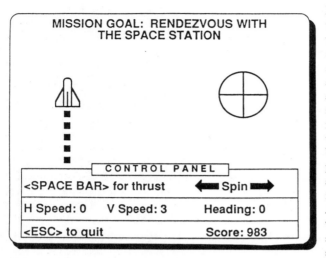

Figure 1. A representation of the computer screen during an episode of "Mission 5: Rendezvous." The animated "shuttle" is under student control. Arrow keys rotate the shuttle in 90-degree increments, and the space bar gives the shuttle a "kick" or thrust in the direction it is pointing. The goal of this mission is to maneuver the shuttle to the space station. (Not drawn to scale.)

Missions are the two units upon which SSC is designed. Each mission is a stand-alone microworld, and each flight lesson represents one interpretation of the knowledge base of motion principles which are represented in its related mission. An instructivist would probably see the flight lessons as the focus for sequential learning, starting with flight lesson 1, followed by its related mission as practice, then moving on to the next lesson. A constructivist would focus on the missions, but not in any particular order, and would either disavow the flight lessons entirely or use them as one of many possible "reference materials." While a more specific

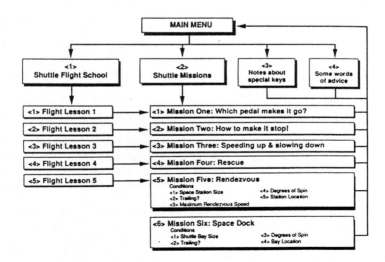

Figure 2. A course flowchart of *Space Shuttle Commander*. (Copyright 1991 by Lloyd Rieber.)

description of SSC and its related research can be found elsewhere (Rieber 1990a; Rieber 1990b; Rieber, in press-a), snapshots of SSC will be presented here to show examples of a working collaboration of instructivism and constructivism.

CONSTRUCTIVISM

Central to constructivism is the idea that learning involves individual constructions of knowledge and occurs through the natural interaction with one's environment or "culture." Constructivists assert that there is a direct relationship between learning and the degree to which the environment provides a rich source of engaging experiences. Computer enthusiasts such as Papert (1980) feel that the computer offers a powerful medium for the exploration and discovery of many ideas, particularly in science and mathematics, just as a young child might explore the concepts of volume given a sandbox, and mass and momentum with marbles.

Piaget's theory of cognitive development essentially has two parts: stage dependent and stage independent (Mayer 1983). Most of the attention to Piaget's theory is usually given to the stage dependent part or that which refers to the four stages of development that people presumably progress through: sensorimotor, preoperational, concrete operations, and formal operations. However, Piaget's stage independent theory is of most interest in the design of computer microworlds. It concerns two processes: the *adaptation* of an individual to survive and flourish in an ever-changing environment, and the need for *organization*, or stability and coherence of the world knowledge and experiences learned "to date." Piaget referred to this organized set of knowledge and understanding as internal structures or schemata. The processes of adaptation and organization create an intrinsic conflict which provides the platform for lifelong learning. The goals of one process are in contrast to the goals of the other. Life entails a continual balancing between the two that is accomplished through the process of *equilibration. Assimilation* and *accommodation* are the two well-known mechanisms that carry out the process of equilibration. Assimilation subsumes new information under an already understood structure or schemata. A baby who has learned to throw a tennis ball is also likely to throw a rubber ball as well as an orange. Both the rubber ball and the orange fit the structure that one "throws round objects." Accommodation builds new structures from old structures when incoming new information no longer fits an existing structure. The baby soon learns to accommodate the fact that some round objects are to be thrown, and others are to be eaten.

Assimilation and accommodation result from the natural conflict that occurs through everyday encounters with the environment. Learning can only occur when an individual is in a state of disequilibrium, or cognitive conflict. Given new information, an individual intrinsically seeks to assimilate, or incorporate, the information into preexisting structures. Accommodation is triggered when new information no longer fits established structures, necessitating the adaption of new structures.

Constructivism consists of three properties: epistemic conflict, self-reflection, and self-regulation (Forman and Pufall 1988b). *Epistemic conflict* is the Piagetian process of equilibration. Learning occurs when the environment presents a problem that is outside of the individual's repertoire and a problem that the individual seeks to resolve. Even though the conflict may be induced by some external agent such as a teacher, resolution is achieved only by the individual. Each resolution is an individual construction. *Self-reflection* involves an individual's attempt at objectively and explicitly representing reality; it is a response to conflict. *Self-regulation* is the spontaneous restructuring of thought, which is embodied in the Piagetian process of accommodation and assimilation. Old mental structures are refined to be more comprehensive, and new structures are formed as well. Conflict and reflection are necessarily prerequisite to self-regulation.

Proponents of constructivism emphasize the quality of knowledge structures over their quantity. Learning is viewed not so much as the acquisition of knowledge but as the constant reconstruction of what is already known. Individuals do not simply add information to their knowledge "banks." They either revise existing mental structures to accept new information or formulate new structures based on old ones when an existing structure is no longer sufficient. As Forman and Pufall (1988b) note: "Central to constructivism is the assumption that to know is to continually reconstruct, to move from a more to a less intuitive state" (240).

One of the most promising attempts at creating computer environments that foster the construction (assimilation and accommodation) of knowledge are microworlds. Microworlds offer instructional designers two key advantages. First, microworlds present learners with experiences *within* specific boundaries of a domain. Second, microworlds offer learners with "stepping stones" *between* interconnected ideas within the domain by allowing rudimentary ideas to first become established and then transformed into more sophisticated aspects of the domain.

COMPUTER MICROWORLDS

A microworld is a small, but complete, subset of a specific environment or domain. Microworlds are often confused with simulations. Characteristics of the two can heavily overlap, depending on their design and, most important, on how they are used in a learning interaction. However, a microworld has two essential characteristics that distinguish it from a simulation. First, a microworld represents and contains the simplest model of a domain that is judged accurate and valid by an expert. Second, the parameters of a microworld match the cognitive level of the learner, thereby offering an entry point into the domain. For example, Cuisenaire rods act as a microworld for many mathematical ideas. They embody the manipulation of a set of introductory mathematical ideas that are valid from any mathematician's point of view but function at the level of the children who use them. Cuisenaire rods offer not a mathematical simulation, but they permit real mathematics to take place. On the other hand, simulations can be designed that do not offer any significant difference from real-life experiences, such as the sophisticated flight simulators used for training by the military and many major airlines. These simulations would not be considered microworlds for most people because they are designed to represent as many of the variables and factors of the real experience as possible. It is important to remember that microworlds, like simulations, do not have to be computer generated. However, the computational and graphical power of computers provide a powerful arsenal to microworld designers.

Simulations can become microworlds with design changes. A computer flight simulation can be designed to permit only limited control and manipulation with only one part of the aircraft, such as the rudder. In this sense, the simulation becomes a "rudder microworld." Similarly, many microworlds can easily become simulations. Consider a mathematical microworld that involves the estimation of distances, for example, by using the LOGO command FORWARD to move the turtle from one point to another on the screen with as few commands as possible. This microworld becomes a "whale search" simulation simply by changing the turtle into an animated "boat" and the screen target into a "whale." The mathematical microworld has not changed, only the context.

Papert (1980) suggests that microworlds should fulfill four criteria: they should be simple, general, useful, and syntonic. *Syntonic learning* loosely translates as "it goes together with" and, in a sense, this criterion subsumes the other three. Syntonic learning involves connecting new ideas to prior knowledge and engaging the learner in a never-ending pattern of going from the "known to the unknown."

Syntonic learning within a microworld relies on the natural tendency of a learner to seek equilibrium. Successful microworlds actually encourage learning conflicts to arise in order to activate the process of equilibration. It is only through the resolution of these conflicts that learning can take place. The trick is to structure the microworld so that learners have an environment in which conflict resolution is within their grasp. Microworlds offer learners the opportunity to exercise a cognitive or intellectual skill that they would be unable or unlikely to do on their own, either because there is no intrinsic reason to do so or because no sufficient tool is available with which to allow the learners to construct the experience. Constructivists assert that learner control is essential in microworlds, a point that contrasts with research on learner control of direct instruction, research which frequently suggests that learners are often poor judges of their own learning paths (Clark 1982; Steinberg 1977, 1989).

In SSC, each mission acts as a stand-alone microworld that simulates one particular aspect of Newton's laws of motion. One instructivist influence is on the overall design of the *series* of missions. The missions are hierarchically organized from simple to complex; early missions focus individually on the simplest ideas, and later missions combine and extend these ideas. For example, Missions 1, 2, and 3 take the learner through a series of activities that introduce the simplest aspects of Newton's first and second laws. At first, structure is heavily imposed but then reduced as the learner establishes a foothold with these concepts. The first three missions further constrain the learner's experience to one dimension. Mission 4, as illustrated in Figure 3, introduces the effects of two dimensions in a highly structured way so as to make the simplest relationships of two-dimensional motion as explicit as possible. This is accomplished by placing several artificial constraints on the microworld: (a) the amount of force that the learner can exert on the shuttle is limited; (b) the learner is completely prevented from manipulating the heading of the shuttle; (c) the learner has no control over the initial state of the

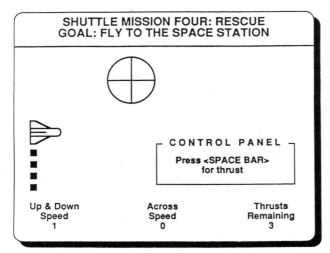

Figure 3. A representation of the computer screen during "Mission 4: Rescue." The goal of the mission is to maneuver a "disabled" shuttle to the space station. Students do not have control over the shuttle's rotation, and they only have enough fuel for three bursts of thrust. These constraints make the relationship of orthogonal forces more apparent. (Not drawn to scale.)

shuttle; as the mission begins, the shuttle is moving slowly in vertical fashion from bottom to top (i.e., toward a zero degree heading); and (d) the shuttle's heading is fixed at 90 degrees (i.e., directly to the right). These conditions limit all subsequent forces to be orthogonal to the shuttle's initial state. These somewhat complicated restraints, though difficult to express here in a pure narrative fashion, are easily presented to the learner through a "rescue" context where students are asked to imagine that their shuttle has been disabled by a collision with a small asteroid. The collision has resulted in limited control of the shuttle.

The final two missions, "Rendezvous," shown in Figure 1, and "Space Dock," shown in Figure 4, involve highly detailed simulations in which students can experience the full range of ideas presented in SSC. Students who attempt these final two missions early on but are not successful are encouraged by on-line coaching (and, hopefully, by the teacher as well) to go through earlier missions or to consult the flight lessons. The use of coaching in this way is an instructivist convention.

CREATING "OBJECTS TO THINK WITH"

LOGO represents the most ambitious attempt to date at creating computer-based microworlds in which learners have ready access to mathematical ideas and heuristic problem solving. Another is BOXER, a computer-based microworld that combines computational and text-processing tools in a programming environment based on interacting windows called "boxes" (diSessa and Abelson 1986). Turtle geometry is one such LOGO microworld that gives learners access to geometric principles through interactive graphics (see Abelson and diSessa 1981; Lockard et al. 1990; and Lukas and Lukas 1986 for further descriptions and discussions of turtle geometry). Research on LOGO is inconclusive (Clements 1985; Palumbo 1990; Walker 1987), perhaps because, as Papert (1987) contends, LOGO is meant to be part of a cultural

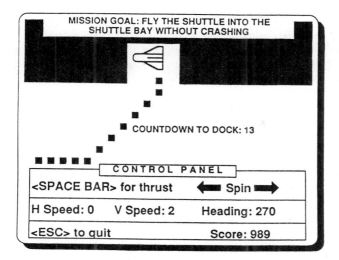

Figure 4. A representation of the computer screen during "Mission 5: Space Dock." The goal of this mission is to maneuver the shuttle to the shuttle "bay" of the space station without touching the walls of the bay or the station. (Not drawn to scale.)

influence on learning and considering its effects on learning in isolation is not defensible. The issue here is not the success of LOGO but the way in which LOGO was developed to achieve constructivist goals.

Successful LOGO learning experiences are founded on many key ideas, many associated with programming. For example, LOGO is procedural, which encourages "top-down" problem solving in which a large problem can be broken down into smaller, more manageable "chunks." However, the turtle embodies the essence of constructivism by being an "object to think with" (Papert 1980). Papert (1988) goes so far as to suggest that the turtle may represent a "mother structure" for differential geometry. Mother structures are the most fundamental ideas or concepts upon which cognition is built.

At the heart of constructivism is a search for good objects to think with. Almost anything can become a good object to think with: pots, pans, mud pies, blocks, Legos, and so on. Some are more flexible and generalizable to a variety of domains than others. Meaningful interaction with objects in the environment liberates and encourages the equilibrium process. The turtle is but one example of an object to think with that is made possible by computer technology. Papert contends that the computational power of the turtle makes certain ideas from the world of mathematics and problem solving accessible to children, ideas that were previously considered too "formal" or abstract to be understood. Several characteristics of the turtle, in particular, help make this possible: the turtle as a transitional object and the turtle as an aid to "debugging."

Young children often begin using LOGO for self-guided learning within minutes of encountering the turtle. This is largely achieved by the role of the turtle as a transitional object between themselves and the computer. The turtle is body syntonic with a child in that both share two important characteristics: a position and a heading. This simple fact has powerful consequences. From the start, even a young child has something in common with the turtle. This commonality immediately provides the bridge to new ideas. Young children quickly anthropomorphize the turtle, creating an ego-syntonic relationship with the turtle. This encourages the Piagetian concept of decentering in which young children begin to see the world from several perspectives. The anthropomorphization of the turtle also encourages the articulation and communication of mathematical ideas. Children begin to acquire the "vocabulary" of turtle geometry through their communications with the turtle. Since the language of the turtle is LOGO and the language of LOGO is mathematics, children have a way in which to verbalize mathematics (Papert 1980).

The second important characteristic of the turtle is as an aid to "debugging," which is the identification and correction of errors within a computer program. In most forms of instruction, errors are something to be avoided; errors imply failure. Constructivists accept the fact that errors are a natural consequence of interactions with the environment. Errors, instead of having negative connotations, are useful so long as they provide a rich source of information for subsequent interactions. Successful error handling drives adaptation. The informational feedback that errors provide is very potent, especially when a learner has a strong commitment to the action that triggered the error (i.e., Kulhavy 1977). Error detection is made intuitively obvious in LOGO with the turtle's role as a graphical tool. The turtle's animated graphics provide almost instantaneous graphic feedback to a learner. This rapid turnover of learner action and feedback encourages risk taking and hypothesis testing. The forming and testing of hypotheses through graphical feedback becomes more important as a learner's computer program becomes more complex.

SSC is a direct application of a physics microworld designed by Andy diSessa (1982) and uses LOGO, which involves a dynamic turtle, or "dynaturtle." The dynaturtle has one more characteristic beyond position and heading—velocity. The dynaturtle acts as a free-floating Newtonian particle whose velocity can be changed by giving it a "kick" in whichever direction it is pointing. Other applications of the dynaturtle microworld include White (1984) and White and Horwitz (1987). Like its cousin the dynaturtle, the animated space shuttle is under the learner's control and acts as a transitional object between the learner and Newtonian physics. However, there are instructivist influences on each of the SSC microworlds as evidenced by the term *mission*. A clear and simple goal is made overt to each student in each mission. The goal provides a simple tool for students to evaluate their interaction during the mission. All of the continuous feedback received from the microworld—the motion of the shuttle, the "trail" left by the shuttle, and the verbal information from the "control panel"—can be used to compare the shuttle's current state against the desired state (i.e., the goal). For example, the goal from mission 5, as shown in Figure 1, is "fly the shuttle to the space station." Norman (1988) referred to this use of feedback as bridging the gulf of evaluation, which is the amount of effort that a learner must exert to determine whether the intentions and expectations of an action have been met. Successful completion of the goal also provides a means to evaluate across a group of students, similar to the use of performance objectives in most instructivist models. Research conducted thus far has shown the missions to be an effective practice strategy (Rieber 1990a, b).

MENTAL MODELS

Daily activities require all of us to interact with a complex environment. Many researchers and theorists suggest that people form mental models of the physical world and the objects and activities within it (see Gentner and Stevens 1983, for a review). A mental model is an individual's conceptualization of a specific domain or "system." The purpose of mental model research is to lay out as precisely as possible how people understand a certain domain. The study of mental models combines the fields of cognitive psychology and artificial intelligence. An area of computer-based instruction related to mental model research is the development of intelligent tutoring systems that attempt to diagnose student errors and misconceptions within a domain and then prescribe instruction from an "expert" model through the use of a carefully designed computer interface (Clancy 1984; Wenger 1987).

Students develop and use mental models to help explain and solve general classes of models. Far from formal and static entities, mental models are loosely organized and forever changing as new interactions with the environment suggest adaptations. Mental model research has concentrated on technical domains, like physics or electricity, because they are far more normative and are more easily made explicit than most other domains such as "parenting." Although researchers may use highly specific domains, theorists suggest that people form mental models of a large number of systems, ranging from the kitchen stove to Newtonian mechanics (see Norman 1988 for a discussion of ways in which people form mental models of everyday things).

Mental models serve explanatory and predictive purposes. Survival demands that we be able to predict everyday events with a high degree of success. The routine needed to cross a street is a prime example. Beyond all of the perceptual requirements (such as estimating the width of the street and the speed of oncoming cars) is a need to understand the many "street" systems that operate together. Just a few of these systems include the workings of automobiles, traffic lights, and physics. Our understanding of each of these systems is crucial as we decide when is an appropriate time to cross the street, as well as if we can casually stroll across or should attempt an "Olympic sprint." Any misunderstanding of one of these systems could be as deadly as any misjudgment of distance or speed. For example, consider your mental model of an electric "walk" sign at an intersection and what it means when it begins to flash. Can you still initiate the crossing? What should those already part way across do? Obviously, each interpretation can have dramatically different consequences. Mental models of everyday things usually form through interactions with the environment. However, some systems, such as the physical sciences, are difficult to understand through a wide range of random interactions. Microworlds offer a platform for appropriate conceptualizations of such systems to form and develop.

The application of mental models to the instructional design of microworlds involves the consideration of three things: the target system, the conceptual model of that target system, and the user's mental model of the target system (Norman 1983). The *target system* is the actual system that a learner is trying to understand. For example, Newtonian mechanics or thermodynamics are typical target systems. A *user's mental model* describes each individual's conception of the target system; a person describes how the target system works and predicts its behavior based on her mental model of the target system. A user's mental model may not be an accurate reflection of the target system, as evidenced by the growing body of research which indicates that people frequently hold a wide range of scientific misconceptions (Eylon and Linn 1988; Perkins and Simmons 1988). For example, research has shown that many people hold a mental model of motion problems based on Aristotelian, rather than Newtonian, principles (Champagne et al. 1980; diSessa 1982). Hence, many students assume that a ball comes to a stop soon after being thrown because it "runs out" of force. Some (such as McCloskey 1983) have suggested that the mental models people form constitute their own personal "theories" of the physical world with which formal instruction often conflicts. Others propose that people's mental models are no more than a collection of knowledge "pieces" (diSessa 1988). Either way, a person's mental model is likely to exert a strong influence on all problem solving in a particular domain.

Even as adults, we all have and form mental models of everyday objects and events, such as kitchen stoves, bathroom shower controls, and light switch panels (Norman 1988). A particularly good example is the thermostat of a home furnace. Many people erroneously believe that the higher you set the thermostat, the faster your home gets heated. A personal example of how mental models form and affect my daily behavior concerns a broken television I have in my bedroom. When I turn on the set, its power quickly goes on and then off. If I tap the set, it will quickly turn on and off again. In order to get the set to turn on and stay on, I have to carefully tap it until the "connection catches." I have learned that there are better places to tap the set than others. I have become particularly adept at how to turn the set on, and I frequently coach my wife and daughter on the "technique." Now, I know nothing about television repair or about electrical devices, yet I have developed my own "loose connection" theory, or mental model, to define the set's electrical problem (assuming it is an electrical problem). So far, this mental model allows me to predict future instances of the set's performance and thereby governs my actions. There are some interesting connections between mental models and superstitious behaviors. For example, does pressing the already lit elevator button in the lobby help ensure that the elevator will really come? Does repeated pressing of the button make the elevator come to your floor faster? If confronted with either question, most people would probably not argue forcibly that such behaviors really work, yet our actions persist.

The formation and refinement of mental models over time can be explained well by the equilibration process. For example, I enjoyed water play when I was a very young boy. I particularly liked to experiment with the action of running water from the kitchen or bathroom faucet. I discovered that I could "power" certain toys with this stream of running water. I would take a toy car or truck and hold it under the faucet so that the stream would be running against the side of the wheel to make the wheel spin. I quickly learned that I could make the wheels go fast or slow by adjusting the rate of flow. I could also make the tires spin clockwise or counterclockwise, depending on which side of the tire touched the water. I then generalized these principles to real automobiles by accommodating the "facts" I observed in traffic. I noticed how exhaust pipes were always found at the back of a car on one side. I also noticed that some cars had two exhaust pipes, one on each side of the car. I knew that the car's motor was in the front, so I deduced that the flow of exhaust acted like the stream of water and made the car's wheels move by flowing over one or both of the wheels as it traversed from the engine to the back of the car. All of the evidence seemed to support this hypothesis. When a car's engine revved, the flow of exhaust was heavy and powerful. I also figured that a car with two exhaust pipes was more "powerful" than a car with only one. In winter, the visual effect of "smoke" coming from the condensation of warm exhaust in the cold air strongly reinforced this notion.

A conceptual model is often presented to a learner as an aid to understanding the target system. *Conceptual models* act as both "bridges" between the target system and a user's model and "anchors" upon which a user's model can grow and develop (Mayer 1989). Conceptual models are usually invented by teachers, designers, or engineers. For all practical purposes, a microworld is largely synonymous with an *interactive* conceptual model. It embodies the simplest working model of a system in which an individual can begin to understand the target system. Each of the missions of SSC easily fit this definition. A conceptual model can often be metaphorical to the target system, such as suggesting that the flow of electrons in a wire is "like water running through a pipe." Conceptual models, like microworlds, offer a temporary "doorway" to a set of larger ideas. For example, Papert (1980) has recounted the way in which his fascination with gears as a young boy offered him a beginning conceptual model of mathematical ratios and proportions. For Papert, gears became a personal microworld that helped make tangible many abstract mathematical ideas. This does not imply that the gears metaphor should be introduced into the mathematics curriculum, but, instead, suggests how a microworld can provide an introductory conceptual transition into a learning domain.

In order for an interactive conceptual model to truly become a microworld, one more condition must be met—students must find the experience personally satisfying and rewarding. Designing a microworld in such a way so that students choose to engage in the activity involves the issue of motivation.

INTRINSIC MOTIVATION

Constructivists make the assumption that learners must find an activity, within a microworld or otherwise, intrinsically motivating in order for the process of equilibration to occur. Lawler (1982) has suggested that microworlds, like those presented in LOGO, are successful because they produce "neat phenomena," or "phenomena that are inherently interesting to observe and interact with" (141). However, constructivists offer little guidance on this issue to designers of microworlds. Turtle geometry, for example, may capture an innate human interest in the visual appeal of graphics, although this is totally speculative.

What motivates an individual to initiate and complete a task? A strict behaviorist would derive this empirically by trying a variety of reinforcements and observing which best sustains the desired response. School environments typically use a variety of reinforcements, such as praise, rewards, and grades. All of these are examples of extrinsic motivators, in that an activity is engaged in order to get the promised incentive, whether it be a star pasted to a school paper or a good grade on a report card. We all need extrinsic motivators at times, such as a paycheck, in order to get an important job done which we may otherwise choose not to complete.

In contrast, people tend to complete other tasks without the promise of a reward. These tasks or activities have their own inherent appeal and people do not need coaxing to participate. Typical examples include leisure activities such as reading, playing cards, cooking, or crafts. In each of these cases, participation in the activity itself is reward enough. Such activities are intrinsically motivating in that the nature of the activity or domain itself produces significant reason to participate, and external reinforcement or pressure to participate is not required (Kinzie and Sullivan 1989; Maehr 1976). Of course, an activity that is intrinsically motivating to one person may require extrinsic motivators for another.

Instruction that is intrinsically motivating relies on incentives that are student-centered, rather than on external lesson reinforcement. In fact, the addition of extrinsic incentives, such as grades or other rewards, can often undermine or destroy the intrinsic appeal of many activities for students, and, in effect, turn "play into work" (Condry 1977; Greene and Lepper 1974; Lepper 1985; Lepper et al. 1973). Several models of motivation in instruction have been proposed. Probably the most widely known and applied model in ISD is the Attention, Relevance, Confidence, and Satisfaction (ARCS) model as proposed by Keller (1983; Keller and Suzuki 1988). This model describes and prescribes motivation as the interplay of students' perceptions of the model's four elements. Similarly, Malone (1981) has suggested a framework of intrinsically motivating instruction based on *challenge, curiosity,* and *fantasy*. Malone's model has been specifically applied to the design of computer games, which often share characteristics of microworlds.

Challenge and curiosity are closely related, and both must be optimally maintained to be effective. Tasks that are too easy can be tedious and boring, and tasks that are too difficult are frustrating and intimidating. In either case, it is unlikely that a student would choose to engage in the activity for even short periods of time. Both *challenge* and *curiosity* often result when tasks are novel, moderately complex, or when tasks produce uncertain outcomes. An element of surprise results when the expected and actual results for an activity are different. Such events trigger disequilibrium, thereby offering opportunities for learning. Such events have been termed *"conceptual conflicts"* by Berlyne (1965) and *"critical confusion"* by Norman (1978). Completion of challenging tasks can elicit feelings of confidence and competence, which lead to a feeling of control over one's success (Weiner 1979).

The element of *fantasy* is used to encourage students to imagine that they are completing the activity in a context in which they are not really present. Inducing fantasy relies on mental imagery of contexts that are very meaningful for a student. Fantasy is evident in the intense play of children, especially very young children. Play is a very important cognitive exercise; Piaget (1951) explained children's play as a mechanism for assimilation. Fantasy is best applied to instructional activities when the learning outcome is intrinsic to the imagined context. This means that the skills to be learned are integrated into the fantasy, such as learning about how to use a compass to "rescue a party of lost hikers." When effective, the fantasy becomes a meaningful context in which all subsequent instruction can be "anchored," or situated (Brown, et al. 1989; The Cognition and Technology Group at Vanderbilt 1990).

My experience has shown that imagining to be an astronaut appears to be a common fantasy in which children and adults often indulge. The successes and failures of the world's space flights during the past several decades offer an intriguing context for people's imaginations to go on fantastic voyages in outer space. Beyond all of the fantasy is the very real practicality and necessity of working in space, which is continually documented in the media. SSC tries to capitalize on all of this by encouraging students to fantasize about being the commander of the space shuttle. In SSC, students are not vicariously watching someone else's experience in outer space, such as a video documentary about space travel, instead, through realistic fantasy, SSC hopes to provide students with an imaginary "trip" aboard the shuttle. In SSC, students are not along only for the ride, but they are in control. Preliminary research has supported the contention that SSC contains intrinsically motivating characteristics for elementary school children (Rieber, in press-b).

A FINAL WORD

Given the widely different philosophical positions, do the instructivist and constructivist positions have anything in common when it comes to suggesting what or how to teach? Each would probably agree, at least, that establishing needs and goals for education is important. There comes a point, however, when you cross either the line of excessive or inadequate specificity. Instructivists tend to impose instructional order, but this can lead to narrow and self-limiting instructional designs that are apt to encourage passive or "sponge-like" learning. Conversely, constructivistic designs are largely based on discovery learning and treat a learner as a "candle to be lighted." Unfortunately, this can lead to designs where learners may become either bored or frustrated early on, thus either preventing the necessary "sparks" or prematurely extinguishing the "flame."

Papert has argued that the current education infrastructure is diametrically opposed to constructivism. Certainly, attempts to integrate LOGO into existing scope and sequences have certainly been unsettling, and LOGO is often criticized for not "fitting in" to the mathematics curriculum. However, this is a little like asking what curriculum areas are served by pencil and paper. Computer microworlds, like LOGO, offer a powerful computational medium for children to explore ideas and themselves. LOGO and its successors will not have their proper impact until the educational system, in terms of its goals and organizational and physical structure, are reconsidered and reconceptualized. Introducing LOGO or its counterparts into the schools is like introducing a Ferrari to late nineteenth century America: you can appreciate its futuristic shape and appearance, and you can listen to its powerful motor, but you can not go anywhere useful or practical on the roads provided.

As an instructional technologist, I look not for divisions but for good ideas. I think a working compromise can be reached, in part, through the study of successful microworlds, especially ones that are computer-based. I contend that microworlds which are carefully designed with appropriate structure offer a way to incorporate some of the ideals of constructivism into an educational system based on instructivism. So far, constructivistic interpretations of instructional technology have adopted computer-based microworlds as the principle vehicle for realizing goals within instructional technology. However, when the boundaries of a microworld are appropriately limited and constrained, the range of possible learning outcomes is also constrained, making the achievement of prescribed learning objectives possible and probable, which is the goal of instructivism. This view is similar to (as well as supported by) research indicating that students often need some structure, order, or organization, especially when they are novices in the area (Mayer 1979), a point which has recently been reinforced by initial efforts on applying hypertext principles to instruction (Jonassen 1986; Tripp and Roby 1990). SSC offers some structure to learners, but not excessively so; it lets them stay in the "driver's seat," but with a "learner's permit."

Table 1 provides a summary of the main ideas presented here in the form of a small set of principles or design guidelines. These principles are a first attempt at distilling elements common to constructivism and direct instruction that can be practically applied. These are the principles that guided my development of *Space Shuttle Commander* (see Rieber, in press-a, for an elaboration of each of the principles). Applying these principles has helped me to design computer activities that try to achieve some of the goals of constructivism, within the constraints of many instructional environments. For example, Mission 5, which is represented in Figure 1, allows learners to change and experiment with a variety of mission "conditions," such as target size and increment of shuttle rotation, as shown in Figure 5. The conditions are initially set to use the easiest combinations but can

Table 1. Some considerations in the design of computer-based microworlds to support direct instruction.

- provide a meaningful learning context that supports intrinsically motivating and self-regulated learning

- establish a pattern where the learner goes from the "known to the unknown"

- emphasize the usefulness of errors

- provide a balance between deductive and inductive learning

- anticipate and nurture incidental learning

be changed to both increase the level of challenge and difficulty, as well as to broaden the range of physics principles to be encountered.

The last two principles summarize much of the pedagogical controversy between instructivists and constructivists over the years (Slavin 1988). For example, instructivists generally follow traditional instructional methods based on a deductive approach where general concepts are introduced first, followed by specific examples (Ausubel 1968; Gagné et al. 1988). Constructivists, on the other hand, generally advocate discovery learning based on inductive strategies where learners must induce and interpret (construct) principles from a set of specific examples or occurrences of the principles (Bruner 1966). Instructivists also seek to take learners through a learning sequence based on predetermined learning objectives and to discourage any learning which is incidental to these objectives (Reigeluth 1983). Advocates of pure discovery learning suggest that any learning path taken by a student is the one to be encouraged, and leading the

```
SHUTTLE MISSION FIVE: RENDEZVOUS

Here are your current mission conditions:

<1> Size of Space Station: 9
<2> Trailing? YES
<3> Maximum Rendezvous Speed: NONE
<4> Degrees of Spin: 90
<5> Station Location: RANDOM

Press the number of the condition you
wish to change...

...or press ➡ to begin the mission.
...or press ⬅ to go back to the
instructions.
```

Figure 5. Sample screen showing the various mission "conditions" of "Rendezvous" which can be changed to increase the difficulty of the mission. A similar set of conditions are given for "Mission 5: Space Dock." (Not drawn to scale.)

student to any one specific outcome should be avoided (see Klauer 1984 for a review of research associated with intentional and incidental learning). Thus, the last two principles are meant as a reminder to instructivists to not get carried away with their own good intentions.

Finally, it is not the intention of this chapter to try to sway "hardcore" instructivists or constructivists away from their ideals, but, rather, it is to express the opinion that active collaborations of philosophies are not only possible but are generally the aim of instructional technology. To that end, I assert that the best examples of instructional technologists are also our best teachers, even if they have never, as yet, used a computer. Good teachers, like good instructional technologists, use the best of whatever resources and ideas are available to "get the job done."

REFERENCES

Abelson, H., and A. diSessa. *Turtle Geometry: The Computer as a Medium for Exploring Mathematics.* Cambridge, MA: MIT Press, 1981.

Ausubel, D. *Educational Psychology: A Cognitive View.* New York: Holt, Rinehart, and Winston, 1968.

Berlyne, D. *Structure and Direction in Thinking.* New York: Wiley, 1965.

Brown, J.S., A. Collins, and P. Duguid. "Situated Cognition and the Culture of Learning." *Educational Researcher,* 18(1), (1989): 32–42.

Bruner, J. *Toward a Theory of Instruction.* New York: Norton, 1966.

Butts, R.E., and J.R. Brown (Eds.). *Constructivism and Science.* Norwell, MA: Kluwer Academic Publishers, 1989.

Case, R., and C. Bereiter. "From Behaviourism to Cognitive Behaviourism to Cognitive Development: Steps in the Evolution of Instructional Design." *Instructional Science,* 13, (1984): 141–158.

Champagne, A., L. Klopfer, and J. Anderson. "Factors Influencing the Learning of Classical Mechanics." *American Journal of Physics,* 48, (1980): 1074–1079.

Clancy, W.J. "Methodology for Building an Intelligent Tutoring System." In W. Kintsch, J. Miller, and P. Polson (Eds.), *Methods and Tactics in Cognitive*

Science. Hillsdale, NJ: Erlbaum, 1984. 51–83.

Clark, R. "Antagonism Between Achievement and Enjoyment in ATI Studies." *Educational Psychologist*, 17, (1982): 92–101.

Clements, D. "Research on LOGO in Education: Is the Turtle Slow but Steady, or Not Even in the Race?" *Computers in the Schools*, Summer/Fall, (1985): 55–71.

Condry, J. "Enemies of Exploration: Self-initiated Versus Other-initiated Learning." *Journal of Personality and Social Psychology*, 35, (1977): 459–477.

Dede, C. "Empowering Environments, Hypermedia and Microworlds." *The Computing Teacher*, 15(3), (1987): 20–24, 61.

diSessa, A. "Unlearning Aristotelian Physics: A Study of Knowledge-based Learning." *Cognitive Science*, 6, (1982): 37–75.

diSessa, A. "Knowledge in Pieces." In G. Forman and P. Pufall (Eds.), *Constructivism in the Computer Age*. Hillsdale, NJ: Erlbaum, 1988. 49–70.

diSessa, A., and H. Abelson. "Boxer: A Reconstructible Computational Medium." *Communications of the ACM*, 29(9), (1986): 859–868.

diSessa, A., and B. White. "Learning Physics from a Dynaturtle." *Byte*, 7, (1982): 324.

Eylon, B., and M. Linn. "Learning and Instruction: An Examination of Four Research Perspectives in Science Education." *Review of Educational Research*, 58, (1988): 251–301.

Forman, G., and P. Pufall (Eds.). *Constructivism in the Computer Age*. Hillsdale, NJ: Erlbaum, 1988a.

Forman, G., and P. Pufall. "Constructivism in the Computer Age: A Reconstructive Epilogue." In G. Forman and P. Pufall (Eds.), *Constructivism in the Computer Age*. Hillsdale, NJ: Erlbaum, 1988b. 235–250.

Fosnot, C.T. *Enquiring Teachers, Enquiring Learners: A Constructivist Approach for Teaching*. New York: Teacher's College Press, 1989.

Gagné, R., L. Briggs, and W. Wager. *Principles of Instructional Design* (3d ed.). New York: Holt, Rinehart, and Winston, 1988.

Gagné, R., and W. Dick. "Instructional Psychology." *Annual Review of Psychology*, 34, (1983): 261–295.

Gagné R., and R. Glaser. "Foundations in Learning Research." In R. Gagné (Ed.), *Instructional Technology: Foundations*. Hillsdale, NJ: Erlbaum, 1987. 49–83.

Gentner, D., and A. Stevens (Eds.). *Mental Models*. Hillsdale, NJ: Erlbaum, 1983.

Goodman, N. *Of Mind and Other Matters*. Cambridge, MA: Harvard University Press, 1984.

Greene, D., and M. Lepper. "How to Turn Play into Work." *Psychology Today*, 8, (1974): 49–54.

Hannafin, M., and L. Rieber. "Psychological Foundations of Instructional Design for Emerging Computer-based Instructional Technologies: Part I." *Educational Technology Research & Development*, 37(2), (1989): 91–101.

Jonassen, D. "Hypertext Principles for Text and Courseware Design." *Educational Psychologist*, 21(4), (1986): 269–292.

Keller, J. "Motivational Design of Instruction." In C.M. Reigeluth (Ed.), *Instructional-design Theories and Models: An Overview of Their Current States*. Hillsdale, NJ: Erlbaum, 1983. 383–434.

Keller, J., and K. Suzuki. "Use of the ARCS Motivation Model in Courseware Design." In D. Jonassen (Ed.), *Instructional Designs for Microcomputer Courseware*. Hillsdale, NJ: Erlbaum, 1988. 401–434.

Kinzie, M., and H. Sullivan. "Continuing Motivation, Learning Control, and CAI." *Educational Technology Research & Development*, 37(2), (1989): 5–14.

Klauer, K. "Intentional and Incidental Learning with Instructional Texts: A Meta-analysis for 1970–1980." *American Educational Research Journal*, 21, (1984): 323–339.

Knirk, F., and K. Gustafson. *Instructional Technology: A Systematic Approach to Education*. New York: Holt, Rinehart, and Winston, 1986.

Kulhavy, R. "Feedback in Written Instruction." *Review of Educational Research*, 47, (1977): 211–232.

Lawler, R. "Designing Computer-based Microworlds." *Byte*, 7(8), (1982): 138–160.

Lepper, M. "Microcomputers in Education: Motivational and Social Issues." *American Psychologist*, 40, (1985): 1–18.

Lepper, M., D. Greene, and R. Nisbett. "Undermining Children's Intrinsic Interest with Extrinsic Rewards: A Test of the Overjustification Hypothesis." *Journal of Personality and Social Psychology*, 28, (1973): 129–137.

Lockard, J., P. Abrams, and W. Many. *Microcomputers for Educators* (2d ed.). Glenview, IL: Scott, Foresman/Little, Brown High Educ., 1990.

Lukas, G., and J. Lukas. *LOGO: Principles, Programming, and Projects*. Monterey, CA: Brooks/Cole Publishing Co., 1986.

Maehr, M. "Continuing Motivation: An Analysis of a Seldom Considered Educational Outcome." *Review of Educational Research*, 46, (1976): 443–462.

Malone, T. "Toward a Theory of Intrinsically Motivating Instruction." *Cognitive Science*, 4, (1981): 333–369.

Mayer, R.E. "Twenty Years of Research on Advance Organizers: Assimilation Theory is Still the Best Predictor of Results." *Instructional Science*, 8, (1979): 133–167.

Mayer, R.E. *Thinking, Problem Solving, Cognition*. New York: Freeman, 1983.

Mayer, R.E. "Aids to Text Comprehension." *Educational Psychologist*, 19(1), (1984): 30–42.

Mayer, R.E. "Models for Understanding." *Review of Educational Research*, 59, (1989): 43–64.

McCloskey, M. "Naive Theories of Motion." In D. Gentner and A. Stevens (Eds.), *Mental Models*. Hillsdale, NJ: Erlbaum, 1983. 299–324.

Merrill, M.D., Z. Li, and M.K. Jones. "Second Generation Instructional Design (ID$_2$)." *Educational Technology*, 30(2), (1990): 7–14.

Norman, D. "Notes Toward a Theory of Complex Learning." In A. Lesgold, J. Pellegrino, S. Fokkema, and R. Glaser (Eds.), *Cognitive Psychology and Instruction*. New York: Plenum, 1978. 39–48.

Norman, D. "Some Observations on Mental Models." In D. Gentner and A. Stevens (Eds.), *Mental Models*. Hillsdale, NJ: Erlbaum, 1983. 7–14.

Norman, D. *The Psychology of Everyday Things*. New York: Basic Books, 1988.

Palumbo, D.B. "Programming Language/Problem-solving Research: A Review of Relevant Issues." *Review of Educational Research*, 60, (1990): 65–89.

Papert, S. *Mindstorms: Children, Computers, and Powerful Ideas*. New York: Basic Books, 1980.

Papert, S. "Computer-based Microworlds as Incubators for Powerful Ideas." In R. Taylor (Ed.), *The Computer in the School: Tutor, Tool, Tutee*. New York: Teacher's College Press, 1981. 203–210.

Papert, S. "Computer Criticism vs. Technocentric Thinking." *Educational Researcher*, 16(1), (1987): 22–30.

Papert, S. "The Conservation of Piaget: The Computer as Grist to the Constructivist Mill." In G. Forman and P. Pufall (Eds.), *Constructivism in the Computer Age*. Hillsdale, NJ: Erlbaum, 1988. 3–13.

Piaget, J. *Play, Dreams, and Imitation in Childhood*. New York: Norton, 1951.

Perkins, D.N., and R. Simmons. "Patterns of Misunderstanding: An Integrative Model for Science, Math, and Programming." *Review of Educational Research*, 58(3), (1988): 303–326.

Reigeluth, C. (Ed.). *Instructional-design Theories and Models: An Overview of their Current Status.* Hillsdale, NJ: Erlbaum, 1983.

Reiser, R. "Instructional Technology: A History." In R. Gagné (Ed.), *Instructional Technology: Foundations.* Hillsdale, NJ: Erlbaum, 1987. 11–48.

Rieber, L. "Animation in Computer-based Instruction." *Educational Technology Research & Development*, 38(1), (1990a): 77–86.

Rieber, L. "Using Computer Animated Graphics in Science Instruction with Children." *Journal of Educational Psychology*, 82, (1990b): 135–140.

Rieber, L. *Space Shuttle Commander* [Computer Software]. Washington, DC: NASA Educational Technology Branch, 1991.

Rieber, L. (in press-a). "Computer-based Microworlds: A Bridge Between Constructivism and Direct Instruction." *Educational Technology Research & Development.*

Rieber, L. (in press-b). "Computer Animation, Incidental Learning, and Continuing Motivation." *Journal of Educational Psychology.*

Romiszowski, A.J. *Designing Instructional Systems.* New York: Nichols Publishing Co., 1981.

Saettler, P.L. *The Evolution of American Educational Technology.* Englewood, CO: Libraries Unlimited, 1990.

Shuell, T.J. "Cognitive Conceptions of Learning." *Review of Educational Research*, 56, (1986): 411–436.

Skinner, B. *Technology of Teaching.* New York: Meredith Publishing Co., 1968.

Slavin, R. *Educational Psychology: Theory into Practice.* Englewood Cliffs, NJ: Prentice-Hall, 1988.

Steinberg, E.R. "Review of Student Control in Computer-assisted Instruction." *Journal of Computer-Based Instruction*, 3, (1977): 84–90.

Steinberg, E.R. "Cognition and Learner Control: A Literature Review, 1977–1988." *Journal of Computer-Based Instruction*, 16, (1989): 117–121.

Suppes, P., and M. Morningstar. *Computer-assisted Instruction at Stanford, 1966–1968: Data, Models, and Evaluation of the Arithmetic Programs.* New York: Academic Press, 1972.

The Cognition and Technology Group at Vanderbilt. "Anchored Instruction and its Relationship to Situated Cognition." *Educational Researcher*, 19(5), (1990): 2–10.

Tripp, S., and W. Roby. "Orientation and Disorientation in a Hypertext Lexicon." *Journal of Computer-Based Instruction*, 17, (1990): 120–124.

Walker, D.F. "LOGO Needs Research: A Response to Papert's Paper." *Educational Researcher,* 16(5), (1987): 9–11.

Watzlawick, P. (Ed.). *The Invented Reality.* New York: Norton, 1984.

Weiner, B. "A Theory of Motivation for Some Classroom Experiences." *Journal of Educational Psychology*, 71, (1979): 3–25.

Wenger, E. *Artificial Intelligence and Tutoring Systems.* Los Altos, CA: Morgan Kaufman, 1987.

White, B. "Designing Computer Games to Help Physics Students Understand Newton's Laws of Motion." *Cognition and Instruction*, 1, (1984): 69–108.

White, B., and P. Horwitz. *ThinkerTools: Enabling Children to Understand Physical Laws* (Report No. 6470). Cambridge, MA: BBN Laboratories, 1987.

PART 3

Teacher Learning and Change

13

Constructivist Perspectives on Teacher Learning

Kenneth Tobin

The focus for the research described in this chapter is on teacher learning and curriculum change. We were interested in factors associated with teachers making changes in their classrooms and institutionalizing those changes. The program of research has been in progress since 1984 and continues today (e.g., Tobin 1990a, Tobin and Espinet 1989; Tobin and Fraser 1987; Tobin and Gallagher 1987; Tobin et al. 1990; Tobin and Ulerick 1989). Throughout the entire research program, we have focused on the teacher and the rationale for teaching practices. Our research questions have involved teacher beliefs and other cognitive factors such as metaphors. It is now clear that a prerequisite to understanding the change process is to understand the culture in which teaching and learning are embedded.

A teacher's experience is sensory and is given meaning by reflection, which involves the construction of images, and, in some cases, the assignment of language to images, which can be thought of as dynamic reconstructions of experience (Clandinin 1986; Paivio 1974). Sanders and McCutcheon (1986) and Elbaz (1983) reported how teachers used images as they thought about teaching. Kennison (1990) described how the teachers in her case studies used images in the process of imagining lessons in their classrooms, developing and maintaining innovations, and planning for learning. In our own research, we observed how Marsha perceived her roles in terms of metaphors and associated visual images (Tobin and Ulerick 1989). Even more graphic was the example of a science teacher who envisioned himself as a swashbuckling captain of a ship, barking orders to his crew to keep them under tight control (Tobin 1990b).

MAKING SENSE OF SALIENT CONCEPTS

An important aspect of language and knowing is referred to by Lakoff (1987) as *metonymy*. This occurs when a person uses a part of a concept to give meaning to an entire concept. For example, Marsha, and most of her colleagues, defined teaching predominantly in terms of management, which, in turn, was defined in terms of control of students. Solutions to teaching problems were, therefore, sought in terms of management and control. Another way to think of metonymy is to identify the central concepts used to give meaning to a concept. This way of framing metonymy recognizes that some ideas are central and others are peripheral to conceptualization. Thus, in relation to science teaching, one might reflect on the salient roles used by a teacher to define teaching. At the beginning of the study (Tobin and Ulerick 1989), Marsha used three roles to conceptualize science teaching: management and assessment, which were central, and facilitating learning, which was peripheral. At the end of the study, the central concepts associated with Marsha's conceptualization of teaching were numerous, interwoven, and included the following: facilitating learning, constructivism, management for learning, assessing what students know, teacher as researcher, teacher as mentor, and teacher as curriculum designer.

Metaphors can be used to make sense of concepts associated with science and the teaching of science (Johnson 1987; Lakoff 1987; Marshall 1990; Munby and Russell 1990; Tobin 1990c). These metaphors are personal referents and are not necessarily shared by teachers in general. However, metaphors serve to organize sets of beliefs. Consider Marsha (Tobin and Ulerick 1989), who used the metaphor of teacher as comedian, to make sense of her management and facilitation of learning roles. (Marsha described and named her own metaphors. The words used to depict particular metaphors in this chapter are Marsha's.) Marsha had numerous beliefs associated with these roles, and each was consistent with the teacher as comedian metaphor. Similarly, beliefs about how students ought to think and behave were associated with the teacher as comedian metaphor. When Marsha was unable to control the students' misbehavior or facilitate their learning, she had few alternative metaphors to guide her actions. Marsha had to be the comedian or an effort miser. As an effort miser, she focused entirely on management, keeping students controlled and busy. Learning ceased to be a concern.

Marsha had been a social director many times in her life and understood what it meant to be a good hostess. As Marsha talked about teaching and learning with members of the research team, she made sense of the discussion in terms of being a social director. She thought of the classroom as a party. She could invite students to the party, but whether or not they came was up to them. If they declined, she had a responsibility to make the invitation more attractive. If students decided to accept the invitation, they had to respect the hostess and other guests at the party. In addition, no guest should disrupt the fun of others. Marsha believed the teacher as social director metaphor could work for her and began to plan her teaching accordingly.

Marsha also used other referents to make sense of the science curriculum. For example, Marsha had made a commitment to constructivism (von Glasersfeld 1989) and thought of learning and the curriculum from a constructivist perspective. When she thought about her new metaphor of teacher as social director, therefore, it and the associated beliefs needed to be consistent with optimizing student learning from a constructivist perspective.

Our research suggests that the beliefs, metaphors, and metonymic models of teachers and students are associated with curricular actions. To change the curriculum, therefore, it is necessary for teachers and students to reconceptualize the manner in which they make sense of their salient roles. The implications of this assertion are that teacher education programs ought not focus on learning how to implement projects as intended by writers and designers; rather, an analysis of materials ought to focus on identification of the beliefs, metaphors, and metonymic models on which suggested activities have been built. Teachers can then relate what they find to their personal mind frames, thereby identifying areas of potential conflict and making cognitive adjustments that will influence the manner in which the materials will be deployed if the decision is made to incorporate them as a learning resource in the curriculum.

One of the metaphors Shirley used when she taught elementary science was a movie director. In the past two years, she changed from a traditional science teacher who was reliant on the textbook and focused on students' rote-learning of facts and algorithms to get correct solutions, to a teacher who facilitated learning based on problem solving and cooperative learning whereby students worked together to arrive at consensus solutions to problems. Shirley used the director metaphor to make sense of her teaching role in an environment with which she had no previous experience. The director provides actors (i.e., the students) with a script, but it is for the actors to create their own parts within the confines of the script. The actors cannot succeed unless the director provides them with a good script and guides them as they work together to create the film (i.e., learning). The director is in charge and manages the schedule; however, the quality of the film is dependent upon the participation of the actors and the director. Other metaphors used by Shirley to guide her teaching included co-investigator and concerned parent.

Diana used three metaphors to describe her teaching role in different contexts. Usually, she managed her class as a policewoman, in some circumstances, she was a mother hen, and, on other occasions, she was an entertainer. Her mode of behavior (i.e., the metaphor she used to make sense of what she ought to do) depended on the context in which learning was to occur. And each conceptualization of her role as manager was associated with a discrete set of beliefs.

The following excerpts from an interview indicate how Diana used imagery to establish a mind-frame in which she could explain how the teaching of a former teacher influenced her own teaching.

Diana: I remember a social studies teacher who taught tenth grade history. He was a real energetic type person who did a lot of activities in his classroom to get the whole class excited. He did different games and contests that were really fun. But the games got everyone involved.

Interviewer: Can I ask you a question about what was going on in your mind as you were telling me that?

Diana: I was picturing myself sitting in that classroom.

Interviewer: Was the teacher there?

Diana: Yes. The whole class was there. And I was sitting among the group. And he was sitting in the front of the room directing the class.

Is it possible that images are used metaphorically to make sense of a new role to be adopted by a teacher? The metaphorical use of images would bypass the use of language and guide actions unconsciously. For example, the decision to use a policewoman as a metaphor for managing a classroom is probably not done consciously. Something in the context might result in the construction of an image associated with a role which is well understood (e.g., a police officer) and provide a basis for subsequent behavior. Basing actions on images associated with another role is analogous to using a verbal metaphor to make sense of a new concept. The interview with Diana suggests that she has based her teaching on images associated with her favorite teacher. It is apparent that she did not sit down and meticulously describe in words those behaviors she would adapt and adopt; rather, the association appears to have been more direct, involving re-presented images of her former teacher. Similarly, Diana's decision to be an entertainer is probably associated with images of entertainers she has experienced and occasions when, in other roles, she has been entertaining.

Metaphors are used to make sense of teaching roles (i.e., beliefs associated with phenomena that are familiar are applied to a teaching role). If beliefs are found to be viable, they are retained, and if not, they are modified or discarded so that the belief set enables the teacher to satisfactorily fulfill her role. Thus, beliefs associated with a metaphor are consistent with one another in that they lead to thoughts and actions that make sense to the teacher in the contexts in which they are applied. Beliefs within a set usually are consistent because inconsistencies are likely to produce discrepant practices or thoughts from the teacher or students. The beliefs associated with discrepant actions and thoughts can then become foci for reflection and change.

MYTHS AS REFERENTS

So many of the traditional classroom practices and curricular resources have been built on the myth of objectivism. Accordingly, teachers, students, and policy makers make sense of what they do in terms of objectivism. Innovations are planned and implemented within that traditional framework without questioning the myth. The potential problem of adopting objectivism as a referent for educational activities is that the focus is not on the learner as much as it is on aspects of the milieu in which learning is assumed to be embedded. From a constructivist perspective, the learner becomes the central part of all decisions and concerns. It is the learner who has to learn, not only about the substantive content of the science curriculum, but also about the milieu. The mind frames of students will determine how the context in which learning occurs is framed, as well as what is learned. Accordingly, the roles that are of greatest salience for learners and teachers are different under a referent of constructivism. The problems, and hence solutions, are likely to be different from constructivist and objectivist perspectives.

A puzzle that needed to be solved involved a teacher who maintained a personal epistemology of being a constructivist, yet the classroom practices suggested strong facets of objectivism. The clue to this problem lies in the complexity of classroom life. In order to cope with a high degree of complexity, teachers are required to do most things in a routine, or nondeliberative, manner. In such situations, teachers are "guided by experience," which is akin to being guided by an image of past experience. Within that image of a routine practice are embedded an epistemology and metaphors that are consistent with the epistemology. Thus, when a teacher gets involved in a question-and-answer routine, the teacher's deliberative actions might focus on the content of the interactions and the selection of students to be involved. However, at the same time, many other events are occurring. The types of questions asked and the reactions to student responses might be handled as a routine based on extant experience. This opens the door to frustrations of the type experienced by Rod, who has made a transition to constructivism only to find that when he gets involved in questioning routines, he looks for correct answers to convergent questions, provides immediate feedback on the response adequacy, and quickly searches for a student who can provide the correct answer to his question. Rod's routines belie the attempts he is making to incorporate collaborative learning based on constructivist principles.

When Rod reflects on re-constructed images of his classroom practices, he recognizes that his routines have embedded within them an objectivist epistemology and that the metaphor which is guiding his actions is teacher as quiz master. This realization frustrates him and makes him wonder if he ever will be an effective teacher. However, recognition of the images, metaphors, and epistemology embedded within his questioning routine, seems to be a substantial step forward. If Rod continues to reflect on his questioning routine, it is likely that verbal interactions will become a focus for deliberative actions in the classroom (i.e., a focus for reflection), and opportunities will be taken to adopt new practices and develop new routines that are consistent with constructivism. Thus, Rod's reflection on re-constructed images of routine practices raises the possibility of him becoming a reflective practitioner in the sense that, while he is teaching, he will deliberate on his actions in relation to verbal interaction.

The idea of a master switch for changing teaching practices had its genesis in the story of Peter, who switched from being captain of the ship to being an entertainer. We wondered whether teachers could change their metaphors in a deliberative way by reflecting on existing metaphors and the beliefs embedded within them, and by constructing new metaphors that have meaning in the contexts in which learning is to occur. Thus, the process of reflection was seen as a means to change metaphors, embedded beliefs, and associated classroom practices. But what would be the objects of reflection? A study involving Marsha provided us a key to the problem. Marsha had severe management problems and had only one metaphor (teacher as comedian) and set of beliefs that she could use to manage the class. When it would not work, Marsha was bereft of ideas. At that time, Marsha was involved in intensive discussions with a research team and discussed her teaching practices, beliefs, metaphors, and images of teaching and learning. Of considerable importance was continued discussion on constructivism (von Glasersfeld 1989). Marsha was becoming a radical constructivist and had made sense of learning from a radical constructivist perspective. Marsha knew what she wanted to do to facilitate learning, if only she could manage her classes. Her saintly facilitator role was being adjusted in accordance with her beliefs about knowledge (i.e., constructivism).

Marsha decided to reconceptualize her role as manager in terms of being a social director. She understood what it meant to be a social director as she had been a social director on many occasions in the past. Her application of this role to teaching was metaphorical and resulted in rejection of many of her previous beliefs about managing a class. Her social director metaphor was associated with beliefs that were compatible with constructivism. The social director managed the class in a manner that supported learning. According to the metaphor, the teacher invites students to the party of learning. Students decide whether or not to come, and the teacher's role is to create opportunities for them to learn. If students decide not to accept the invitation, the teacher has the responsibility to make the invitation more attractive. Only two rules applied. Guests (i.e., students) should be courteous to their host (i.e., the teacher) and to one another, and guests should not disrupt the fun (i.e., learning) of others. Guests who violated these rules would be invited to leave the party. Student misbehavior, which was previously widespread, almost disappeared overnight. With management less of an issue, Marsha pursued her roles as facilitator of learning and assessor of students. Numerous changes occurred in the practices of teacher and students.

So what was involved in the change of metaphor? The regular reconstruction of images of Marsha's classroom in discussions with the research team undoubtedly resulted in an environment that was conducive to changing her conceptualizations (i.e., beliefs, images, and metaphors for science teaching and learning). Marsha was able to construct a vision of what science teaching and learning could be like, project herself and her students into that vision, develop a strong commitment to change selected aspects of her teaching (i.e., the metaphor for management), implement the changes, and then reflect on reconstructed images of what happened in her classes. Being able to refer what happened in her class to a vision that was based on constructivism facilitated the change process. Marsha was not content with simply managing her students and keeping them under control. She had practical and emancipatory interests at the heart of her restructured curriculum.

The referent for Marsha's meaning-making with respect to teaching and learning was her personal epistemology. As Marsha changed from an objectivist to a radical constructivist perspective, she was able to relate the components of her vision to constructivism, and, through reflection, make adjustments to her personalized vision of what science teaching and learning could be like. Furthermore, as Marsha developed beliefs that were consistent with her new metaphor for management, she was less able to relate the beliefs to constructivism as well. Thus, Marsha's metaphors for her roles were consistent with one another not so much by relating one to the other, but by relating each of them to her personal epistemology. Thus, constructivism became the most important ingredient in the process of reflection.

The traditional culture of schools seems to be permeated by school as work place and teacher as controller of student myths. The two myths are interconnected. Adherence to the myth of the school as work place is evident in the management of schools within the traditional culture. Teachers manage so as to maintain control of student thinking and behavior. Rather than placing responsibility for behavior and learning on the student, teachers arrange students so that they cannot interact with one another and assign tasks that keep students busily engaged in activities of the teacher's choosing. Keeping busy and task completion are regarded as desirable and often are rewarded in the evaluation procedures that are implemented in the traditional culture. Just as control is emphasized in the sense that employers and shop stewards control employees, teachers control students in traditional classrooms. Management is seen as the first consideration, and initiates into the culture are soon exposed to the conventional wisdom of gaining control in the first few days. "First gain control over the students. If you don't manage to do this in the first few days, the remainder of the year will be a disaster." So goes the conventional wisdom of the teaching profession. As a consequence, the initial focus is on management, and students are arranged and supervised with control and suppression as a major goal. Only when management concerns have been addressed are the learning needs of students given consideration. The teacher decides at what pace students will cover content, assessment tasks emphasize engagement in and completion of tasks rather than learning, and extrinsic rewards are used to increase time on task. This type of approach to the curriculum has been described by Grundy (1987) as emphasizing technical interests.

Observations of Jim, a middle school science teacher, provide support for the assertion that the school as work place metaphor is a cultural phenomenon. Jim endeavored to maintain strong control over his students. He kept them quiet and ensured that they engaged in their work. Technical interests were emphasized in his curriculum, and evidence of the work metaphor pervaded his speech and practices in the classroom. The main concern in Jim's classes appears to be content coverage and completion of the work in time for the test. Time is viewed as a major constraint. There is very little evidence that Jim is concerned that students understand what they are to learn. Indeed, the environment in Jim's classes is similar in many respects to the traditional practices described in case studies by Stake and Easley (1978) and Tobin and Gallagher (1987).

Janice, a middle school mathematics teacher, used a number of metaphors to describe the roles of her students (e.g., robots, Indians, ants, and light bulbs). Each of these metaphors was consistent with the teacher having control over the curriculum and the learning of students. There are several significant aspects of Janice's use of metaphors to describe her teaching and learning roles. First, as Janice described her experiences in terms of metaphors, she consciously accessed images of her classroom and the school environment. For her, identifying metaphors involved assigning language to re-constructed images of school life. Second, it was apparent that the metaphors were an active part of her endeavors to make sense of what happened at school. Janice wanted to be a co-worker, but saw herself mainly as a dictator and the students as robots, Indians, ants, and light bulbs. Although these metaphors were used in Janice's description of what happened in her classroom, it is not clear if the metaphors had an active role in constraining the practices that she adopted in her classroom. What does seem apparent is that the metaphors were embedded within the re-constructed images of teaching and learning. The metaphor was not just a verbal device used to communicate teaching and learning practices. When Janice used the metaphor of teacher as dictator, she was intending to communicate the meaning that the role of the teacher in a classroom was akin to the role of a dictator in a state. She elicited the metaphor by referring to an image of her teaching her class. The metaphor has associated with a set of beliefs and practices. What a dictator would do to his subjects is used to define what a teacher would do to her students. Thus, in this instance, the metaphor is an organizer of beliefs and practices.

An examination of some of Janice's metaphors provides insights into the personal epistemology of the teacher. The metaphors used to define student roles suggest that the teacher has an objectivist perspective of knowledge and learning. Robots have no internal control but are assigned specific tasks for which they are programmed. Students were seen as robots in the sense that the teacher decided what students were to do and when to do it. The teacher had control at all times and could instruct the robots when to work and when to stop work.

The Indian metaphor was used in the stereotypic sense of a bunch of painted warriors following one another mindlessly into a battle. Each Indian had little control over his actions; he simply followed his peers into the battle with little thought given to his actions. The metaphor was used to communicate an idea associated with peer pressure. Students were seen as having little control over their behavior and did not have a responsibility for deliberative actions.

The light bulb metaphor was related to student learning. The teacher's role was seen as illuminating light bulbs. Once again, it was apparent that control is with the teacher, rather than the students. Light bulbs do not switch themselves on. Thus, the metaphors used by Janice to describe student roles do not assign responsibility for learning to students. Instead, the teacher has control, and students are assigned subservient roles consistent with an oppressed population in a dictatorship. The metaphors used by Janice were all consistent in the actions they implied in that the teacher having control over students was a common element. The need for control is consistent with a curriculum that gives emphasis to technical interests (Grundy 1987). The belief that teachers should control the curriculum therefore emerges as a third myth associated with the traditional culture of schools. According to Grundy, the epistemology that underpins a curriculum emphasizing technical interests is objectivism. Is it possible that Janice uses objectivism to organize her metaphors, beliefs, and practices?

A contrast to the school as work place metaphor is the school as learning place (Marshall 1990). One would assume that adherence to such a myth would result in an initial focus on learning. Students would be arranged so as to facilitate learning, and practical and emancipatory interests (Grundy 1987) might be of primary importance. It is conceivable that the design of classrooms, furniture arrangement, and perhaps the organization of students into classes, and the allocation of teachers to students might be characteristically different in schools where the myth was the school as learning place.

BELIEFS, METAPHORS, ACTIONS, AND CONTEXT

Our first clue that a role might be defined by different metaphors on different occasions was evident in a study involving a science teacher known as Peter (Tobin et al. 1990). Peter used the metaphor of entertainer to make sense of his management role on some occasions, while at other times, he was the captain of the ship. The classroom environment appeared to be very dependent on the metaphor Peter used to define his role in different contexts. When Peter decided it was appropriate to be captain of the ship, he maintained strict discipline and was a hard taskmaster. The captain of the ship delivered content, explained areas that were potentially difficult to understand, and ensured that the content was covered in timely fashion. Knowing the right answers was a high priority in this instructional mode. However, the highest priority was keeping the students on task and ensuring that they obeyed the teacher's orders. There were harsh penalties for not following orders. The teacher's beliefs, associated deliberative actions, and routine practices, embedded within an image and metaphor of being captain of the ship, constrained the roles of students. The students engaged in a range of teacher-centered, whole class activities that were prescribed by the teacher (e.g., listening to the teacher, answering questions when called upon, providing written responses to assignments).

As an entertainer, Peter focused on humor. He was intent on projecting an image of being friendly, amenable to interactions with students, and having interests that were not unlike those of the students in the class. The entertainer role was evident in whole class, small group, and seatwork activities. In this mode, the teacher told stories about the world outside of the classroom, his personal life outside of education, and other personal, humorous anecdotes. The major difference in the entertainer and captain of the ship metaphors was that Peter demonstrated that he was amenable to off-task behavior, and students could be humorous at the teacher's expense. It was safe to take risks with the entertainer. It was also in this mode that gender-related differences were most apparent in the interactions between students and the teacher. With some male students, Peter demonstrated macho tendencies, and, with the more attractive females, he engaged in social discourse. The entertainer was not in a rush to cover content or to keep students on task. It was a time to enjoy school, to relax, and to wander off task. Engagement was not controlled by the entertainer. As long as students did not become too disruptive, they were free to work at their own pace, usually completing exercises from a workbook or the textbook.

There were similarities in the two metaphors as well. Each demonstrated an emphasis on technical interests (Grundy 1987). The captain of the ship controlled what was to be learned, and the entertainer interacted with students while their tasks were controlled by the workbook or textbook. Both metaphors were embedded within an objectivist epistemology and a myth of the school as work place. Thus, Peter switched between roles when he perceived the context to be appropriate. Peter defined his role in the way that made sense to him. Each guiding metaphor or image appears to have been a routine for Peter. Referring to the metaphor was not a conscious or deliberative process as he taught. However, when he reflected on his teaching, Peter used the terms *captain of the ship* or *entertainer* to describe what he was doing and to justify his thoughts and practices. Furthermore, there was ample evidence of both metaphors in the actions of Peter as he taught and reflected on teaching.

A second example was evident in Frank's teaching of marine biology and chemistry. For the first year in which we observed Frank teach, he was a dictator in his class, and his class was perceived as a dictatorship. He was clearly in charge and technical interests were evident in terms of the type of assessment used, the emphasis on learning content by rote from the textbook, and lectures that focused on content coverage.

In a marked change of approach to teaching, Frank implemented his marine biology class in a radically different fashion at the beginning of our second year of observation. The students were assigned projects with a problem-solving emphasis. Each person became a part of a team that would work together to undertake research. Frank's teaching role was conceptualized in terms of a metaphor of tour guide. He described his role as "reminiscent of Father Time, just flowing, pointing out directions here and there," giving the students an indication, and then allowing them to do what they decide to do. Not only had Frank changed his metaphor (i.e., dictator to tour guide), he had also changed the epistemology associated with learning and the cognitive interests represented in his curriculum. Underlying the tour guide metaphor was a more constructivist epistemology than had been the case when Frank was the dictator. In the marine biology class, Frank emphasized practical and emancipatory interests. Students were in control of what to investigate, what materials to use, where to find the materials, what resources to consult, what products they would produce, and how they would demonstrate that they had learned. In marine biology, we had seen a significant change in classroom practices. Yet Frank argued that he had always believed he ought to teach that way. His insistence that he had not changed his beliefs about teaching or learning was a puzzle to us. However, we were more puzzled by his revolutionary approach to marine biology while maintaining a very traditional approach to chemistry. Nothing had changed in his chemistry class. This was a paradox.

Frank assisted us in understanding how he was making sense of his roles in the two classes. Lower-ability students studied marine biology, and Frank believed it was important for them to enjoy what they were doing and to learn by doing. However, the chemistry class contained college-bound students, and he believed that they had to know specific facts of chemistry prior to their college studies of the subject. In previous years, he had always felt constrained to teach in a traditional manner and prepare students for standardized tests. However, in the past year, he had observed his colleagues trying different student-centered approaches to learning and realized that he too could do what he believed he ought to do. As the semester progressed, it was apparent that Frank was changing his beliefs about learning and what the emphasis in the curriculum ought to be. From an epistemological perspective, Frank was making sense of potential teacher and learner roles in terms of constructivism, and, as he implemented his new roles, he was making sense of constructivism. In the second semester, Frank changed his chemistry class to a project-oriented approach, too. Throughout the semester, the emphasis shifted from an emphasis on technical interests to an emphasis on practical and emancipatory interests (Grundy 1987). By the end of the second year, Frank was a strong advocate of project-oriented learning in both chemistry and marine biology.

Frank maintains to this day that he has not changed his beliefs; yet it is apparent that his classes have continued to evolve throughout the third year of study. Emancipatory and practical interests characterize the curriculum, and most of Frank's routine practices are consistent with constructivism. It would appear that a change in the culture of the school initiated Frank's thoughts about possible change. From the time he first became a teacher, Frank apparently believed that students ought to be involved in laboratory activities, and he favored an approach that involved students in solving problems with societal implications. When he perceived the context in the school and the state to be such that he could implement a curriculum to provide the emphasis in which he believed, he seized the opportunity. Reflection on practice and his beliefs about constraints, the nature of science and knowledge, and his most salient roles precipitated additional changes in his beliefs and practices. The changing culture of the school and the context in which education was occurring facilitated the change of beliefs and practices that led to a restructuring of the science curriculum in Frank's classes.

IMPLICATIONS

Teachers are professionals who endeavor to implement the curriculum in a manner that makes sense to them. What they do in their classes is usually considered to be the best thing to do in the circumstances. And what is considered best depends not only on the teachers' personal beliefs, but also on the culture of teaching. Accordingly, consideration needs to be given to the myths associated with teaching and learning. In this chapter, I have discussed two of these, the school as work place and the teacher as controller of students. At the very least, teachers ought to reflect on these myths and consider each in relation to what they believe and value about teaching and learning. There is little payoff in the traditional approach to teacher enhancement, whereby teachers are asked to change to a different approach without a strong rationale for doing so. If a teacher adheres to an objectivist epistemology, it makes no sense at all to change from a traditional curriculum that emphasizes technical interests and dissemination of the truth from teacher to students. Why place students in groups for discussion and problem solving? From an efficiency perspective, it makes sense to keep them quiet, tell them the truths they are to learn, give them opportunities to practice, and then test what they have learned. Assisting teachers to change their epistemology so that they understand learning from a constructivist perspective is empowering in the sense that teachers can then analyze what they do in classrooms and make adjustments that make sense to them. As teachers use constructivism to make sense of their roles and those of their students, the reform of the curriculum evolves.

Reflection involves deliberating on re-constructed images of past thoughts and practices. What emerges as having important implications for teacher education is the range of objects for reflection. It is possible that teachers can identify and effect salient changes by reflecting on images, metaphors, beliefs, and values in relation to what they perceive to be happening in their classrooms. Since metaphors are used to make sense of what teachers do in classes, it is important for teachers to reflect on their images of teaching and the language they use to describe teaching and learning. If metaphors constrain the actions of teachers, it is imperative that teachers are conscious of their metaphors, and consider alternatives and the consequences of adopting alternative conceptualizations of teaching roles. Reflecting in this manner enables teachers to modify their visions of what the curriculum could be like and compare what is happening in their classes to the vision of what they would like to happen.

REFERENCES

Clandinin, D.J. *Classroom Practice: Teachers Images in Action*. London: Falmer Press, 1986.

Elbaz, F.L. *Teacher Thinking: A Study of Practical Knowledge*. London: Croom Helm, 1983.

Grundy, S. *Curriculum: Product or Praxis?* London: Falmer Press, 1987.

Johnson, M. *The Body in the Mind: The Bodily Basis of Meaning, Imagination, and Reason*. Chicago: University of Chicago Press, 1987.

Kennison, C.W. *Enhancing Teachers' Professional Learning: Relationships Between School Culture, and Elementary School Teachers' Beliefs, Images, and Ways of Knowing*. Unpublished Educational Specialist thesis, Florida State University, 1990.

Lakoff, G. *Women, Fire, and Dangerous Things: What Categories Reveal about the Mind*. Chicago: University of Chicago Press, 1987.

Marshall, H.H. "Beyond the Workplace Metaphor: The Classroom as a Learning Setting." *Theory into Practice*, 29(2), (1990): 94–101.

Munby, H., and T. Russell. "Metaphor as an Instructional Tool in Encouraging Student Teacher Reflection." *Theory into Practice*, 29(2), (1990): 116–121.

Paivio, A. "Images, Propositions, and Knowledge." In M. Nicholas (Ed.), *Images, Perception, and Knowledge*. Dordrecht, The Netherlands: D. Reidel Publishing Co., 1974.

Sanders, D.P., and G. McCutcheon. "The Development of Practical Theories of Teaching." *Journal of Curriculum and Supervision*, 2(1), (1986): 50–67.

Stake, R.E., and J. A. Easley. *Case Studies in Science Education* (Vols. 1&2). Urbana, Il: Center for Instructional Research and Curriculum Evaluation and Committee on Culture and Cognition, University of Illinois at Urbana-Champagne, 1978.

Tobin, K. "Social Constructivist Perspectives on the Reform of Science Education." *Australian Science Teachers Journal*, 36(4), (1990a): 29–35.

Tobin, K. "Teacher Mind Frames and Science Learning." In K. Tobin, J. B. Kahle, and B.J. Fraser, *Windows into Science Classrooms: Problems Associated with High Level Cognitive Learning in Science*. London: Falmer Press, 1990b. 33–91.

Tobin, K. "Changing Metaphors and Beliefs: A Master Switch for Teaching. *Theory into Practice*, 29(2), (1990c): 122–127.

Tobin, K., and M. Espinet. "Impediments to Change: An Application of Peer Coaching in High School Science." *Journal of Research in Science Teaching*, 26(2), (1989): 105–120.

Tobin, K., and B.J. Fraser (Eds.). *Exemplary Practice in Science and Mathematics Education*. Perth, Western Australia: Curtin University of Technology, 1987.

Tobin, K., and J.J. Gallagher. "What Happens in High School Science Classrooms?" *Journal of Curriculum Studies*, 19, (1987): 549–560.

Tobin, K., J.B. Kahle, and B.J. Fraser. *Windows into Science Classrooms: Problems Associated with High Level Cognitive Learning in Science*. London: Falmer Press, 1990.

Tobin, K., and S.J. Ulerick. "An Interpretation of High School Science Teaching Based on Metaphors and Beliefs for Specific Roles." Paper, annual meeting of the American Educational Research Association, San Francisco, 1989.

von Glasersfeld, E. "Cognition, Construction of Knowledge, and Teaching." *Synthese*, 80(1), (1989): 121–140.

14

Staff Development and the Process of Changing: A Teacher's Emerging Constructivist Beliefs about Learning and Teaching

Francine P. Peterman

The question of why school reforms recur, following a pattern of small successes and overwhelming failures, has been addressed from a range of macro- to micro-perspectives. From the broadest view, Cuban (1990) explained these recurring reforms as political reactions to the ebb and flow of public optimism and pessimism about schools' abilities to accommodate shifts in societal values. Sarason (1971) explained that a firmly embedded culture of the school—its history, routines, values, attitudes, and beliefs—inhibits lasting change from occurring. While Doyle and Ponder (1977) explained that the ecology of the school and, furthermore, the ecology of the classroom, impact the teacher's reconciliation of the intended reform and the perceived classroom realities, Olson (1980, 1981) identified the similarities and differences in innovator's and teachers' conceptions of the innovation and its implementation in functioning school and classroom systems as the keys to success and failure. Further, Cohen (1988) suggested that changing practice to reflect current constructivist notions of learning is unlikely because social organizations, expectations, and roles inhibit the risk taking, ambiguity, and inquiry required for "adventurous," constructivist teaching and learning.

Staff development is frequently used as a process to promote significant and worthwhile changes. To promote the implementation of constructivist notions in classroom practices through staff development, this chapter was designed to address questions regarding (1) why change is so difficult to accomplish, (2) how constructivist interventions might promote change in practice, (3) how a teacher expresses changes in beliefs throughout the process of changing, and (4) how staff development interventions might be designed for more effectively implementing constructivist practices. For worthwhile and significant change to occur, Richardson (1990) identified an interactive approach to staff development that empowers teachers through reflection upon personal, practical, and theoretical frameworks and the activities of teaching. Fenstermacher and Richardson (1991) suggest that, for change to occur, teachers must express and reconstruct their beliefs privately and publicly.

Fullan (1985) described research agendas about staff development as dealing with either "theories of change" or "theories of changing." Research questions related to theories of change asked, "What factors explain change (or the failure to change)?" These were addressed by researchers during the 1960s and 1970s. Questions related to theories of changing asked, "How do changes occur?" and "How is new knowledge used in the process of change?" These questions were raised by Doyle and Ponder (1977), echoed by Olson (1980, 1981), and examined during the 1980s. As reviewed, research regarding the implementation of constructivist innovations and teacher beliefs includes evidence of a process of changing through which teachers reconstruct beliefs. Like several of Olson's (1980, 1981) subjects, teachers frequently reconstruct the beliefs implicit in the innovation to match their own beliefs. Then, the mutual adaptation of the innovation becomes something more familiar or practical to the teacher. In other cases, where the teachers' beliefs conflict with those implicit in the innovation, the teachers may reconstruct their own beliefs. Thus, studying the process of changing involves studying the process of changing beliefs.

STATEMENT OF THE PROBLEM

To inform the development of a theory of changing, a longitudinal case study of one teacher's changing beliefs as she participated in a staff development project was completed. The project involved her in discussion about, reflection upon, and evaluation of both ideas and practices—*what* and *how* beliefs. Changes in beliefs were examined throughout the process and at the beginning of the school year after the project had ended. This study focused on the following questions:

- After participating in a particular type of staff development process, how are a teacher's beliefs about learning and teaching different from those she expressed prior to her involvement in the project?

- In the teacher's language, how is the change in beliefs exhibited throughout the process of changing?

The research methodology includes components used to elicit and analyze beliefs from a cognitive science perspective, to implement a staff development project designed to change not only *what* the teacher believes but *how* she teaches, and it functions to retell the story of the teacher's process of changing.

CONCEPTUAL FRAMEWORK: DEFINING BELIEFS

This study focused on a teacher's process of changing, specifically, the changes in her beliefs after and throughout her participation in a particular staff development project. Debbie, the subject of this study, has taught senior high school science for 16 years. Debbie said she participated in the staff development project about thinking skills in order to learn how to implement the Hilda Taba teaching strategies and to improve her questioning skills. She was involved in a process of changing her beliefs about learning and about questioning. According to Sigel (1985), these would be her *what* beliefs and her *how* beliefs; both must be addressed to predict a change in her behavior.

Beliefs: Concepts, Truth, and Action

Beliefs have been defined in a variety of ways. In their extensive review of the literature, Eisenhart et al. (1988) found many definitions of beliefs upon which current research has been based. As represented in the studies these researchers reviewed, the inconsistencies in the definitions they noted may be attributed to the foundations of the research agendas cited. Some rely upon anthropological studies, others upon sociological, psychological, or philosophical agendas. These differences are highlighted in additional reviews conducted by Clark and Peterson (1986), Hamilton (1989), Kagan (1990), and Richardson and Anders (1990).

Despite their differences, most definitions of beliefs and belief systems contain referents to concepts or other linguistic representations, to truth, and, in some instances, to action. In many explanations of the differences between beliefs and knowledge, the veracity of the proposition distinguishes beliefs from knowledge (Abelson 1979; Black 1973; Lehrer 1990; Sigel 1985). Sigel (1985) includes this differentiation in his definition of beliefs, which will provide a foundation for this research. Sigel (1985) proposed the following definition:

> Beliefs are knowledge in the sense that the individual knows that what he (or she) espouses is true or probably true, and evidence may or may not be deemed necessary; or if evidence is used, it forms a basis for the belief but is not the belief itself.... In sum, beliefs are constructions of reality. They may incorporate knowledge of what and knowledge of how, but do not necessitate evidential propositions. Beliefs are considered as truth statements even though evidence for their veridicality may or may not exist. (348–349)

Definition

For purposes of this research, *beliefs* are defined as an individual's "mental constructions of experience—often condensed and integrated into schemata or concepts" (Sigel 1985, 351) that are held to be true and may guide personal action.

Assumptions

If beliefs are mental representations integrated into schemata and concepts, the following assumptions about schemata and concepts may apply:

- Beliefs may be held as semantic networks similar to concepts and schemata (Abelson 1979; Donald 1987; Howard 1987; Sigel 1985).

- Contradictory beliefs may exist within different knowledge domains (Abelson 1979; Donald 1987; Sigel 1985).

229

- Certain beliefs may be core beliefs, and, like core schemata, these core beliefs may be difficult to change (Abelson 1979; Green 1971; Howard 1987; Nisbett and Ross 1980).

These assumptions frame the methodological design of this study. Comparisons of semantic maps of Debbie's initial and final interviews yielded distinct similarities and differences in her beliefs about learning and teaching. The similarities may represent Debbie's core beliefs—those that form the basis for most of her practical arguments—the primary premises of her explanations for her actions (Fenstermacher 1986; Green 1971, 1976). The differences may represent changes in her peripheral beliefs, those most easily examined, discussed, and changed (Green 1971; Howard 1987; Nisbett and Ross 1980). Semantic maps of Debbie's intermediate interviews were analyzed to examine changes in Debbie's beliefs throughout the staff development project. A narrative of these changes was written to combine semantic maps and field notes in a representation of the ideas and actions that provoked Debbie throughout the process of changing.

METHODOLOGY

As part of a larger study at a suburban high school in the Midwest, this case study is the story of one science teacher whose beliefs about how students learn began to change during her participation in a staff development project. The nine-month research project required that the 12 teachers engage in all of the staff development activities, including attending monthly structured classes, preparing at least three audio-taped model lessons and their analyses, and participating in five audio-taped, structured interviews. Because they involved some questions about the content of the classes and the teachers' lessons, and about the teachers' beliefs, these interviews were integral to both the staff development and the research processes. The subject of this study, Debbie, participated in each class, completed each of the taped model classes and follow-up analyses, and participated in each structured interview. The researcher, who also served as the staff developer, recorded and transcribed each interview and maintained field notes regarding her interactions with her subjects and her impressions of the staff development project and the process of changing. Thus, the extensive data set included (1) the transcriptions of five structured interviews that took place across nine months, (2) transcriptions of additional conversations between the researcher and participants, (3) transcriptions of each class session, (4) field notes about the class sessions, such as what was written on the blackboard, where participants sat, or what may have happened or been said that indicated change was occurring, and (5) field notes about the researcher's impressions of the participants and the changes they may have been experiencing.

Intervention and Data Source: Staff Development

In accordance with Taba's (1962) theory of curriculum development, the staff development intervention included a theory of changing teachers' theories of learning and a method of changing teachers' classroom practices. The theory formed the basis of the *what* beliefs addressed; the teaching strategies formed the basis of the *how* beliefs addressed. As presented in the staff development project, the Taba Teaching Strategies engage students in a structured discussion that consistently moves the students from a focus on concrete examples to their inferences of broad, abstract generalizations about the relationships among the concepts discussed. The teacher plays the role of director, questioning students to provide evidence and reasoning and to draw conclusions along the way. Because the strategies can simply model the use of prior knowledge in the construction of knowledge, and because they complement the content regarding cognition and constructivism, they were incorporated into the staff development project by the researcher. The Taba Strategies challenged Debbie's *how* beliefs.

During the staff development project, participants attended seven class sessions—three centered on model discussions framed as Taba Teaching Strategies: Concept Development, Interpretation of Data, and Application of Generalizations. The content generalizations which focused these three model discussions framed the content of the staff development project. The model Concept Development lesson focused on learning, Interpretation of Data on the effects of cognitive dissonance on learning, and Application of Generalizations on the difficulties and complexities involved when school change occurs. The researcher designed these lessons both as models of the strategies the participants were to learn and as thought-provoking discussions to challenge the participants' current beliefs about how students learn. Unlike some staff development projects in which the coordinator models lessons that are student-oriented and lessons that the teachers might use in their classrooms, these lessons were intentionally teacher-oriented and included content about schooling to engage the teacher as learner (Driver 1988; Taba et al. 1964). Thus, the content of these model discussions challenged Debbie's *what* beliefs.

Intermittent class sessions

Each month, class sessions included the introduction of a strategy, the analysis of the previously introduced strategy, common misconceptions and errors identified in teachers' taped trial lessons, and the discussion of concepts and schemata, generalizations, and concept mapping. In the seven class sessions they attended, the 12 teachers discussed their own theories of how students learn and those theories presented by Bransford and Johnson (1972), Howard (1987), Neisser (1976), Novak and Gowin (1988), Resnick (1987), Rumelhart (1980), Taba (1962, 1966), Taba, Durkin et al. (1971), Taba and Elzey (1964), Taba et al. (1964), and von Glasersfeld (1987).

Data

Each class session was audio-taped. Transcriptions of the audio-taped staff development classes, participants' trial lessons and their analyses, discussions between the researcher/staff developer and the participants, structured and informal interviews, and field notes served as data for this study.

THE SUBJECT: DEBBIE

The subject of this study, Debbie, is a 37-year-old science teacher who had been teaching for 16 years at the onset of the staff development project.

Debbie was sponsor of the senior class and involved in both academic and extracurricular activities at the school. Recently, she began a master's degree program in administration and supervision at a prestigious midwestern university, hoping to secure an administrative certificate and to move into a school-site or district-level administrative position that involves working in the area of curriculum. As an undergraduate, Debbie was a science major who interned at the school where she now teaches.

"The thing that I am constantly amazed about teaching is that I like it more and more as I go on," Debbie said as an addendum to her first interview, expressing the enthusiasm that underlies her personal beliefs about teaching and learning. Debbie was selected as the subject of this case study because she was one of only a few subjects to attend each session, submit and analyze each taped lesson, and participate in each interview. Further, she was one of the few subjects who designed her next semester plans to include the strategies and ideas she learned.

Interviews

Debbie was interviewed in January at the end of the first semester of the 1989–1990 school year (and prior to participation in any staff development activities, at monthly intervals throughout the project, at the beginning of the 1990–1991 fall semester, four months after the project ended, and briefly before this paper was written). Questions included in the interviews were written to elicit both declared and private beliefs (Goodenough 1963; Richardson-Koehler 1988; Richardson-Koehler and Hamilton 1988). To elicit responses about the subject's declared beliefs that were recorded in the subject's own language, the pre- and post-interviews included questions based on the Kelly Repertory Grid (Kelly 1955; Munby 1982, 1984; Olson 1980, 1981) as revised by Munby (1982). In addition, questions to elicit the subject's private beliefs about learning were included, as Kelly (1955) and Richardson-Koehler and Hamilton (1988) recommended. These questions were based on the Heuristic Elicitation Method (Eisenhart et al. 1988; Richardson-Koehler 1988; Richardson-Koehler and Hamilton 1988). Intermediate interviews throughout the study involved discussion about the researcher's and the teacher's analyses of the model lessons and an ongoing dialogue about how students learn, what the teacher was learning, and how she knew that she and her students had learned something. To complement categorical data elicited in the pre- and post-interviews, several questions in each of these interviews were designed to elicit *critical incidents* (Flanagan 1962). Critical incidents are those that the subject can recall with the most detail possible. This detail allows for a categorical analysis of the interviews to determine the characteristics of the core concept, in this case, learning.

Data Analysis

Interviews and semantic maps

The transcribed data from each interview was used to develop semantic maps, which graphically represent a subject's understanding of the concepts.

Many researchers use semantic maps to represent a subject's construction of concepts and the relations she sees among them (Anderson 1983; Donald 1987; Fisher et al. , in press; Novak and Gowin 1988), as mental models are key to theories of cognition that include schemata as the building blocks (Neisser 1976; Rumelhart 1980).

To develop conceptual maps, or semantic maps, from interview data, Novak and Gowin (1988) suggest framing interview questions to include the types of relationships among concepts to be analyzed. For example, to elicit responses about the relationship between what students do and learning, the researcher asked, "Think of a classroom where learning is taking place. What are the students doing?" Similarly, questions regarding the relationships between learning and what the teacher does, what students do to demonstrate it, what the teacher does when she is doing it, and what she does to demonstrate it were included in the interviews. The responses to these questions were mapped to develop a semantic network of the teacher's understandings of these relationships.

Using the subject's words and phrases as concepts, but using the relations implicit in the questions asked, a semantic map with learning as the central concept was developed for each interview. The extent and content of each map, however, were dependent upon the subject's responses. Each map includes similar sets of relations among learning, activities, and assessment, determined by the pre- and post-interview questions. The words that Debbie used to name and describe learning, activities, and assessment were represented as *concepts* in the semantic maps and appear in the squares and ellipses. Verbs and adverbs from the interview questions were represented as the *relations* among these concepts and appear on the lines connecting concepts. An *instance* is any combination of concepts and relations stated by the subject.

Domain and descriptive analyses: Toward defining Debbie's beliefs.

Two analyses were used to determine the ways in which Debbie's beliefs changed after participating in the staff development process. The semantic maps developed from the pre- and post-interviews provided a framework to facilitate a domain analysis, as suggested by Spradley (1979) and corroborated by Flanagan's (1962) categorical analysis. A descriptive analysis provided a framework to identify cultural themes and to write a narrative to express the findings.

The domain analysis included three categorical analyses of the data collected as responses to the initial question of the pre- and post- interviews. This question was similar to those Kelly (1955) used to create a repertory grid and later revised by Munby (1982, 1984). The quantitative and qualitative categorical analyses involved the following procedures:

- Analyses of the number and types of activities that occur when learning takes place involved counting the activities that Debbie named and noting those in which the teacher, the students, or combinations of the two were included

- Analyses of the number and types of categories into which Debbie placed the activities involved counting and identifying the function of the emergent categories (Spradley 1979)

■ Analyses of the number and types of instances or connections among the activities and categories that Debbie stated involved counting these connections and identifying the nature of activities and the participants included in each category

By outlining annotations of Debbie's responses to these questions, counting and describing the types of activities, categories, and connections were simplified.

These domain and descriptive analyses were examined to determine qualitative differences in the subject's beliefs about learning. These qualitative analyses of the teacher's responses were used to explicate the subject's beliefs as to what Spradley (1979) called cultural themes, which he defined as "any cognitive principle, tacit or explicit, recurrent in a number of domains and serving as a relationship among subsystems of cultural meaning" (186). These principles, or propositions, and their supporting descriptive analyses framed a case study of the ways in which Debbie's beliefs about learning changed.

To address the question of how these changes were exhibited throughout the process of changing, the same analyses of the intermediate interviews were completed and augmented by data from the staff development classes and the researcher's field notes. The researcher synthesized all of these analyses to create a narrative of Debbie's process of changing, that is, Debbie's story (Connelly and Clandinin 1990). The narrative provides a sense of the whole process and does not imply any cause-and-effect relationship but rather "brings theoretical ideas about the nature of human life as lived to bear on educational experiences as lived" (Connelly and Clandinin 1990, 3). The validity of this case, as Wolcott (1990) suggested, lies in the telling of Debbie's story in as many of her own words as possible.

Throughout the process of changing, Debbie struggled with her *what* beliefs about how students learn and her *how* beliefs as she practiced new teaching strategies. By comparing and analyzing semantic maps of Debbie's responses to the same interview question prior to and after the staff development project, differences between Debbie's prior beliefs about learning and teaching and those she expressed later were determined. By reviewing and analyzing field notes, taped class sessions, and interviews with Debbie and other teachers at the site throughout the project, the manner in which this change in beliefs was exhibited throughout the process of changing was reconstructed in narrative form.

DEBBIE'S CHANGE IN BELIEFS

The First Interview

Debbie's initial proposition about learning might be summarized simply as *through activity, students learn*. This proposition is supported by a domain analysis (Spradley 1979) of the categories of activities she derived. Each category Debbie created named the function of or included activities within it. For example, "applying information students find" names the function of building models, doing well on quizzes, explaining concepts, and identifying the unknown.

Because she talked about "traditional kinds" of activities in which the teacher played a "traditional" role of giving information and the students "find" answers, Debbie also proposed initially, *When they learn, students are given or find knowledge.* In this interview, Debbie seemed to project the teacher as an interloper, a one-way communicator. In her examples, while the students were frequently engaged in interaction, working in small groups, and teaching each other, the teacher delivered information in "traditional" activities such as demonstrating, answering student questions, lecturing while students take notes, and reading. These "traditional" activities" "are always quick and easy ways, possibly the most efficient ways to get information across, although not necessarily the longest lasting," Debbie said. "I still think that the most efficient way to get information across is short lectures."

The Last Interview

After the staff development project ended, summer vacation passed, and the new school year began, Debbie talked about learning differently. As in the first interview, Debbie focused on a laboratory experience when she was asked to recall a classroom where learning was taking place. In the final interview Debbie proposed, *By interpreting what they experience and interacting with others about what they see and know, students learn.* Debbie indicated there were cause-and-effect, means-to-an-end, and sequential relationships among activities. In fact, the language Debbie used to describe these activities indicated the students' and teacher's mindfulness about the activities in which they were engaged. Rather than just doing, helping, talking, working, and identifying, students were deciding, interacting, observing, discussing, and hypothesizing. They record, graph, and translate data, comparing it to what has been done before and to what happens in the real world.

Although social interaction was an integral activity in learning during her first interview, new insights about the significant role of interaction in learning was supported by a change in Debbie's personal learning style. During her summer in graduate school, she had learned about case studies by participating in "group work, which I've never been big on doing personally." Further, when asked how she knew she had learned something, Debbie said, when I "know I can convince someone else of it." By thinking and interacting in a group about what they do, see, and know, students learn. In this final interview, Debbie projected the teacher as an active participant whose questions and directions guide students to put the puzzle together and to synthesize—not just to apply information. She implied that *Learning is making sense of the world by organizing information about experiences.*

Summary of the Change

Although the domain analyses of Debbie's initial and final interviews resulted in slight quantitive differences, they indicated qualitative differences that shift the teacher from the passive role of interloper to the more interactive role of director. In her last interview, Debbie named two more activities than she had initially named in which the students and teacher engage when learning takes place. Further, she named three more activities in which the teacher participated, included the teacher in two more categories of activities, and made 13 more connections among activities and categories than she had during the first interview. Based on Spradley's (1979) delineations, the categories changed from those that name the function of, or characteristics included in the first interview, to those that name cause-and-effect, means-to-an-end, and sequential relationships.

As interpreted by the researcher and corroborated by Debbie, Debbie's beliefs about how students learn shifted from a belief that students learn simply by doing classroom activities to one where students learn by mindfully engaging in experiences inside and outside of the classroom. In both the initial and final interviews, Debbie talked about students doing laboratory experiments and interacting with other students, and about the teacher answering questions. Only one category in each interview included activities involving the teacher and students: "traditional kinds" from the first interview and "students and teacher interacting" from the second interview. The difference between the names and contents of these two categories mirrors Debbie's theoretical shift from a behaviorist view that doing is learning to a constructivist view that sense making is learning.

Traditional kinds of activities including, in Debbie's words, the teacher's "answering questions, demonstrating the problem," and lecturing while students were "asking questions and taking notes," reflect a behaviorist view that the teacher tells students what they should know and that students learn by recording what was said. Activities in which "students and teacher [were] interacting," including the teacher's "answering questions, asking questions, bringing students back, and pointing to the wrong answer and asking" while the students were "interacting" and "discussing," reflect a constructivist view that, through social interaction, the students are making sense of their world and their activities. In the latter case, the teacher interacts with the students, facilitating the sense-making activities with open-ended questions about what students are doing and thinking. Thus, Debbie's propositions shifted from *When they learn, students are given or find knowledge* to *Learning is making sense of the world by organizing information about experiences.*

Because Debbie related *students' doing* and *students' interacting* to learning throughout the project, these concepts may be key elements of Debbie's core beliefs, formulated in her own classroom experiences. These were key concepts in her introductory through graduate coursework in science education, she explained. Further, the science curriculum she followed was laboratory-based, involving students in doing experiments and interacting with other students about their findings. Perhaps because these beliefs were indoctrinated throughout Debbie's career in schooling—as a student, as a student of teaching, and as a teacher—these core beliefs remained constant. Green (1971), Howard (1987), and Nisbett and Ross (1980) discussed the difficulty in changing beliefs that are formed without evidence. Further, Sigel (1985) stated that the likelihood of change "depends on whether or not they are evidentially based" (348). Perhaps Debbie's more peripheral beliefs about her role in the learning process changed because she trusted the new evidence with which she was consistently confronted in her readings, discussions, and experiences.

THE PROCESS OF CHANGING

As she struggled with *what* about learning and *how* to teach, Debbie engaged in the process of changing. Throughout the staff development project, she followed a path of resisting, questioning, and assimilating new information and new practices. Initially, Debbie resisted *what* she was learning about concepts and schemata from the content of the staff development activities. "We do need more concepts [in science]," she said in February, "[but] students don't really have a handle on them." She resisted *how* she might implement new teaching strategies, as well. "You can't do that in science," she said in February. "Schools can't be like that," she echoed in April. Yet, as she prophesied in her initial interview, "You get involved whether you want to or not," and Debbie "got involved," reading, discussing, practicing, analyzing, and reflecting.

"How do I play the game?" Debbie asked at the onset, and throughout the process of changing, Debbie focused both on *what* the game was and *how* she played. First, the game was defined by school district policies that outlined what was to be taught in each science course and that mandated annual, district-level assessments of what students knew about this content. Debbie explained that she played the game by working with science teachers from the other four high schools in the district to plan her classroom activities, relying upon what they agreed had worked and what she knew had worked before. By April, however, when Debbie found that students' "grades on the test were not as good as they normally are, but I think they have more of a feeling for what the subject is," she asked, "Now, how do you explain that?" Debbie continued to ask questions about district assessments, time, and how students organize what they learn. She began to change the game: "Do what I know is right and be a risk taker," she said. She determined that she would take the time she needed to cover more than just "trivial waste" and asked, "Am I teaching the right thing?"

Throughout the staff development project, Debbie focused on "concepts," a key construct guiding the development of the teaching strategies that she learned and examined in the model discussions and supplementary readings. In the process of changing, Debbie assimilated her understanding of concept, representing her iterations of the idea throughout the project. In February, concepts were what she taught, like "diffusion" and "electromagnets." Even though "students don't really have a handle on them," she said early in the study, "We do need more concepts [in science]." In April, "concepts" were "trivial pursuits knowledge" in the science curriculum, where "everything is at a pretty concrete level." She talked about the complexity and abstraction of concepts, notions that framed the development of the teaching strategies that she practiced. By May, she explained "concepts" as fitting into a "logical picture" that students construct by organizing what they know with the teacher's help. A year after the project began, Debbie continued her study of concepts by enrolling in a course called "Cognition, Learning, and Instruction." When asked to review the findings of this study, she reported, "I was really intrigued by 'concepts and schemata,'" noting that the findings did not address her new understanding. Perhaps, by constructing a "concept" schemata, she formed a belief system that substantiated her proposed changes in practice.

When the researcher explained the findings expressed in this paper, Debbie said that she agreed with the differences in her thinking about her role in the teaching process and with her having core beliefs related to science teaching, beliefs that defined students' learning as doing and interacting about what is done. Further, she said that she takes more time to be "a better teacher. It reminded me that I need to listen to what [students] are saying . . . [realize] what prior knowledge they have . . . what connections they make." Debbie thinks about the "whole idea of schemata and building a structure . . . to hang things on . . . every day."

"I'm turning into being more like the kids instead of like the teacher," Debbie said. The game changed. Ten months after the project ended, Debbie continued planning with the district's science teachers, deciding *how* and *what* to teach. As the teachers skipped from topic to topic without making connections among concepts previously studied, activities students had done, and/or students' prior knowledge, Debbie asked, "How does that make sense?"

By the end of the study, Debbie's beliefs were more constructivist, as represented by her putting together a puzzle as a metaphor for teaching. In May, Debbie said, "Teaching is like helping kids put the pieces of a puzzle together, because they do have a lot of random knowledge, but they just don't know where it fits yet." She extended her metaphor as follows:

> You have to, starting at the beginning, find out where they are, what their bits of knowledge are. Then you try to help them put it together into a logical picture by showing them that you turn the picture this way and that. . . . Turn the pieces around the other way; flip them over; and arrange them. Show them how to group their information. All the pieces that are green may be grass. Again, try to help them organize what they know into some logical picture. . . . I am really more aware. . . I spend a lot more time talking about it, introducing it, laying some groundwork, and making it real for them before I try to extend what they know about the subject. . . . I'm more aware that they have data . . . it's just helping them take their database and organize it, and add to it. . . . I am really trying to help students make connections and thinking through where I want them to end up at the end. . . . You start reminding yourself that you're not just teaching subject matter, you're teaching a kid. And, there's something going on in the kid's head. . . .

Thus, Debbie's role as the director began to emerge. In September, Debbie recalled her metaphor: "Teaching is helping students manipulate puzzle parts, helping them put it together, without doing it for them." She explained as follows:

> I don't think you can put a puzzle together for someone . . . whenever you put a puzzle together for even a little kid, they always want to dump it right out again, because they didn't do it themselves. (Teaching) is the same thing . . . I think you just have to help them, guide them to do it. . . . Otherwise, they are going to shuffle (the pieces) and try again anyway.

Further, she explained that everyone's puzzle was not the same, "including colors . . . the blue that I see isn't necessarily the blue that you see."

Unlike the subjects upon whom Guskey (1986) based his findings that changes in beliefs follow changes in practice, in the process of changing, Debbie discussed ongoing interactions with *what* she learned and *how* she might teach differently. Throughout, and long after, this study, Debbie was constructing new understandings of "concepts," "teaching," and "learning." Debbie personified the beliefs that framed this intervention—that knowledge is constructed within the individual and that meaning is socially constructed.

Conversations as Constructivist Interventions

In this study, as in Taylor's (1990, 1991) studies, the intervention played an important role throughout the process of changing; and, although the intervention was clearly defined as the teaching strategies—their content and process—the interviews must be considered as an intervention, as well. Both the model discussions and the interviews engaged Debbie in public and private conversations, challenging her *what* and *how* beliefs about learning and teaching. Further, these conversations were an integral part of the developing relationship between Debbie and the researcher—me.

As I reviewed the data and shared my findings with Debbie and other colleagues, they frequently asked, "But what about the relationship between Debbie and you?" "What relationship?" I replied quickly. Yet, by listening carefully to the taped class sessions and interviews, I heard and noted my own beliefs about the emerging teacher-student relationship and how they were enacted throughout the staff development project.

Debbie and I engaged in a continual conversation—publicly, in six half-day workshops, and privately, in five hour-long interviews. These conversations were predicated on Debbie's interest in and willingness to participate in the planned activities and on my reputation as the expert on the Hilda Taba Teaching Strategies and thinking skills. Two important informants at the school—my friend Jennifer, a well-respected English teacher there, and the assistant principal, who, based upon Jennifer's recommendation and my research proposal, had hired me to conduct the inservice training—advertised my expertise to the faculty. As the participants remarked during their initial interviews, they were recruited by these informants, both of whom shared a high regard for my expertise.

As I reviewed the audio-taped lessons and interviews, I found Debbie's and my emerging relationship modeling a cognitive apprenticeship. According to Collins et al. (1989), a *cognitive apprenticeship* involves the teacher and learner or a group of learners in an expert-novice relationship. Sometimes changing roles, the teacher and learners change implicit practical knowledge to explicit theoretical and practical knowledge through shared reflection on the details of expert performance in problem-solving situations. Through a dialectic about an experienced problem-solving situation, the learner constructs knowledge and skill, in context. Similarly, Debbie and other participants in this staff development project engaged in many conversations about the model lessons. As if engaged in a cognitive apprenticeship, after modeling in each Taba Strategy for the teachers, the participants and I discussed and analyzed my planning the lesson, my focusing intentions and guiding generalizations, my decision making throughout the enacted lesson, and my reasoning along the way.

Within this cognitive apprenticeship, I expressed and enacted my personal beliefs about teaching and learning, and about the dynamic relationships among teachers and students. Trust, playfulness, wonder/reflection, and dissonance were concepts that I frequently expressed and principles upon which I based my relationships. "Trust they will get there," I told the class one day. "Am I getting there?" I asked Debbie during an interview in February. "You know this better than I." "Let's play with that for a second. . . . Does somebody else want to take another one, just to play?", I asked in the March class. "I wonder what you can do with them," I asked Debbie, after suggesting a plan and hearing myself say, "Listen to me: I'm telling you what to do . . . I wonder what would happen if . . . ?"

Throughout our interactions, Debbie and I developed a relationship of mutual respect and trust. In addition to the time that we spent together in class, we had the luxury of private interviews, in which my questions focused Debbie's reflections. "It's wonderful," she said in April. "It helps me reflect. It's things I haven't really thought about." In a way, I became an "other" for Debbie. That is, in the dialectic, I became the person Fenstermacher and Richardson (1991) call the "other," a person who elicits another person's beliefs and provokes the normative reconstruction of beliefs. Since our cognitive apprenticeship was both public and private, both the descriptive task of elicitation and the normative task of reconstruction were possible. During the interviews, I elicited Debbie's beliefs; but among her peers in class discussions, Debbie reconstructed her beliefs. In class, for example, Debbie shared her general beliefs about cognitive dissonance:

> Cognitive dissonance provides an uncomfortable situation that must be handled by the student in order to return to an equilibrium state when comfortable and learning. And, I have in parentheses, "Thinking must occur for this to happen." My three reasons are that the student has to reflect to clarify what he really believes. He has to give thought to this, that it is something that is dissonant to him. So, he has to have a fairly clear belief structure somewhere. And, he must then be able to justify that position, and he has to have the flexibility to resolve within his own mind how he is going to resolve these two dissonant ideas in order to return to a more comfortable state.

Debbie and I had time together to plan and reflect, resolve and create dissonance, and playfully engage in teaching and learning, and over a period of time, we would "return to a more comfortable state." Thus, throughout the continual dialectic process of the staff development project, Debbie's beliefs about learning and teaching changed. Consistently, and both privately and publicly, Debbie was confronted with new ideas and practices. Although she initially resisted using ideas such as prior knowledge, concepts and generalizations, and cognitive apprenticeships in teaching and learning, she continued to engage in conversations about these issues. She talked to her colleagues, publicly in the staff development classes and privately in the workroom for scheduled interviews. She expressed her frustrations regarding time and expectations, yet she continued to plan new teaching strategies and express changed beliefs about the role of these social interactions in learning and teaching.

How Might Change Include Constructivist Practices?

This research addressed questions about the process of changing (Fullan 1985). The findings indicate that changing practice and changing beliefs occur throughout the staff development process and that these changes represent interactive processes rather than causal ones, as suggested by Guskey (1985, 1986) and Fullan (1982, 1985). Distinguishing Debbie's changes in practice and changes in beliefs as ends of each other misinterprets the iterative process through which Debbie made sense of the learned strategies and concepts. As she engaged in learning by reading about, reflecting upon, discussing, practicing, and analyzing new practices and ideas, Debbie seemed to experience *praxis*—a synergistic change in both beliefs and practice (Goffmann 1973). Therefore, this study raises important questions about the nature and content of staff development projects and about the methodology necessary to identify and change core beliefs.

Constructivism and Staff Development

This close look at the process of changing provides evidence of the need for constructivist staff development projects that address the teacher as a learner and that involve the teacher in *praxis*—doing, reflecting, learning, changing (Goffmann 1973). As constructivist notions about students' making sense in classroom settings impacts instruction, similarly, teachers' sense making must be considered in the design of staff development. Therefore, in deference to Cohen's (1988) notion that change will always be the same, we must begin to address the complexity and ambiguity issues that cloud the implementation of constructivist notions in classroom practice. Perhaps a critical application of Doyle's (1983) notion of academic work and the tensions created by more ambiguous and complex academic tasks would inform the design of more successful staff development projects. For example, in the context of a cognitive apprenticeship, Debbie seemed to deal more comfortably with the ambiguity, risk taking and difficulty involved with this type of adventurous, constructivist teaching and learning. The features of cognitive apprenticeships that encourage learning and changing may be examined more closely.

Further, those projects that focus on the implementation of procedures (the *how*) but not on the beliefs that support the design and implementation of these procedures (the *what*) may always result in mutual adaptations, whether they are desirable or not. However, as Sigel (1985) suggests, the chances of changing beliefs are increased when the individual trusts the new evidence provided, and the structure of the belief system is permeable. If core beliefs are indoctrinated throughout teachers' careers in schooling, staff development projects must include rich evidential arguments that challenge participants' beliefs (Green 1976; Howard 1987; Nisbett and Ross 1980) and provide a variety of practices for participants to analyze (Richardson 1990; Richardson and Anders 1990). As Sigel (1985) and Howard (1987) explained, changing beliefs involves providing repeated evidence to change or to complete the invalid belief structure. This is complementary to Fenstermacher (1986) and Green's (1976) proposition that, to change practice, one must change or complete the teacher's or student's practical argument and complementary to Richardson and Anders' (1990) findings about how teachers change beliefs and practice through staff development that includes research and practices as the content.

This case provides evidence to support the notion that staff development requires components that require the descriptive elicitation of beliefs as well as the normative reconstruction of beliefs, as argued by Richardson (1990) and Fenstermacher and Richardson (1991). One cannot assume causal relationships among the interviews, the intervention, and the change in Debbie's beliefs; however, one cannot deny the evidence of Debbie's ongoing process of changing as she reconstructed her beliefs about learning and teaching. The demonstrated change in her beliefs appeared throughout the process as iterations of learning about, reflecting upon, and enacting new ideas. These iterations were the results of private and public conversations about learning and teaching; conversations shared, revisited, and renewed.

CONCLUSION

Based upon the process and results of this study, staff development projects should address the following concerns:

- What are the prior beliefs that participants bring to the staff development process?

- In what ways might these beliefs be elicited and reconstructed, publicly and privately, throughout the staff development process?

- What are the beliefs embedded in the recommended practices emphasized by the staff development process?

- How might the beliefs embedded in practice be made more explicit?

- How might both *what* and *how* beliefs be integrated in the design of the staff development process?

Further, future research studies may include similar intervention strategies and a larger number and broader variety of subjects; examine the relationships among prior beliefs, changed beliefs, and those implicit in the implemented innovation; investigate the interactive role of elicitation and reconstruction of beliefs throughout the process; and explore the use of cognitive apprenticeships in the practice of teaching.

For research to impact practice, an imminent goal of many staff development projects (Fenstermacher 1986; Richardson and Anders 1990; Richardson-Koehler 1987), questions about the interactive nature of beliefs and practice and how they change will continue to be raised. Methods for identifying and analyzing core beliefs—those held most strongly and those most difficult to change—may assist researchers in more carefully examining these beliefs throughout the process of changing. Further, examining the constructivist nature of learning and changing can inform the development of successful staff development projects in which teachers are empowered by their own sense making.

REFERENCES

Abelson, R.P. "Differences Between Belief and Knowledge Systems." *Cognitive Science,* 3, (1979): 355–366.

Anderson, J.R. *The Architecture of Cognition*. Cambridge, MA: Harvard University Press, 1983.

Black, M.B. "Belief Systems." In J.J. Honigmann (Ed.), *Handbook of Social and Cultural Anthropology*. New York: Rand McNally, 1973. 509–577.

Bransford, J.D., and M.K. Johnson. "Contextual Prerequisites for Understanding: Some Investigations of Comprehension and Recall." *Journal of Verbal Learning and Verbal Behavior*, 11, (1972): 717–726.

Clark, C.M., and P.L. Peterson. "Teachers' Thought Processes." In M. C. Wittrock (Ed.), *Handbook of Research on Teaching*. New York: Macmillan Publishing, 1986. 255–296.

Cohen, D.K. *Teaching Practice: Plus a Change*. (Issue paper no. 88-3). East Lansing, MI: Michigan State University, The National Center for Research on Teacher Education, 1988.

Collins, A., J.S. Brown, and S.E. Newman. "Cognitive Apprenticeship: Teaching the Crafts of Reading, Writing, and Mathematics." In L.B. Resnick (Ed.), *Knowing, Learning, and Instruction: Essays in Honor of Robert Glaser*. Hillsdale, NJ: Erlbaum, 1989. 453–494.

Connelly, F.M., and D.J. Clandinin. "Stories of Experience and Narrative Inquiry." *Educational Researcher*, 19(5), (1990): 2–14.

Cuban, L. "Reforming Again, Again, and Again." *Educational Researcher*, 19(1), (1990): 3–13.

Donald, J.G. "Learning Schemata: Methods of Representing Cognitive, Content and Curriculum Structures in Higher Education." *Instructional Science*, 16, (1987): 187–211.

Doyle, W. "Academic Work." *Review of Educational Research*, 53(2), (1983): 159–199.

Doyle, W., and G.A. Ponder. "The Practicality Ethic in Teacher Decision-making." *Interchange*, 8(3), (1977): 1–12.

Driver, R. "Theory into Practice II: A Constructivist Approach to Curriculum Development." In P. Fensham (Ed.), *Development and Dilemmas in Science Education*. London: Falmer Press, 1988. 133–149.

Eisenhart, M.A., J.L. Shrum, J.R. Harding, and A.M. Cuthbert. "Teachers' Beliefs: Definitions, Findings, and Directions." *Educational Policy*, 2, (1988): 51–70.

Fenstermacher, G.D. "Philosophy of Research on Teaching: Three Aspects." In M.C. Wittrock (Ed.), *Handbook of Research on Teaching*. New York: Macmillan Publishing, 1986. 37–49.

Fenstermacher, G.D., and V. Richardson. "The Elicitation and Reconstruction of Practical Arguments in Teaching." Paper, annual meeting of the American Educational Research Association, Chicago, 1991.

Fisher, K.M., J. Faletti, H. Patterson, R. Thornton, J. Lipson, and C. Spring (in press). "Knowledge Networks: Theoretical Considerations." *Journal of College Science Teaching*.

Flanagan, J.C. *Measuring Human Performance*. Palo Alto, CA: The American Institutes for Research, 1962.

Fullan, M.G. *The Meaning of Educational Change*. New York: Teacher's College Press, 1982.

Fullan, M.G. "Change Processes and Strategies at the Local Level." *Elementary School Journal*, 85(3), (1985): 390–421.

Goffmann, E. *The Presentation of Self in Everyday Life*. Woodstock, NY: Overlook Press, 1973.

Goodenough, W.H. *Cooperation in Change*. New York: Russell Sage Foundation, 1963.

Green, T. *The Activities of Teaching.* New York: McGraw-Hill, 1971.

Green, T. "Teacher Competency as Practical Rationality." *Educational Theory,* 26, (1976): 249–258.

Guskey, T.R. "Staff Development and Teacher Change." *Educational Leadership,* 42(7), (1985): 57–60.

Guskey, T.R. "Staff Development and the Process of Teacher Change." *Educational Researcher,* 15(5), (1986): 5–12.

Hamilton, M.L. *The Practical Argument Staff Development Process, School Culture and Their Effects on Teacher's Beliefs and Classroom Practices.* Unpublished dissertation, University of Arizona, Tucson, AZ, 1989.

Howard, R.W. *Concepts and Schemata: An Introduction.* London: Cassell Educational, 1987.

Kagan, D.M. "Ways of Evaluating Teacher Cognition: Inferences Concerning the Goldilocks Principle." *Review of Educational Research,* 3, (1990): 419–469.

Kelly, G.A. *A Theory of Personality.* New York: W.W. Norton, 1955.

Lehrer, K. *Theory of Knowledge.* San Francisco: Westview Press, 1990.

Munby, H. "The Place of Teachers' Beliefs in Research on Teacher Thinking and Decision Making, and an Alternative Methodology." *Instructional Science,* 11, (1982): 201–225.

Munby, H. "A Qualitative Approach to the Study of a Teacher's Beliefs." *Journal of Research in Science Teaching,* 21(1), (1984): 27–38.

Neisser, U. *Cognition and Reality.* San Francisco: Freeman, 1976.

Nisbett, R., and L. Ross. *Human Inference: Strategies and Shortcomings of Social Judgment.* Englewood Cliffs, NJ: Prentice-Hall, 1980.

Novak, J.D., and D.G. Gowin. *Learning How to Learn.* New York: Cambridge University Press, 1988.

Olson, J.K. "Teacher Constructs and Curriculum Change." *Journal of Curriculum Studies,* 12(1), (1980): 1–11.

Olson, J.K. "Teacher Influence in the Classroom." *Instructional Science,* 10, (1981): 259–275.

Resnick, L. "Learning in School and Out." *Educational Researcher,* 16(9), (1987): 13–20.

Richardson, V. "Significant and Worthwhile Change in Teaching Practice." *Educational Researcher,* 19(7), (1990): 10–18.

Richardson, V., and P. Anders. *Reading Instruction Study* (Grant No. G008710014). Washington, DC: Office of Educational Research and Improvement, 1990.

Richardson-Koehler, V. "What Happens to Research on the Way to Practice?" *Theory into Practice,* 26(1), (1987): 38–43.

Richardson-Koehler, V. "Teachers' Beliefs about At-risk Students." Paper, annual meeting of the American Educational Research Association, New Orleans, LA, 1988. (ERIC Document Reproduction Service No. ED 306–343.)

Richardson-Koehler, V., and M.L. Hamilton. "Teachers' Theories of Reading." Paper, annual meeting of the American Educational Research Association, New Orleans, LA, 1988.

Rumelhart, D.E. "Schemata: The Building Blocks of Cognition." In R.J. Spiro, B.C. Bruce, and W.F. Brewer (Eds.), *Theoretical Issues in Reading Comprehension.* Hillsdale, NJ: Erlbaum, 1980. 3–26.

Sarason, S.B. *The Culture of School and the Problem of Change.* Boston: Allyn and Bacon, 1971.

Sigel, I.E. "A Conceptual Analysis of Beliefs." In I.E. Sigel (Ed.), *Parental Belief Systems: The Psychological Consequences For Children*. Hillsdale, NJ: Erlbaum, 1985. 345–371.

Spradley, J.P. *The Ethnographic Interview*. New York: Holt, Rinehart, and Winston, 1979.

Taba, H. *Curriculum Development: Theory and Practice*. New York: Harcourt, Brace, and World, 1962.

Taba, H. *Teaching Strategies and Cognitive Functioning in Elementary School Children*. (Cooperative Research Project No. 2404.) San Francisco: San Francisco State College, 1966.

Taba, H., M.C. Durkin, J.R. Fraenkl, and A.H. McNaughton. *A Teacher's Handbook to Elementary Social Studies: An Inductive Approach*. Reading, MA: Addison-Wesley, 1971.

Taba, H., and F. Elzey. "Teaching Strategies and Thought Processes." *Teachers College Record*, 65, (1964): 524–534.

Taba, H., S. Levine, and F. Elzey. *Thinking in Elementary School Children*. (Cooperative Research Project No. 1574.) San Francisco: San Francisco State College, 1964.

Taylor, P.C.S. "The Influence of Teacher Beliefs on Constructivist Teaching Practices." Paper, annual meeting of the American Educational Research Association, Boston, 1990.

Taylor, P.C.S. "Collaborating to Reconstruct Teaching: The Influence of Researcher Beliefs." In V. Richardson and K. Tobin (Chairs), *The Emerging Role of Constructivism in Changing Teachers' Beliefs*. Symposium, annual meeting of the American Educational Research Association, Chicago, 1991.

von Glasersfeld, E. "Learning as a Constructive Activity." In E. von Glasersfeld (Ed.), *The Construction of Knowledge: Contributions to Conceptual Semantics*. Salinas, CA: Intersystems, 1987. 307–333.

Wolcott, H.F. "On Seeking—and Rejection—Validity in Qualitative Research." In E.W. Eisner and A. Peshkin (Eds.), *Qualitative Inquiry in Education: The Continuing Debate*. New York: Teacher's College Press, 1990. 121–152.

15

Learning to Teach Science: Constructivism, Reflection, and Learning From Experience

Tom Russell

The decade of the 1980s saw research on the learning of science concepts dominated by the perspective of constructivism. At the same time, teacher educators generally became increasingly attentive to the perspectives of reflection and metaphor as aids to the process of learning to teach. The process of learning to teach science is a significant meeting ground for the perspectives of constructivism and reflection, and this chapter explores the potential value of bringing the two perspectives to bear on each other. This task requires consideration of both content and process in the two domains of science teaching and science teacher education.

Constructivism and *reflection* are two terms that have joined the science education vocabulary in the last decade from quite different sources. Constructivism is supported by studies of how children learn and driven by perspectives on the nature of learning. Reflection is related to analyses of the nature of professional knowledge and the ways it is acquired, held, and renewed, and it is driven by perspectives on the relationship between thought and action.

These two terms share several significant features. Both are adaptable to and suggestive of a broad range of meanings, so that any two people using either term assume at their peril that they are using the term in the same way. Both terms are also indicative of potential challenges to the traditions of teaching and learning that most people experienced in their own schooling. Linked together, the two terms gain power, as the argument of the chapter reveals.

The argument opens with accounts of two fundamental dilemmas: the science teacher must come to terms, both practically and theoretically, with the tension between "the curriculum," as content to be covered, and "constructivism," as a process by which students deal with content. People learning to teach science confront an even greater dilemma, perhaps masked by the ease with which we often speak of "putting theory into practice." The most basic dilemma is that generated by the contrast between their personal actions of teaching in schools and the teaching they experience in teacher education classes, and this dilemma must be considered with reference to both content and process. Against the background of these dilemmas, identified as ones brought into focus by issues of constructivism and reflection, the argument turns to the debate about what needs to be changed in the teaching of science: content or strategies. The debate applies both to our schools and our programs of teacher education. The opportunities offered by reflection, constructivism, and metaphor are then identified, and the argument closes by describing the challenges arising for programs of science teacher education. A personalized style is adopted at points where the argument bears closely on my own science teacher education teaching experiences.

THE SCIENCE TEACHER'S DILEMMA: CURRICULUM VS. CONSTRUCTIVISM

When I participated in the Science Council of Canada's development of a set of case studies of science education in Canadian schools (Olson and Russell 1984), the discussions among the authors of the seven cases left a profound impression. I realized more fully just how deeply teachers care about the subjects they teach and about the task of helping students understand, in some significant way, the subject being taught. Yet in the final analysis, teaching students in classes of 25 to 30 or more generates an even higher "governing variable": maintaining equality of access to further education. Quite simply, teachers lack the time and resources to bring all children in a class to the same level (as in "mastery learning"). The compromise that is often most palatable is to teach so that one can always assure oneself that all children were provided with equal opportunity to demonstrate levels of achievement that would enable them to continue their schooling.

To illustrate, it is common for a teacher to be asked, "What will be on the test tomorrow?" on the day before a test is scheduled, and it is not uncommon for the teacher to reply by describing the types of questions and the content areas to be examined. These familiar interactions have little to do with fostering deeper understanding of the content and much to do with guiding the review practices of those children who are prepared to invest time in further study before the test. When society places virtually exclusive emphasis on test scores, and when parents transmit this emphasis to their children, it becomes more and more difficult for children to tolerate a teacher's interest in giving less attention to assessment and more attention to the learning process. Consequently, the curriculum (to be taught and tested) competes with and usually takes precedence over the goal of meaning and understanding that is implicit in constructivism.

Constructivism speaks to fostering better understanding of a subject, not to the pressure to provide equal opportunity to all students to succeed in evaluated work, though we tend to assume (and do so at our peril) that those who understand will display that understanding in their test responses. Thus, concern for understanding competes with concern for covering the curriculum and testing what has been "covered." The dilemma for the science teacher is fundamental, profound, and unavoidable. Every student who presents a signal that some aspect of the content has not been understood is a potential source of frustration and further effort.

To sum up the dilemma, constructivism represents a plausible perspective on learning and a valuable addition to traditional views of the nature of learning. The perspective is appealing to teachers because it speaks to the longstanding goal of helping others understand. Yet because covering a curriculum and testing children on what has been taught are fundamental to the practical realities of teaching in schools, constructivism can be seen as generating issues that interfere with the traditionally fundamental goal of teaching.

THE BEGINNING SCIENCE TEACHER'S DILEMMA: RELATING THEORY AND ACTION, TO LEARN FROM EXPERIENCE

Learning to teach has never been regarded as easy. The most obvious task is to develop confidence with the countless everyday practical skills of the classroom. At the same time, there is a great deal to be accomplished in conceptual terms, involving the underlying theories and rationales for teaching practices as well as the inevitable tensions between aspirations and accomplishments in teaching. From the perspectives of constructivism and reflection, the beginning science teacher (taken here as "student teacher" or "first-year teacher") faces a subtle but profound dilemma with facets of both content and process. In terms of content, a discipline studied initially for purposes of mastery must be reconstructed for purposes of pedagogy, and this involves deepening one's personal understanding of the subject while also learning how the subject presents conceptual difficulties to the students one is teaching. The content dilemma can be resolved only in response to personal opportunities to teach. In terms of process, the content of teacher education programs typically advocates teaching strategies that are not consistent with those being used by experienced teachers, with those expected by students, or with those being used by teacher educators. All too often this very complex dilemma is resolved by writing off teacher education as unrealistic and theoretical and by using strategies very similar to those one experienced personally as a student: "We teach as we were taught."

The process side of this learning-to-teach dilemma can be illustrated by a recent personal example. A first-year science teacher from my 1989–1990 class made the effort to describe to me in some detail his initial teaching experiences and found himself "apologizing" for the fact that he had settled on alternating sessions of "lecture" and "solving problems" as his initial compromise in terms of how he taught.

> As for actual classroom teaching style, I have found that, believe
> it or not, lecture style is the most effective and the most popular.
> I can get straight to the point and give them time to try the
> problems. Sometimes lessons have (heaven forbid!) taken up
> most of a period, but then I devote the next class to seatwork and
> taking up problems. The communication I have with students
> helps tremendously as I can warn them about what is to come.
> (R. Hypher, personal communication, January 25, 1991).

The apologetic tone, signaled by the phrases "believe it or not" and "heaven forbid!" is in response to a teaching strategy that I adopted when I began to work with preservice teachers in 1977. At the beginning of every year, I announce my intention to reduce the amount of talking I do as teacher, and I appear to succeed in doing less talking during classes than most of my colleagues. The message is "felt" by my students but not understood fully; it cannot be fully understood without significant experience, which is not yet available to the teacher in the first year of teaching. As I better understand the confusion my behavior generates for the preservice students in my science methods class, I have improved my explanation and justification for my own behavior. In this task I was helped by a student who asked, with frustration, but without seeing the irony in her words, "Why didn't you tell us you weren't going to tell us?" When we modify teaching strategies from the familiar and expected, we are dealing with influences that are very powerful in potential and yet also poorly tolerated and even actively resisted by students.

Since 1984 I have had the opportunity to include in my teaching a research perspective in which I gather interview data on several students in each year's group. It is difficult to recall one student teacher who did not acutely feel the tension between teacher education's ideals and schools' realities. Yet in program terms, teacher education offers little support to the beginning science teacher who realizes how central is this dilemma of practice vs. theory, in both content and process. Constructivism and reflection both speak directly to this dilemma implicit in learning to teach.

CHANGING CONTENT VS. CHANGING STRATEGIES, IN SCHOOLS AND IN TEACHER EDUCATION PROGRAMS

Sarason (1982, 1990) is my favorite author on the topic of innovation and change in schools. He offers little in the way of the much-desired recipe, but his explanations are deep and profound and show why recipes are likely to be misleading. Existing practices are embedded in such a complex web of values and compromises that each and every attempt at change generates a new set of problems that are usually not addressed by those seeking change. Again, there are two dimensions: first, content can be changed and ordered in different ways, and this is the familiar domain of curriculum development. Second, new strategies can be advocated, and these are often attached to new curricula, profoundly complicating any innovation.

The argument here asserts that curriculum development is not the central issue facing science education, whether it proceeds on constructivist principles or not. How we teach and learn, in science classes and in programs of science teacher education, is the issue of importance. Process, not content, is the untested domain. Progress is incredibly difficult because we have so little useful experience to guide us in understanding the effects of teaching strategies. The range of strategies that we have actually experienced in our own schooling is but a highly restricted portion of possible strategies. Teachers have accepted being told by others to change the textbooks they teach from the topics they teach, and the sequence in which they teach the content of their subjects. The myth persists that the teacher's professionalism resides in selecting the most appropriate strategy for a specific group of children. (Implicit is the assumption that any strategy will do because strategy does not influence the content learned.)

Harmin and Simon's (1965) parody of curriculum change was so subtle that many missed the point of their superficially absurd suggestion that children could be taught the telephone directory. Teachers appeared to be praised for being clever enough to use a variety of teaching methods to make the ultimately dull content more palatable. My personal reading was that the authors were pointing to the sad state of affairs in which teachers had no say about what their students should be taught and no understanding of the significance of the process by which students are taught. The annual race to beat the school-year calendar ensures that students' test scores will continue to decline and that politicians will continue to criticize those who teach and those who teach teachers. Society wants genuine understanding and students who love to learn and value life-long learning, yet it imposes conditions that make those goals unattainable.

The last 30 years in particular (since PSSC, BSCS, Chem Study, and the "new math" (Hurd 1969)), and the last 80 years in general (Roberts 1982), have shown us that curricula can be and have been written in different ways with resulting variations in messages to students. In secondary science, Harvard Project Physics was shown to be a clear alternative to the "grand intellectual view" of physics in the Physical Science Study Committee (PSSC) course:

> Clearly, then, the content of a physics course can have marked effects on the students' view of physics. Harvard Project Physics was able to draw on the completed work of the Physical Science Study Committee and, going off in a different direction, apparently succeeded in restoring the social, humanitarian, and artistic aspects that had been lost in PSSC, and succeeded in augmenting the historical and philosophical aspects. (Ahlgren and Walberg 1973, 189)

Reducing the emphasis on mathematics and increasing the historical and humanistic content did make a difference to students, but as Walker and Schaffarzick (1974, 94) reported, innovative curricula appear to teach content better than those they replace only when the criteria for comparison of achievement match the innovative curricula. There is little to suggest that we will ever find the one best curriculum for a science subject (Munby and Russell 1983), and the risks associated with publishing an unusual approach to content are so high that the likelihood of shifting science education by that route seems very low, whether one's interest is in a constructivist perspective or in some other important perspective on teaching and learning.

Yet the pursuit of content to the exclusion of process persists among those who understand and pursue constructivism. Millar (1989) distinguishes the "alternative conceptions movement" from the constructivist view of learning and then concludes that "a constructivist model of learning does not, however, logically entail a constructivist model of instruction" (589). He then argues that construction of meaning is a learner's activity that "occurs whenever any successful learning takes place and is *independent of the form of instruction*" (589, emphasis added). Millar goes on to conclude that teachers should engage students in ways that will lead them to reconstruct meaning, ending with the suggestion that "this contribution [of the constructivist research programme] should be seen more in terms of improving the sequencing and pacing of science curriculum and less in terms of changing teaching styles and approaches" (590). What advantages accrue by ignoring teaching styles is not readily apparent.

Within his argument, Millar cites as support the fact that many people who now understand science arrived at that understanding by nonconstructivist teaching methods. This view identifies a crucial line of demarcation among those who would improve teaching. Some would say, with Millar, that we must continue to rewrite curricula in hopes of better supporting the requirements of fostering the reconstruction of meaning by students. Others, among whom I count myself, would argue that those who achieved an understanding of science by nonconstructivist teaching strategies did so in spite of the strategies used by their teachers. Many others who experienced those strategies "fell by the wayside" and abandoned any initial interest they may have had in the content of the natural sciences, perhaps more in response to the process of being taught science than to the content of science. Adherents of this view might go on to argue that rewriting curricula continues at a significant pace without profound effects.

We have long known that content can be changed and reordered, but making significant changes in teaching-learning strategies used by teachers is a promising but largely untested domain that deserves greater attention. Barnes (1976) made the fundamental point that communication in schools conveys curricular messages. He argued, with convincing illustrations, that when students work in small groups with data and without a teacher, interesting and important changes happen. These changes are related to the constructivist assumptions about learning. Barnes also contrasted "transmission" with "interpretation" and noted that science teachers tend to be at the transmission end of the teaching spectrum while English teachers tend to be at the interpretation end. There is a great deal of research suggesting that "telling" (or transmission) has been the dominant strategy around which teachers may develop individual variations. Telling wins, as a strategy, whenever "covering information" is the fundamental objective. And telling persists as a dominant strategy in teacher education, in university classrooms of course, but also in discussions of practice with individual student teachers during practice teaching.

Barnes issued a challenge that has received little attention within teacher education, yet that is the domain where the messages of communication have double impact—on the nature of learning to teach and on the nature of the beginner's subsequent teaching. Both constructivism and reflection offer insights into long-standing issues of teaching and learning to teach science, and they may contribute to long-awaited and long-overdue developments in the way we teach, in science classrooms, and in science teacher education programs.

REFLECTION, CONSTRUCTIVISM, AND METAPHOR

Reflection may be the most misunderstood word in teacher education. In common sense terms, reflection refers to thinking about one's teaching, when one has the time and environment for thought about one's work (or intended work). It is assumed that reflection is desirable and productive, that there is not enough of it, and that it is separate from action. It is also assumed that schools do not encourage or support reflection by teachers, but that universities somehow *do* have the resources for nurturing reflection. At the same time, reflection is something one does outside of lectures and seminars, perhaps in the library or in one's favorite chair; reflection is thought to be completely removed from action. Also, reflection may be seen as related to the issue of bringing theory and practice to bear on each other, though in keeping with the specialty of the university, reflection is likely to refer to using theory to improve practice rather than to taking practice as the starting point for developing personal understanding of theory.

My basic inclination is to dismiss all of these as "everyday" and "common sense" meanings for a term that is the best we have for a much more important process: learning from experience. As Edwards and Mercer (1987) have argued, we tend to assume that learning from experience is straightforward and automatic. There is a growing recognition within science education that students of science do not relate laboratory experiences directly or easily to the science concepts that the experiences are meant to illustrate or stimulate thought about (Driver 1983). Similarly, I have come to believe that experiences of teaching (in the practicum or throughout one's career) are not directly or easily related to one's personal beliefs and assumptions about how students learn and why one teaches as one does. This is important because our programs of preservice and inservice teacher education encourage the everyday assumption that action follows from and is consistent with one's beliefs.

Unlike *reflection, constructivism* is virtually free of common sense meaning, although it may be associated with the term *construction* and the familiar meanings of physical (rather than mental) activity. As a term, it must seem as strange to beginning teachers as reflection seems familiar. In both instances, those learning to teach are likely to have difficulty coming to grips with the ways in which science teacher educators are using the terms. *Constructivism* is closely related to an image of reflection that develops new frames for experience, leading to new possibilities for action. Driver et al. (1985) offer the following account of a constructivist perspective on learning:

> Individuals internalize their experience in a way which is at least partially their own; they construct their own meanings. These personal "ideas" influence the manner in which information is acquired. The personal manner of approaching phenomena is also found in the way in which scientific knowledge is generated. Most philosophers of science accept that hypotheses or theories do not represent so-called "objective" data but that they are constructions of the human imagination. In this way of thinking, observations of events are influenced by the theoretical frameworks of the observer. The observations children make and their interpretations of them are also influenced by their ideas and expectations. (2-3)

My perspective for reflection continues to be that expressed in Schon's (1983, 1987) outline of a process of "reflection-in-action" that he offers as an alternative to the underlying epistemology of the university. My attempts to understand this process through my own teaching and research since 1983 have led me to see that reflection-in-action involves reframing of experience in response to messages from the practice setting, followed by new actions in the practice setting, actions that express the new frame and test its viability (Munby and Russell 1989). Thus, teachers' practical knowledge is seen as developing from experience as thought prompted by action develops a new dimension that returns to action. Neither of these steps—reframing and returning to action with the new frame—can be separated from the immediacy of events of practice. Indeed, they require attention both within and outside practice. But before one can recognize such reflection, one must have a perspective that recognizes the possibility and the importance of being attentive to experience and opportunities to learn from it. The similarities between this view of reflection and the general account of constructivism are striking and significant. Reflection-in-action is applied by Schon to the on-the-job learning of professionals. Constructivism is applied by researchers to children attempting to understand concepts of science. In both instances, the focus is on action and the subsequent understandings of action for how we see the world we live and act in. One major difference involves the source of direction for the learning process. Reflection is seen as directed by the individual adult professional, while constructivism is seen as guided and mediated by the teacher. The essential point is the implied similarity between how children and adults learn, between learning of science by children and learning of professional knowledge by beginning science teachers.

The concept of metaphor spans both constructivism and reflection. In our studies since 1985 of the development of teachers' professional knowledge, Hugh Munby and I have found the concept of metaphor to be valuable in the analysis of what teachers say about their development of practical knowledge (Munby and Russell 1990; Russell and Johnston 1988). We believe that teachers' development involves coming to see teaching experiences and the goals of teaching in new frames and that the metaphors apparent in talk about teaching can be useful guides to identifying and understanding the development of practical knowledge of teaching. Carter (1990) has explored the concept of metaphor in the context of constructing case studies:

> Cases with metaphor may be useful as conversational tools for deliberate reflection about what it means to teach and as currency for conveying knowledge about teaching. Case based conversations using metaphor may also be one way to allow teachers to confront their own conceptions of teaching and to critically examine their views about accomplishing the work of teaching. (114)

Marshall (1990) has made similar suggestions for teacher education, but the words could also be interpreted from the perspective of teaching science:

> One path toward heightening awareness of their implicit belief systems involves focusing on the metaphors and images they use as they describe their teaching. (128) Being attuned to the language student teachers use to describe their teaching may provide clues concerning the cause of their problems in teaching. A first step in helping student teachers develop may be bringing into awareness inconsistencies between stated beliefs and the implicit framework on which their actions are based. (131)

Just as teacher educators are interested in exploring the role of metaphors in teacher education, so the concept of metaphor has practical implications for a constructivist approach in the science classroom and laboratory. When students' language for discussing or explaining phenomena shows signs of development, then the teacher gains valuable clues about the nature and extent of that development.

CONSTRUCTIVISM'S CHALLENGE TO TEACHER EDUCATION

These perspectives on constructivism and reflection and the associated views of dilemmas facing experienced and beginning science teachers lead me to the conclusion that the untested but compelling domain that awaits our attention is the arena of classroom teaching and learning strategies. (Baird, in press, has made important beginnings in his research in Australia.) This aspect of schooling has always been accessible to us, but we often turn away from it because it is so difficult to make headway, let alone to know where to start. Teachers have been so strongly criticized for the folly of "discovery methods," "new math," "open classrooms," and other variations from traditional and familiar approaches that they are understandably wary of changing their strategies. Of particular interest is the question of whether "constructivism" and "reflection" will come to be seen as two more unrealistic fads whose time has passed.

The environment of teaching—begun in the preservice teacher education program and continued within the schools where one teaches—can persistently undermine the importance and significance of attention to the reactions of students as individuals and as groups. We are encouraged to put theory into practice, but I am much more inclined to the view that we must see practice as the medium by which we come to understand existing theory—theory induced tacitly by years of experiences in schools. We already have too much theory that has never been understood against the background of experience. Theory is taken in from lectures and texts, but is also deeply embedded in the images we hold for what is "proper and possible" in the classroom setting. When we come to act in the classroom, images rather than theory are much more likely to guide our teaching. Both constructivism and reflection-in-action stress the importance of messages back from action, particularly the messages that are seen to have been unexpected because they do not match the way one has been thinking about the situation.

Teacher educators have long been criticized for their "irrelevant theories" about how to improve education. There is little doubt that moving from the school to the university or college context generates a shift of perspective. A similar shift occurs when teachers move from the classroom to the summer curriculum writing project. How else can we explain the recurrent phenomenon that new curricula generated by teachers always include more content than any individual teacher can feel comfortable "covering" in the available time? When a teacher moves from school classroom to teacher education classroom and is thereby removed in part from the daily routines and commitments of life in a school, it is difficult to limit one's scope and vision to "how to be a firm disciplinarian" or "15 ways to improve overhead transparencies," even though good discipline and effective transparencies are self-evidently desirable when teaching. And so teacher educators are well known for their criticisms of practices in schools, exhorting beginning teachers to repair the problems that have defied solution for decades (Sarason 1990).

After almost 15 years as member of a team of more than 60 in a formal teacher education program preparing 650 teachers a year, I have a great deal of respect for what teacher educators manage to accomplish in the space of eight or nine months. At the same time, I accept that teacher education needs to reorder its own house. As school practices remain the same, so too do those of teacher education. Teacher education continues to be seen as idealistic and theoretical by its clients and by society for at least two reasons. First, the unresolved tension between theory and practice is quite predictable when preservice teachers have such strong expectations of how one learns to teach and have such a short program of initial preparation. Second, teacher education has failed to practice its own recommendations for others and needs to examine its own routines of content and process. We always chuckle when someone refers to the stereotypical professor of education who gave a lecture on how to teach by the discovery method. We laugh because the contradiction is so apparent, but much more we laugh because the problem is so serious and intractable.

Recently, I have sensed that teacher educators are beginning to realize how important it is to scrutinize our own practices, not take the superficially easier route of lecturing so that we can "get on with our research." When McIntyre (1988) wrote about plans for an "internship" program of teacher education at the University of Oxford, he offered an unusual account of learning to teach, one that is consistent with the growing interest in constructivism and reflection:

> What I conclude then from our limited knowledge about how people learn to teach is, first, that they have their own extensive repertoires and their own agendas; and that we as teacher educators, if we are realistic, need to accept that we can only help them in their efforts, not define the enterprise in which they are engaged. Second, I conclude that even if we did believe that we had "the answers," reliable knowledge about how best to teach, student-teachers would not accept it but would want to test it for themselves in various ways; we can probably exert more influence by encouraging this process of testing than by pretending it is not necessary. And third, we can have some confidence that if we do not put student-teachers into situations which overwhelm or seriously threaten them, we have good reason to believe that they will explore the problems of teaching with a high degree of objectivity about their own performances and rationality in their investigations. (104–105)

Reflection-in-action and constructivism are approaches to learning from experience that gain power when their usually separate literatures are combined. Learning to teach science is a complex and confusing exercise, and teacher education's traditional program and teaching approaches extend, rather than reduce, the complexity and confusion. If the strategies of constructivism and reflection were practiced and discussed with some consistency in teacher education, then those learning to teach science would be better prepared to approach and sustain the long-term challenge of developing similar teaching strategies for teaching science.

Learning from experience, rather than from spoken or written words, is central to both constructivism and reflection-in-action. Speech and writing play significant but not exclusive roles, and the role of action and experience is the unique common feature of these perspectives. Schooling makes both children and adults better at learning by words than at learning from experience. If science teacher education is to encourage learning from experience as a way of linking thought and action more closely, then teacher education itself must develop learning from experience as a feature of the process of learning to teach.

I have always been skeptical of suggestions that a paradigm shift is underway in a particular field of inquiry; it is difficult to know how one could recognize a Kuhnian revolution from the inside. Yet one can sense that many people are trying to develop new rationales for teacher education, and science teacher education is an obvious venue for attention to constructivism and reflection. While our practices may appear to evolve by explaining away anomalies in current rationales, our research and our thoughts about our work are often more focused on the development of new theories that will apply to both teacher education and science teaching. The resolution is far from clear, but the activity is encouraging.

REFERENCES

Ahlgren, A., and H.J. Walberg. "Changing Attitudes Towards Science Among Adolescents." *Nature*, 245, (1973): 187–190.

Baird, J.R. (in press). "Collaborative Reflection, Systematic Inquiry, Better Teaching." In T. Russell and H. Munby (Eds.), *Teachers and Teaching: From Classroom to Reflection*. London: Falmer Press.

Barnes, D. *From Communication to Curriculum*. Harmondsworth, UK: Penguin Books, 1976.

Carter, K. " Meaning and Metaphor: Case Knowledge in Teaching." *Theory into Practice*, 29, (1990): 109–115.

Driver, R. *The Pupil as Scientist?* Milton Keynes, UK: Open University Press, 1983.

Driver, R., E. Guesne, and A. Tiberghien. "Children's Ideas and the Learning of Science." In R. Driver, E. Guesne, and A. Tiberghien (Eds.), *Children's Ideas in Science*. Milton Keynes, UK: Open University Press, 1985. 1–9.

Edwards, D., and N. Mercer. *Common Knowledge: The Development of Understanding in the Classroom*. London: Routledge, 1987.

Harmin, M., and S.B. Simon. "The Year the Schools Began Teaching the Telephone Directory." *Harvard Educational Review*, 35, (1965): 326–331.

Hurd, P.D. *New Directions in Teaching Secondary Science*. Chicago: Rand McNally, 1969.

Marshall, H.H. "Metaphor as an Instructional Tool in Encouraging Student Teacher Reflection." *Theory into Practice*, 29, (1990): 128–132.

McIntyre, D. "Designing a Teacher Education Curriculum from Research and Theory on Teacher Knowledge." In J. Calderhead (Ed.), *Teachers' Professional Learning*. London: Falmer Press, 1988. 97–114.

Millar, R. "Constructive Criticisms." *International Journal of Science Education*, 11, (1989): 587–596.

Munby, H., and T.L. Russell. "A Common Curriculum for the Natural Sciences." In G. Fenstermacher and J. Goodlad (Eds.), *Individual Differences and the Common Curriculum, 82nd Yearbook of the National Society for the Study of Education, Part 1*. Chicago: University of Chicago Press, 1983. 160–185.

Munby, H., and T.L. Russell. "Educating the Reflective Teacher: An Essay Review of Two Books by Donald Schon." *Journal of Curriculum Studies*, 21, (1989): 71–80.

Munby, H., and T.L. Russell. "Metaphor in the Study of Teachers' Professional Knowledge." *Theory into Practice*, 29, (1990): 116–121.

Olson, J., and T.L. Russell (Eds.). *Science Education in Canadian Schools, Vol. 3, Case Studies of Science Teaching*. Ottawa: Science Council of Canada, 1984.

Roberts, D.A. "Developing the Concept of 'Curriculum Emphases' in Science Education." *Science Education*, 66, (1982): 243–260.

Russell, T.L., and P. Johnston. "Teachers Learning from Experiences of Teaching: Analyses Based on Metaphor and Reflection." Paper, meeting of the American Educational Research Association, New Orleans, LA, 1988.

Sarason, S.B. *The Culture of the School and the Problem of Change* (2d ed.). Boston: Allyn and Bacon, 1982.

Sarason, S.B. *The Predictable Failure of Educational Reform*. San Francisco: Jossey-Bass, 1990.

Walker, D.F., and J. Schaffarzick. "Comparing Curricula." *Review of Educational Research*, 44, (1974): 83–111.

16

Transitioning into Constructivism: A Vignette of a Fifth Grade Teacher

Kenneth L. Shaw and Mia Lena Etchberger

Jessica, a self-contained fifth grade teacher, is considered by her district to be a successful elementary teacher. She received the District Science Council's Outstanding Elementary School Science Teacher Award in 1990 and the Teacher of the Year Award from her school for the 1988–1989 school year. Yet, while the awards were encouraging, Jessica realized something was lacking in the students' abilities to conceptualize or formulate knowledge. During the 1989–1990 school year, she realized she and her students were but playing a "game" and calling it school. She contemplated her past two years of teaching and realized she taught quite traditionally; she felt she was no different from the ordinary teacher. She became aware of a disturbing pattern emerging in her students. They would memorize certain bits of information, sets of facts, or operational procedures and algorithms, and repeat them back on a test. When she asked her students what meaning they had for the information, sets of facts, or operational procedures and algorithms, most were unable to adequately explain what they had learned. With few exceptions, a few weeks or months later, they were unable to converse about a previously studied concept. They could not or would not relate that information and use it in any meaningful way. The game essentially was as follows: teacher dispenses information; student receives information; student transmits information back to the teacher. Most students played the game well, as evidenced by their grades, but Jessica began asking the question, "Were they learning?"

Jessica was very perplexed by what was taking place in her classroom and by what was taking place in classrooms within her school and county. What could she do to make a difference? This perplexity was heightened because of her colleagues' perceptions that she was a successful teacher. This chapter will endeavor to capture the epistemological and methodological struggles of Jessica and will describe the resulting reconceptualizations of how she made sense of her own instruction and her students' learning as she came to understand constructivism.

GRADES AND KNOWLEDGE

The realization that good grades do not necessarily indicate students' understanding of concepts caused Jessica to begin a process of rethinking her position, ideology, role, and purpose in the field of education. She was thoroughly disturbed that children did not "know." As a result of this perturbation, she decided she must do something to help children know. Jessica resolved that she must not continue in her old pattern of teaching, which was strongly bound to textbooks and scope-and-sequence curriculum plans. She realized she needed to change, but did not know where to begin.

This initial overriding and ever-present perturbation led Jessica to begin searching for a new way to teach. This led her back to school to complete her master's degree in Elementary Education. What she found instead of new ways to teach was a new way of viewing the entire learning process. She began to formulate a vision of possibility which was nonexistent prior to her return to college.

Realizing something had to change was not enough for Jessica. Committing to change empowered her to search for an alternative. She could not envision an alternative, but knew it must be "out there" somewhere. She did not have a clear vision, just a general hope that there was something better.

When Jessica began reading and looking into case studies, she discovered that teachers found it more helpful to focus on helping students learn by enabling them to form their own knowledge, rather than to focus on learning new teaching methods. Instead of focusing on teaching and better ways to teach, Jessica began to focus on learning and better ways to learn. Constructivism became the alternative through which she was able to generate an initial vision of how she could change what she was doing to help children come to know. She perceived that she needed to change her metaphor of teaching and learning from that of teacher-dispenser, student-receiver to that of teacher-enabler, student-constructor. This led her to another dilemma.

TEACHING FROM A CONSTRUCTIVIST PERSPECTIVE

Jessica asked herself, "How do I take concepts, information, and other people's knowledge, and arrange them in such a way as to allow students to construct their own knowledge?" Jessica wanted a quick answer to this question. She was not aware of any textbooks, programs, or people that could show her how to take her next week's science lessons and make the activities constructivist in nature. She began at ground zero, with much fear and trembling. She was very comfortable with her textbook and felt safe as she answered questions, showed films, did preset projects and demonstrations, and reviewed for the test. She was familiar with this approach and so were her students. Furthermore, this was the way she learned when she attended elementary school. Even though she was on unfamiliar ground, she decided to create her first constructivist activity. Just thinking about it raised several questions in her mind: "How can I be sure they will learn or come upon the concept themselves? How much do I need to tell them? How much is too much? Do I tell them anything at all?"

It was a severe struggle, and Jessica lacked confidence in her ability to produce a truly constructivist environment, but her vision propelled her to try. She began in science, the one subject in which she believed students would readily construct knowledge. She arranged her students into groups, believing that cooperative groups were the perfect vehicle for optimum construction to occur. Indeed, her vision at that time saw cooperative groups as necessary and intrinsic to the process of knowledge construction; they were inseparable. She gave students a few ground rules for working together because they were used to working individually and sitting in rows: (1) Be kind to each other, (2) Respect everyone's opinions, (3) Allow all students to have their say, and (4) Everyone is equal—no leaders. Jessica had no experience in cooperative grouping, so this too was uncharted territory for her. She commented that "the only way to begin is to begin, so we all jumped in head first."

Following the scope-and-sequence, they were on the unit for the study of invertebrates. The lesson she chose to use was ants, as those were the next creatures in the book. Jessica's confidence was heightened as a result of a recent group activity she experienced in her college class, the subject being ants. The student groups were told very generally to go outside, find some ants, and watch them. She told the students to "write down everything they could about them, try to copy their motions, draw pictures . . . ANYTHING." Then they would come back to the room and share what they had found. Their energy, excitement, and total absorption in the activity would have been enough to convince Jessica to change forever, but there was much more. They came back to the room bubbling over with information about ants. They filled three chalkboards full of their observations and drawings; Jessica even learned new things about ants. In the process, the class covered all of the information that the book covered (three body parts, six legs, etc.) and much more. The students talked with authority and conviction because of their experience. They KNEW about those ants. That was in October, 1989. The following May, they still knew about ants and had added to their knowledge as they read and interacted further with others and made additional observations of ants.

Experiencing the exuberance of true knowledge construction and innate motivation of her students, Jessica was propelled to further reflection and strengthened commitment to further change. Her vision became broader. Together, she and her students explored concepts and relationships, learned how to work together, and increased their own knowledge. However, the transition from a teacher-dominated environment to a student-centered one was not, by any means, smooth.

WHEN THE NOVELTY WORE OFF

When Jessica's science class began learning in groups, the students were focused and motivated. It was novel; it was something innovative and unique. The activities and situations in which they were involved were stimuli enough to keep them interacting, discussing, arguing, negotiating, and searching for meaning and understanding. But it became apparent that the initial novelty was wearing off after about a week into their new learning environment. Students were fighting and refusing to work with others; they were arguing about things such as who would record information, who would present it, and who was not doing anything. This was taking a lot of class time and detracting from any construction that could be taking place. It was a very real problem to the children and it took precedence over everything else.

This led Jessica to a commitment to try new groupings and to learn more about cooperative groups. She envisioned a room of happy cooperative children constructing knowledge together without conflict. She asked them what could be done and discerned a new way of solving problems: "Let THEM solve the problems!" They made lists on the board of things they did not like about groups and things they did like. The class talked about all of the items and discussed ways to improve the groups.

What they came up with was a highly flexible system of grouping. Jessica had been assigning them to work in groups of four. They wanted to choose their own groups, give themselves jobs to do, and switch jobs for each new activity. They also wanted to switch group members from time to time; so that is what they did. The first time, the class ended up with one group of five, two groups of four, three groups of three, and one set of partners. As time went by, Jessica occasionally saw the need to group them heterogeneously for certain activities and in groups of six or four because of materials constraints. In the meantime, Jessica became familiar with the research on cooperative learning and discovered that her students had constructed much of their own knowledge about successful cooperative groups, information that was reflected in the literature. The class did it through trial, reflection, and re-trial. They continued to have class discussions whenever a problem arose, and they searched for acceptable alternatives.

Jessica learned something. The children were undergoing their own perturbations, commitments, and visions as well. They saw that group learning was a much better way to understand. Realizing that, they committed to change their method of learning and generated a new vision of what school science was all about.

This was not true for all the children. Some of them liked the individual rows and the quiet room; they liked the way it used to be. The first year that Jessica was incorporating constructivist cooperative groups, she did not allow anyone to work by themselves in science class. She was convinced they would not have experiences as enriching as they could have if they were in groups. But as she reflected over the summer and continued her studies, she became disturbed about those few children who just could not function in a group. Her understanding of constructivism at that time involved interaction with others. She did not see how successful negotiation of meaning could occur without group processing.

KNOWLEDGE CONSTRUCTION

What Jessica discovered the following year (1990–1991) was that students do not have to be in a group in order to construct their own knowledge. However, she recognized that students' knowledge would be richer, fuller, and more intricate if they did confer with others. Broader relationships would emerge as a result of interaction.

For example, the following fall, Jessica allowed several students to observe ants individually. They came back with several good observations; however, the number of observations was less, and they tended to see only one aspect (i.e., legs, antennae, body parts, or patterns of movement). Jessica struggled with how to handle those students that suffered in group situations or caused groups to suffer in some way by their lack of involvement and/or behavior problems. Her vision at that time was of successful constructivist situations being group-oriented and problematic in nature, with her role being the questioner and prodder, never quite "giving away" the concept. However, when she began to pull students out of groups or allowed students to work by themselves, she realized that they were constructing knowledge every time they related the problem to something they knew, and they were able to understand and utilize the concept. One of Jessica's students, working with molecules in the form of colored marshmallows and toothpicks, put two blues and one green together to form H_2O, and discerned that the numbers indicated how many atoms of each element a molecule contained. She was able to come up with that relationship herself. But without interaction with others, she was not forced to prove or disprove it; discussion, amplification, clarification, and consensus were not available to her. But Jessica did see that she grasped a major concept in molecular construction for herself.

Because of her realization that individuals construct their own knowledge, Jessica began to relax somewhat about her role as information source. She realized that along with activities and situations, books and printed material, audio and video, she, too, could be a source of information for the students. Initially, she desperately attempted to remove herself from the position of information-giver but realized if she could provide some information they could not "come upon" themselves in order to help them solve some problem, then she should provide it for them. This proved to be further enriching for them, as long as she guarded against moving into the familiar lecture format.

Group learning, which allows rich constructions to occur, was working so well in science that she began to become perturbed about other subjects as well. So, during the second year, she began quite another chapter in her development as a constructivist teacher and attempted to overhaul her mathematics program as well (Etchberger and Shaw, in press).

COMPONENTS OF RECONCEPTUALIZATION

As we noticed Jessica wrestling with her new epistemology, we have noticed several recurring components that have affected her reconceptualization of teaching and learning.

Perturbation

Change cannot occur without some perturbation. This is analogous to Newton's first law of motion, "A body at rest or in uniform motion will remain at rest or in uniform motion unless some external force is applied to it." A teacher will continue teaching in a similar way unless perturbed by something or someone. Perturbations often cause frustration, discomfort, and a great deal of reflection. For Jessica, the perturbation of seeing children listen intently to the teacher, relaying the information back on a test, and then promptly forgetting the material was a significant perturbation that caused her to rethink what she was doing in her classroom, and more broadly, what education should be. Her readings when she began her master's degree of constructivism and cooperative learning also caused a perturbation in how she believed students learn. Prior to her graduate studies, she spent little time thinking of how students learn and a great deal of time thinking about how to teach more effectively. Now she believes students' thinking is vitally important to consider in every aspect of teaching.

Commitment

Commitment is a personal decision to make a change as a result of one or more perturbations. After Jessica made a commitment to change her instruction to a more constructivist approach, many perturbations arose. For example, she asked herself: "What do I do about grades? How do I assess students' knowledge? How do I teach in a constructivist manner? How do students learn?" Shaw and Jakubowski (1991) suggest when a perturbation disrupts a mental state of equilibrium, teachers tend to deal with the perturbation in one of three ways: (1) they block out the perturbation, thus stifling the opportunity to learn or change; (2) they develop a rationale for not dealing with the perturbation; or (3) they plan to make a change. With each perturbation, Jessica committed to make a change.

Vision

For teachers to change, they need to construct a personal vision of what teaching and learning should be like in their classrooms. Jessica needed alternative images to replace the traditional views of teaching. She had constructed a vision: students should be responsible for their own learning; the learning environment should be conducive to students being responsible for their own learning; and students learn best when they negotiate, conjecture, test, evaluate, and justify their thinking with other students. Even though this vision was not totally in focus, it, nevertheless, was a vision. She had a level of confidence in what she wanted in her classroom. This vision became more focused as she experienced changes in her classroom and as she had further dialogue with her colleagues and instructors.

Cultural Environment

The cultural environment for each teacher is different. Geertz (1973) stated that a "man is an animal suspended in webs of significance he himself has spun" (5). For Jessica, she noticed distinctive elements of the culture that affected her change process. These were support, time, resources, and beliefs about taboos and customs. Jessica believed that she was on her own in making changes in her classroom. For teachers to make veritable change, Rogers (1969) stressed that when teachers are making changes, they need support that is genuine and laden with respect, unconditional acceptance, sensitivity, and understanding. When new teaching strategies are incorporated into a traditional classroom, students' routines are sometimes abruptly changed, and dissonance is often the result. This classroom dissonance can easily influence teachers to return to the traditional way of teaching if they have received no other support. Fortunately for Jessica, the opposite was true; students overwhelmingly enjoyed the cooperative activities and thus strengthened her belief that the changes she was implementing were worthwhile.

Her recent readings of constructivism were coupled with the awareness of a need to change, and so Jessica set out to make changes in her classroom. She frequently inquired of her instructors what a fifth grade constructivist science classroom should look like. Her intense desire to have viable alternatives was not adequately met. Nevertheless, she decided to "jump in head first." She looked for curricular materials that would best fit her newly held beliefs about learning, but was unsuccessful, and so she created her own lessons. Resources are essential for teachers to gain alternative images of what can occur in their classrooms. Jessica was quite exceptional in that the constraints of not having what she considered appropriate resources did not overwhelm her into returning to her normal teaching routines.

Within the culture of school, teachers hold certain beliefs about existing taboos and customs. For Jessica, one custom was that teachers should strictly follow their textbooks, and students should sit in rows and remain quiet during class. When a reconceptualization of an epistemology to constructivism occurs, it is necessary that customs and taboos be questioned. It is also necessary to question routines and reflect on if and how they need to be modified. This is not an easy task since routines may be formed and reinforced over many years.

Reflection

Reflective thinking means "turning a subject over in the mind and giving it serious and consecutive consideration" (Dewey 1933, 3). Reflection is integral throughout any change process. Because Jessica's change process was gradual, it required time and understanding, and necessitated that she give serious consideration to her daily decisions. Becoming aware that change was needed was a product of reflection on the congruence between her beliefs and practices. Making a personal commitment to change required Jessica to reflect upon the possible costs and benefits of the change. The creation of a vision required considerable reflection and deliberation of possible viable alternatives. The social negotiation of the vision among other teachers enhanced her reflections. The reflection process was important as Jessica constructed a vision and projected herself into a vision of teaching from a constructivist perspective.

CONCLUSION

For successful and positive change to occur, teachers need to be perturbed, to be committed to do something about the perturbation, to establish a vision of what they would like to see in their classrooms, to develop a plan to implement this vision, and to have the support to carry out their vision. Change is a slow arduous process that requires patience, persistence, and respect. Respect is shown when the teachers are given the ownership of what they are changing.

The implications of the case of Jessica to educational reform is that teachers must be actively involved in the planning phases of any innovation. However, even teachers who are part of the first levels of planning may hold deeply rooted beliefs that may cause them much mental dissonance when they return to their classrooms. But through collaboration with other teachers, they can discuss, deal with, and support each other with their personal dilemmas and successes. The collaboration will also give teachers alternatives. Jessica is now more empowered: she has taken full responsibility for what happens in her classroom in terms of curricular and pedagogical decisions. She is now aware of the importance of the relationship between student empowerment and learning. She realizes that they, too, should take full responsibility for their own learning.

REFERENCES

Dewey, J. *How we Think, a Restatement of the Relation of Reflective Thinking to the Educative Process*. Boston: D.C. Heath, 1933.

Etchberger, M.L., and K.L. Shaw (in press). "Teacher Change as a Progression of Transitional Images: A Chronology of a Developing Constructivist Teacher." *School Science and Mathematics*.

Geertz, C. *The Interpretation of Cultures*. New York: Basic Books, 1973.

Rogers, C. *Freedom to Learn*. Columbus, OH: Merrill, 1969.

Shaw, K.L., and E.H. Jakubowski. "Teachers Changing for Changing Times." *Focus on Learning Problems in Mathematics,* 13(4), (1991): 13–20.

17

Collaborating to Reconstruct Teaching: The Influence of Researcher Beliefs

Peter C. S. Taylor

Recently, I participated in a collaborative research study which investigated the prospects of reconstructing the established belief system that underpinned a didactic, teacher-centered, "transmission-type" pedagogy of a colleague, an experienced high school teacher of science who was teaching "out of field." Ray was teaching Grade 12 mathematics for the first time and had completed the first of four 10-week terms that comprise the school year in Western Australia. Apart from incidental encounters with mathematics in his science teaching, Ray's previous formal experiences with mathematics had been as an undergraduate student, about 20 years ago.

Ray's interest in professional development led him to enroll, on a part-time basis, in a postgraduate science education degree program at the university in which I was teaching. We established a collaborative research partnership with the intention of addressing his concern about the underachievement of his Grade 12 class. Ray believed that his teaching practice might be partly responsible for a series of generally poor student test results and wanted to investigate ways of improving his teaching effectiveness. Ray committed himself to a collaborative research agenda that included the improvement of his teaching practice, interpretation and documentation of data, and reporting of our study.

My main research interest was to investigate the prospects of facilitating Ray's development of a constructivist pedagogy. In particular, I was interested in assessing the viability of a theory of pedagogical change that, I believed, might facilitate the development of "constructivist-related" pedagogical reform amongst teachers of high school science and mathematics. The theory of pedagogical change, discussed below, was based on constructivist-related theories of student conceptual change in science education and theories of teacher cognition.

The pedagogical change methodology was intended to engage Ray in a process of self-critical inquiry from a constructivist perspective. The main aim was to facilitate Ray's development of a constructivist perceptual "lens" that would enable him to newly interpret his classroom teaching experiences, critically reflect on the relationship between his pedagogical beliefs and practice, and reconstruct his pedagogy. I did not intend that Ray adopt a preconceived or prescribed constructivist pedagogy; rather, one of my goals was to determine the nature of possible constructivist teaching practices arising from Ray's interpretation and classroom application of constructivist theory.

During the study, Ray formed constructivist-related beliefs, and his classroom practice became more inquiry-oriented and student-centered. However, the narrow scope of his pedagogical innovations highlighted the significant role of Ray's established "positivist" epistemology in determining boundary conditions for pedagogical change (Taylor 1990b). Having assessed the central role of the teacher's pedagogical beliefs in constraining a constructivist-oriented pedagogical reform process, I now turn to the contribution of the pedagogical change theory that precipitated the teacher's pedagogical reconstruction. Did the theory contain the "seeds" of its own delimitation? Is the theory "naturally" limited to achieving only minor pedagogical reform of a "positivist" nature? If so, what lessons have been learned that might suggest more effective constructivist-oriented approaches to the professional development of teachers?

The purpose of this chapter is to address these questions and to assess the viability of the constructivist-related pedagogical change theory that influenced my role in the collaborative relationship with the teacher. The following sections set out the interpretive framework of the early part of the study, based on an analysis of earlier reports and the interview transcripts and related fieldwork data. The interpretive framework of the discussion section of the chapter draws on more recent theories of both constructivist theory and curriculum theory in an attempt to evaluate the viability of the theory of pedagogical change.

A Constructivist Perspective on Learning

In recent years, the increasing popularity of *constructivism* amongst educational researchers, curriculum developers, teachers, and teacher educators has resulted in a range of interpretations of its nature, purpose, and value. The interpretation of constructivism as a theory of knowledge that I held at the commencement of this study shaped the discourse between the teacher and myself. My constructivist perspective was shaped by an amalgam of constructivist-related theories of knowledge and learning, drawn from cognitive psychology and the history and philosophy of science, and located largely in science education and mathematics education literature. The main constructivist sources of influence on my thinking included (1) Kelly's (1955) personal construct theory, (2) Ausubel's (1968) theory of meaningful learning, (3) Piaget's (1929) theory of genetic epistemology, (4) Kuhn's (1962) theory of scientific revolutions, and (5) Resnick's (1980, 1983) psychological theory of *mathematics instruction.*

Knowledge

What these theories have in common is a view of human knowledge as a process of personal cognitive construction, or invention, undertaken by an individual who is trying, for whatever purpose, to make sense of her social or natural environment. This epistemology is predicated on a *relativist* philosophy that denies the possibility of obtaining an objective, or depersonalized, account of the world and highlights the inevitably subjective nature of human knowledge. At both the individual and professional levels of knowing, the processes of coming to know are constrained by culturally derived, linguistic frameworks that underpin consensus making as the only possible means of establishing truth. Thus, the truth-status of an individual's knowledge depends on its usefulness (or viability) in solving a personally meaningful problem and on its socially determined acceptance by the individual's peer group and mentors. An individual's knowledge, therefore, remains inherently provisional and relativistic; it both moderates and is moderated by the process of personal cognitive construction.

The Principle of Prior Knowledge

Another common feature of these constructivist epistemologies is their concern with the existing conceptual framework that a learner inevitably brings to bear on new learning experiences. The totality of an individual's prior "life-world" experiences constitute a conceptual frame of reference for perceiving and making sense of new phenomena, whether they are physical, social, or imaginative events. New concepts are constructed to deal with cognitive perturbations arising from novel problematic experiences and are assimilated into an individual's existing conceptual framework. The meaningfulness, usefulness, and ultimate viability of a new concept depends on the extent to which it has been integrated into the network of existing concepts and is available for future application. West's (1982) "evolutionary" conceptual change builds onto students' existing conceptual frameworks; the teacher's concern is with the "linkage" of new concepts. However, on occasions when assimilation is unable to restore cognitive equilibrium, a radical restructuring of the conceptual framework, or accommodation, might occur. A major pedagogical challenge is to develop teaching strategies to facilitate radical restructuring of students' conceptual frameworks.

Since the 1970s, research in science education has focused a constructivist perspective on the influence on learning of students' preconceptions of the natural world. A major research thrust has been the identification of students' preinstructional "alternative frameworks" (Driver and Easley 1978; Driver and Erickson 1983). Although this research is strongly influenced by cognitive psychology, it departs from the Piagetian tradition of identifying generalized cognitive structures, or "content independent" forms of thought (e.g., stage theory), and draws upon Ausubel's theory of preexisting conceptual structures (Novak 1986) and Kelly's theory of "man-the-scientist" (Pope and Gilbert 1983). This field of inquiry is concerned with the role of "context-dependent" knowledge on the learning of science and focuses on the personal theorizing and hypothesizing of individual students. By the commencement of this study, this research was flourishing internationally (see reviews by Gilbert and Watts 1983; Pfundt and Duit 1987; Giordan 1987).

Conceptual Change

Of particular relevance to this study were the emerging pedagogical theories for facilitating conceptual change amongst science students. The conceptual change work of Posner et al. (1982) was directly related to the epistemological priorities of Driver's research and drew on the philosophy of science (Kuhn, Lakatos, Toulmin) to explain learning in terms of a process of conceptual change, which they likened to a paradigm shift in the community of scientists. This research group set out specific conditions for facilitating "radical" conceptual change, or accommodation, in students' central, content-related conceptions of the natural world. The main emphasis of their teaching approach was to (1) identify students' preconceptions, (2) organize cognitive conflict learning experiences to assist students to rationally reconstruct their conceptions towards a scientifically acceptable view, and (3) evaluate and monitor students' conceptual change processes. Similar cognitive conflict strategies were advocated by Nussbaum and Novick (1980) in Israel and Osborne (1981) in New Zealand.

The role of classroom discourse in promoting critical, reflective thinking and a more student-centered and participatory classroom environment had been identified as being central to constructivist-related conceptual change teaching. Gilbert and Watts (1983) listed specific classroom discourse strategies in which teachers use language as a "tool" to promote thinking, rather than as a means of transmitting meaning: increasing the proportion of student talk, exploratory use of language by teachers, teachers' use of metaphors and analogies, and teachers and students negotiating the purpose of practical work.

At the same time that science educators were focusing their constructivist perspectives on school science, *mathematics educators* also were advocating constructivist-related teaching approaches and were voicing similar pedagogical concerns. Resnick (1983) characterized learners of science and mathematics as constructors of their own understandings whose "relational" understandings depended on their abilities to meaningfully link "new information" to what they already know. She advocated that teaching should explicitly acknowledge students' prior theories and help them to "confront" their "naive theories."

An Information Processing Approach

A distinctive feature of the field of constructivist-oriented mathematics education, at this time, was its concern with information processing psychology. Resnick (1980) was one of the strong proponents of expressing constructivist-related theory in terms of an information processing framework. Resnick believed that the process of inventing new mathematical knowledge could be facilitated by teachers' elicitation of students' problem-solving strategies. In particular, she was concerned that teachers should present new information which is compatible with students' existing knowledge and which contains "powerful and transparent representations of fundamental mathematical concepts." Resnick advocated teaching strategies that presented carefully tailored mathematical information which would facilitate students' restructuring of their prior knowledge as they struggled to invent new knowledge structures in problem-solving situations. Resnick was not a lone advocate of this approach; most current theories of problem solving in mathematics education are based on information processing models of human cognition (Putnam et al. 1990). Cobb (1987) presents an extensive review and critique of these ideas.

At the commencement of the study, therefore, my constructivist perspective was focused on the adequacy of students' prior mathematical knowledge and its central role in facilitating their construction of new mathematical knowledge. The general problem that needed to be resolved was one of assisting the teacher to take account of students' prior mathematical knowledge by reconstructing her established pedagogy. In practice, this was a problem of overcoming teacher centralism in the classroom.

THE PROBLEM OF TEACHER CENTRALISM

Curriculum reforms introduced into school systems often require teachers to reconceptualize their roles, develop new teaching strategies, and abandon well-established and seemingly effective practices. Research on teacher cognition (Nespor 1987; Clark and Peterson 1986; Eisenhart et al. 1988) has revealed that teachers' theories and beliefs about the teaching-learning process play an important part in determining the nature of their classroom roles. It is not surprising, therefore, that curriculum reform programs associated with innovative student-centered curricula have experienced implementation difficulties attributed to the predominance of teachers' well-established beliefs in their traditional centralist classroom roles (Olson 1980, 1981). The recent trend in curriculum reform towards constructivist-related approaches to teaching and learning science and mathematics (Driver and Oldham 1986; Steffe 1989) has heightened the importance of developing an effective approach for changing teachers' beliefs about their centralist classroom roles.

An Analogous Problem

At the time, my understanding of the problem of changing teachers' beliefs was influenced by an analogous problem experienced by science teachers trying to transform their students' well-established beliefs about the natural world. Research in science education had found that teaching approaches which inform students that their seemingly viable interpretations of the natural world should be abandoned in favor of counter-intuitive, scientifically acceptable alternatives have been far from successful. Despite teachers' rational explanations of the scientific version of a phenomenon, many students' preconceptions, or experiential and intuitive beliefs, tend to persist (Gilbert et al. 1982).

Many students of science, including those in university undergraduate courses, had been found to remain unconvinced about, for example, Newtonian dynamics, when much of their immediate experiences of how things move can be explained to their satisfaction by their well-formed Aristotelian conceptions. Students were also found to be very capable of espousing scientific principles while harboring contradictory theories of "real-life" events (Pines and West 1986). This problem highlights the sometimes tenuous nature of students' formal, or officially sanctioned, scientific knowledge, especially when it conflicts with the intuitive "life-world knowledge" that constitutes the personally viable perspectives of their everyday experiences (Pusey 1987).

Teacher Beliefs

This analogy helped me to understand the implementation problem of curriculum designers that occurs when their reforms require teachers to transform their life-world knowledge of classroom events, especially if a radically counter-intuitive reconceptualization of teacher roles is required. The strength of teachers' commitment to their long-held beliefs in their centralist classroom roles should not be underrated. These beliefs have been formed and reinforced during a lifetime of schooling, commencing with their early childhood classroom experiences and culminating in their teaching careers, and are likely to be so deeply ingrained that they underpin a normative worldview of classrooms and schooling. In other words, a centralist classroom role for many teachers is an unquestioned, "natural" reality.

One of the major constraints to pedagogical transformation is the subconscious nature of the beliefs that shape teachers' centralist classroom roles. Although teachers readily articulate pedagogical beliefs about some classroom-related phenomena, for example, gender-determined cognitive potential, causes of student success or failure, the stable cognitive capability of individual students, and the hierarchical nature of the subject (Ruthven 1987), the teaching routines developed over time by experienced teachers are based on largely subconscious beliefs about teaching and learning (Tobin and Espinet 1988). These partially articulated beliefs form stable perspectives through which individual teachers filter information conveyed by curriculum developers and researchers (Eisenhart et al. 1987; Nespor 1987). Clearly, curriculum reform efforts that are premised on coherent arguments for major changes in teachers' classroom roles are likely to experience very short "half-lives" and to be implemented with much less than the high degree of fidelity desired by curriculum designers.

Laying the blame with teachers for frustrating constructivist-related curriculum reform, however, is not an entirely appropriate, or helpful, approach. The "top-down" rationalist approaches of curriculum reform programs that deliver ready-made curriculum products for implementation have assumed that the rationale of the innovation is universally transparent and, therefore, not susceptible to differential interpretations by teachers. Olson (1980) argued that teachers' translation of curriculum jargon is an inevitable consequence of the cultural differences that exist between curriculum designers and implementers and that the focus of curriculum reform should be on the meaning systems of teachers, rather than on the adoption process.

It seemed to me that a constructivist approach to curriculum reform should address the key role of the teacher, reconceptualize curriculum reform as a *process* of pedagogical transformation, and relocate the locus of control from the external curriculum design team to the teacher in the classroom. The main problem to be resolved for constructivist-related curriculum reform, therefore, is to develop an approach to professional development that enables teachers to engage in a process of reconstruction of their well-established pedagogical beliefs and practice and to develop less centralist classroom roles that embody the principles of a constructivist-related theory of knowledge. The focus of this study was on the development of a constructivist-related pedagogical change methodology for facilitating teachers' reconceptualizations of the centralist nature of their classroom roles.

TEACHER PEDAGOGICAL CHANGE

An Analogous Approach

From the viewpoint of the philosophy of science, major conceptual change for a learner is analogous to a paradigm change, or scientific revolution (Kuhn 1970), or to a change in research programs (Lakatos 1970) for a scientist. When scientists are faced with a challenge to their basic assumptions, they must acquire new theories, or central commitments, and a new way of perceiving the world. However, scientists are unlikely to make major changes to their theories until they believe that less radical changes will not work. Such a belief is precipitated by an accumulation of unsolved problems or anomalies and a loss of faith in the capacity of the current theories to solve these problems.

This analogy was used by Posner et al. (1982) to suggest specific conditions of conceptual change for students of school science. The first condition is that the student must experience dissatisfaction with her existing conceptions. Then the student must be provided with opportunities to develop alternative (scientific) conceptions that she perceives to be intelligible, plausible, and fruitful. The analogy of a paradigm shift in science with major conceptual change in science students serves to highlight the importance of accounting for learners' well-established life-world beliefs in facilitating a process of conceptual reconstruction and suggests useful constructivist-related principles for designing a pedagogical change methodology for teachers. Similar principles for changing teachers' well-established pedagogies have been suggested in research on teacher cognition.

CHANGING TEACHER BELIEFS

Evaluating Established Pedagogy

Perhaps the first step towards pedagogical change is the realization and explication of teachers' pedagogical beliefs, especially the intuitive life-world beliefs that underpin teachers' centralist roles. Fenstermacher (1979, 1986) argues that teachers must become reflexively aware of their "subjectively reasonable" beliefs or "practical arguments" and undertake investigations to confirm or disconfirm them; disconfirmation would constitute compelling rational grounds for change (the experience of dissatisfaction suggested for student conceptual change). Nespor (1987) extends Fenstermacher's ideas by suggesting that because beliefs are self-contained "conceptual systems," rather than logical propositions, the transformation of teachers' beliefs and practice will occur "only if alternative or new beliefs are available to replace the old" (326). The alternative beliefs are necessary to stimulate the cognitive process of transformation that Nespor describes as a "gestalt shift," as distinct from the rational process of argumentation implied by Fenstermacher. These arguments combine to establish the first goal of the constructivist-related pedagogical change methodology:

to enable teachers to assess, from the vantage point of a constructivist-related perspective, the viability of the relationship between their well-established pedagogical beliefs and classroom practices.

Genesis of New Ideas

But what is the genesis of alternative beliefs: from where do the new ideas originate, and how do they achieve the status of beliefs? One answer to this question is suggested by Floden's (1985) argument that teacher educators should try to persuade teachers to make pedagogical changes but not by recourse to their authority. Floden argues in favor of the use of rhetoric by teacher educators, rhetoric that enables a new belief to be held "for reasons that are both sound and are thought by the student to be good reasons and that ... are reasons that the teacher also would use to justify the belief ... the teacher must be ready to answer students' questions about why beliefs should be held." But this process should not be corrupted by resorting to the authority-laden myths of research as a body of universally applicable truths or of the teacher educator as an expert knower. Rather, the teacher should be encouraged to adopt a critical approach to the rhetoric of both the teacher educator and research reports and, for that matter, to the "rhetorical facade" that obscures her own intuitive pedagogical beliefs (Olson 1981).

The teacher educator, therefore, has a number of important roles to play: a stimulus for new pedagogical ideas, an advocate of new ideas, and a negotiator who seeks to establish a consensual viewpoint (the conditions of intelligibility and plausibility of new ideas). Thus, a goal of the constructivist-related pedagogical change methodology is

> to establish a dialectical relationship that recognizes the rationality of the teacher and promotes critical analysis aimed at assisting the teacher to determine the **theoretical viability** of new ideas.

Evaluating New Prospects

A further step in the genesis of new pedagogical beliefs is the opportunity for teachers to assess the practical viability of their new ideas in the context of their own classrooms. Ruthven (1987) argues that because teachers' classroom expertise is largely tacit and grounded in experience, rather than in pedagogical theory, changes in teachers' pedagogical perspectives are likely to proceed changes in their classroom practice. This argument is similar to Nespor's (1987) claim that the transformation of teachers' beliefs cannot be achieved by rational argument alone and suggests that opportunities to experience the success (or fruitfulness) of innovative teaching strategies should be an integral part of a pedagogical change methodology.

However, it is essential that teachers do not assess the efficacy of their innovations in terms of the criteria associated with their well-established centralist pedagogies. For example, there is little point in a teacher judging students' engagement in group discussions in terms of her expectations of students' behavior in whole class teacher-directed discussions. An awareness of constructivist-related criteria is necessary in order that teachers assess their innovative student-centered teaching strategies in terms of their potential to facilitate constructivist-related pedagogical goals. Thus, a goal of the constructivist-related pedagogical change methodology is

> to assist the teacher to establish explicit constructivist-related criteria and to use the criteria reflexively to assess the *practical viability* of the teacher's pedagogical innovations.

Taken together, the constructivist-related principles of the pedagogical change methodology, discussed above, highlight the importance of a *critical discourse* that focuses on the teacher's existing centralist pedagogy and, at the same time, facilitates the development of an alternative constructivist-related pedagogical perspective. The discourse should promote a self-evaluative process of deliberative inquiry, rather than impose a prescription for the teacher's thoughts and actions. An evaluative process requires explicit criteria against which to make judgments about the viability of a new pedagogical idea or strategy.

Constructivist Pedagogical Criteria

My constructivist perspective had a central concern for the prior content knowledge of students. The main pedagogical focus of this perspective was on the teacher's role of investigating student's existing understandings of syllabus-related mathematical concepts and incorporating students' various understandings when planning lessons and interacting with students in the classroom. I readily incorporated the terminology of information processing psychology into the discourse with Ray. Rather than refer to the transmission of the teacher's mathematical knowledge to students, we spoke about the nature of the mathematical information that students interpreted according to their existing conceptual frameworks. This perspective gave rise to three constructivist-related pedagogical criteria that I expressed mostly in the form of questions of Ray's pedagogical beliefs and practices. The form of these questions changed according to daily circumstances and usually focused on specific teacher thoughts and actions. The three evaluative criteria are illustrated below with sample questions expressed in a generalized form.

■ Responsiveness

To what extent do you monitor and evaluate students' learning activities?

To what extent do you believe that your teaching strategies are responsive and relevant to the range of students' learning needs?

■ Focus

To what extent does your teaching focus on students' prior knowledge and provide opportunities for students

— to review and evaluate their prior knowledge?

— to integrate newly constructed knowledge with their prior knowledge?

■ Context

To what extent do you provide opportunities for students to develop their knowledge in a *problem-solving* context?

Having developed a constructivist-related theory of pedagogical change, the next step was to determine the methodology that would guide the study as both a collaborative enterprise and an investigative process of social inquiry.

METHODOLOGY OF THE STUDY

The study comprised a methodological blend of collaboration and intervention within an interpretive framework of social inquiry. Ray and I collaborated to meet both our common research agenda, the improvement of Ray's teaching practices, and my more extensive research agenda, which included an investigation of the prospects of facilitating Ray's development of a constructivist-related pedagogy. During the study, I attempted, with recourse to a pedagogical change theory, to enable Ray to reconstruct his pedagogical beliefs and classroom practices. Making sense of the dynamic nature of the relationship between Ray's pedagogical beliefs and classroom practices was achieved by employing an "interpretive" research approach (Erickson 1986).

Collaboration

A collaborative research partnership was formed between Ray, a high school mathematics teacher and postgraduate student, and myself, a teacher educator involved in educational research and postgraduate teacher education. The common focus of our individual interests was improving Ray's teaching practices, a goal that he had formulated as a research question: "How can the acquisition of mathematical knowledge by students be enhanced by specific teaching strategies?" (Fleming 1988). He was also concerned to report on the study as a postgraduate research project. My additional research interest was to investigate the process of constructivist-related pedagogical change. A collaborative approach to classroom-based research is well recognized for the benefits that can accrue to both educational research and teaching practice (Watt and Watt 1982; Saphier 1982; Kyle and McCutcheon 1984).

Phase One

An initial eight-week period of collaboration focused on 24 consecutive lessons presented by Ray to his Grade 12 mathematics class in a Perth metropolitan, all-girls school. During this period, Ray and I worked closely to document and reconstruct his pedagogy. I attended each of his classes as a participant-observer, making notes on whole-class activities and interacting as a tutor with students during individual seatwork activities. Ray and I discussed his lesson plans prior to each class, and after each class, we critically appraised the lesson, reviewed the fieldnotes from the previous day, and discussed alternative teaching strategies.

Both during this period and later, we conducted a series of audio-tape recorded interviews that explored our perceptions of constructivist theory and appraised our collaborative relationship. The discussions with Ray provided me with an opportunity to implement my pedagogical change strategy by engaging him in critical discourse. Depending on the circumstances, our discussions focused on philosophical issues or more practical teaching-related issues.

Phase Two

In the following school year, our collaboration involved 14 consecutive lessons presented by Ray to his Grade 9 mathematics class in the same school. On this occasion, my goal was to determine the extent to which Ray's recently established "constructivist" pedagogy had persisted over time and had been transferred to another area of mathematics. I was interested in determining the effectiveness of the pedagogical change theory, in terms of the stability and transferability of Ray's reconstructed pedagogy. Again, I attended Ray's lessons as a participant-observer, but our out-of-class discussions were restricted to an exploration of the relationship between his pedagogical beliefs and practices. Although I did not intend to engage him in critical discourse, Ray reported that my presence in his classes caused him to become more self-critical.

Although it seemed that Ray benefited from our collaboration during the second period of fieldwork, I did not feel comfortable with my noninterventionist role. My involvement seemed to be predominantly one of self-interest and, although Ray had agreed to participate, it seemed to me to be a rather inequitable form of collaboration. Was this the guilt response of an "expert-knower" wishing to intervene with advice on how to teach more effectively? I prefer to believe that it was a response evoked by a sense of allegiance to a formerly more equitable partnership that thrived on critical discourse. This experience highlighted, for me, the very real prospects of the collaborative research label being misappropriated by educational researchers concerned only with their own research agenda.

Interpretation

The investigative dimension of the study, with its focus on the relationship between a teacher's pedagogical conceptions and his classroom practices, employed an *interpretive research approach* (Erickson 1986). This approach seeks to generate an understanding about what is happening in a particular social setting, such as a classroom. The approach is based on a relativist philosophy that multiple realities coexist amongst the participants of a social setting and that no single social reality can be claimed to exist for all participants. Thus, the inquirer seeks to establish the nature of those multiple realities while, at the same time, realizing that the best that can be achieved is an interpretation filtered through a perceptual lens formed by the inquirer's own social theories. In order to avoid highly idiosyncratic interpretations, the inquirer is well-advised to explicate her own theoretical assumptions and negotiate with other participants on the nature of their perspectives. Collaborative inquiry involving participant-observation and multiple data sources can enable more reliable higher-level inferences to be made about unobservable phenomena, such as participants' implicit beliefs.

Phase One

Both Ray the teacher and I collaborated as participant-observers and negotiated a daily consensus on what was happening in the classroom. One of my roles was to record daily fieldnotes and to provide Ray with a written analysis of them. These reflections contained descriptions of classroom events, questions about Ray's intentions, and assertions about the nature and impact on student learning of his pedagogical beliefs and practices. These notes provided a shorter-term focus for subsequent classroom observations that sought evidence to support or refute the assertions and served to stimulate Ray's critical reflections on his pedagogy.

Our classroom observations were supplemented with data from other sources, including the *Learning Environment Inventory* instrument (Fraser and Fisher 1983), designed to measure both student and teacher perceptions of the classroom psycho-social environment, semi-structured interviews with students, and extensive interviews with each other. We conducted a series of structured interviews, modeled on the *Repertory Grid Technique* for investigating teachers' implicit beliefs about their classroom roles (Olson 1981; Munby 1984). The interviews were audio-recorded, and we analyzed the transcripts. On completion of the first fieldwork period, Ray and I negotiated an interpretation of the general outcomes of the study in relation to our common research interests of improving his teaching practices (Fleming 1988; Taylor and Fleming 1987).

Phase Two

During the second period of collaboration, undertaken in order to determine the longer term impact of the study on the teacher's pedagogical beliefs and practices, data were obtained by means of classroom observations and teacher interviews. The interpretive process was conducted as before, with the exception that I was no longer attempting to facilitate pedagogical reform. The daily analysis of fieldnotes was not deliberately couched in the type of critical terminology that had previously occurred when my perspective was influenced by the pedagogical change theory. Rather, I was seeking to confirm or refute general assertions about the relationship between the teacher's pedagogical beliefs and classroom practices. These assertions had arisen from a retrospective analysis of data, including written reports on the study, associated with the earlier phase of the study.

DISCUSSION OF THE RESULTS

The relationship between the teacher's pedagogical beliefs and practices was a central empirical focus of the study reported in this chapter. This relationship was assessed in order to determine the influence of the teacher's pedagogical beliefs on the constructivist-related pedagogical reform process. The results of that analysis have been reported in earlier papers (Taylor 1990a) and are referred to in the following discussion. However, the main focus of this chapter is on the influence of the theory of pedagogical change that shaped the teacher's constructivist-related pedagogical reforms. In particular, the constraining influence of the underpinning theory of constructivism is explored from various constructivist-related perspectives. Before this is attempted, however, a brief synopsis of the main pedagogical reforms achieved during the first collaborative fieldwork phase of the study is discussed.

The Teacher's Constructivist Pedagogy: A Synopsis

The key features of the teacher's reformed pedagogy were his decentralized classroom role and the interactive nature of his teaching.

Decentralization

From the viewpoint of the constructivist perspective that shaped the pedagogical reform process, Ray's teaching practices underwent a dramatic transformation. He largely abandoned the formerly daily practice of lecturing to the whole class and adopted a *decentralized* classroom role that enabled him to move about the class and to interact with individual students. Ray designed sets of worksheets to introduce each new topic to the class. The worksheets commenced with a review of salient mathematical issues that, Ray believed, students had encountered in mathematics classes in earlier years. Each worksheet was designed to assist students to construct meaningful linkages between their revised mathematical background knowledge and the mathematical principles and procedures that comprised the new topic. The worksheets contained advice to students on important problem-solving procedures and heuristics, and provided practice exercises to be completed either in class or as homework.

Ray's specially designed instructional worksheets provided daily opportunities for students to engage in the learning activities at a largely self-determined pace. When students experienced difficulties, they sought assistance either from their peer group or, by raising their hands and waiting, from the teacher. Thus, for most of the lesson, Ray acted as a consultant for individual students.

Interactive teaching

During his rounds, Ray's primary pedagogical focus was on assessing students' understandings, detecting students' conceptual difficulties, and providing tutorial assistance. He developed an interrogative questioning style that, he believed, was effective for achieving these goals. Ray's constructivist teaching approach eschewed telling students correct answers; rather, the intention was to prompt students' self-evaluative and self-instructional processes.

Ray continued to use a whole class forum from time to time but usually in a highly interactive mode. When he perceived common difficulties arising amongst the class, he conducted whole class discussions with interested students. During these discussions, Ray employed questioning strategies to probe students' understandings. On the basis of the feedback, he provided spontaneous explanations of mathematical issues. Some students sought clarification of Ray's explanations, while others remained content to listen to the explanations and exchanges. The whole class forum also was used at the commencement of most lessons for "mental tests" of students' topic-related prior mathematical knowledge. Another use of the whole class forum was to revise the mathematical content of the set of topic-related worksheets that the class had been working on for the previous one or two weeks.

Ray seldomly used the whole class forum in a noninteractive mode. This occurred at times when his blackboard-based explanations became protracted, as he spontaneously elaborated on mathematical issues of concern that he had detected during interactions with individual students. Overall, the proportion of teacher talk was greatly reduced in favor of individual student activities. In addition, it assumed a new strategic purpose. Whereas in the past teacher talk was almost entirely oriented towards the transmission of Ray's mathematical knowledge, now it served a mostly transactional purpose. Ray wanted to find out about students' attempts to make sense of their learning activities and to respond with relevant, multimodal teaching strategies that would facilitate both meaningful learning and the development of students as independent (of the teacher) learners. The latter goal was a long-established priority that had underpinned his formerly didactic, lecturing approach and found new expression in his constructivist-related pedagogy.

Conclusion

In terms of the constructivist perspective which shaped the pedagogical reform process, it seems that the teachers' pedagogy had been successfully reconstructed. Not only did Ray articulate constructivist-related pedagogical beliefs, but his teaching was more responsive to students, clearly focused on students' prior mathematical knowledge, and provided a problem-solving context for developing students' mathematical knowledge. It seems reasonable to conclude, therefore, that the pedagogical reform process and its underpinning constructivist perspective evoked a radical reformation in the teacher's pedagogical beliefs and classroom practices. Follow-up fieldwork in the preceding year confirmed that Ray's constructivist-related pedagogy had largely stood the dual tests of time and transfer, and provided grounds for a positive assessment of the long-term influence of the pedagogical reform process and its underpinning constructivist-related theory.

Nevertheless, the problem of teacher centralism persisted throughout the pedagogical reform process. A shift in perspective reveals that the theory of pedagogical change failed to account for important socio-cultural influences that constrained the reform process. An analysis of these influences reveals that the resilience of the teacher's centralist classroom role owes much to his underpinning fundamental beliefs in rationalism and realism and his technical curriculum interests.

The Problem of Pedagogical Rationalism

The constructivist perspective adopted in this study was closely associated with the epistemological position of science educators who, in the late 1970s and throughout the 1980s, explored students' alternative frameworks and developed theories of conceptual change for use in high school science classes. This research was based on theories of cognition proposed by cognitive psychologists such as Ausubel and Kelly and came to be known as the *personal constructivist position* (Solomon 1987).

Overcoming mentalism: A sociological perspective

This popular interpretation of constructivism was attacked by Solomon (1987) who argued, from a sociological perspective (Berger and Luckman 1966), that social influences are a major determinant of all forms of learning, including school knowledge, and that the mentalist assumptions of the personal constructivist approaches to conceptual change failed to address this integral component of the learning process. The undue regard for rationality of the personal constructivist pedagogies had paid only lip-service to the central role of social communication in the construction of children's culturally shared life-world knowing. The persistence of students' alternative frameworks was cited as evidence of the emotionally laden and socially reaffirmed nature of the beliefs that children form during their primary socialization into their life-world schemes. Social psychological perspectives (e.g., symbolic interactionists, neo-Piagetians) and research into interactive aspects of school learning were reviewed to support the assertion that social interactions of groups are a primary determinant of learning outcomes, perhaps more so than rational considerations of empirical data obtained in practical work.

The major pedagogical implication of this sociological perspective is the need for the teacher to focus on the social interactions between students that give rise to their intersubjective beliefs or shared perspectives. Understanding how students have come to hold beliefs about the natural world, or views on the nature of mathematics, or perspectives on the purpose of schooling, is important for facilitating further social interactions amongst students in relation to specific teaching goals. Solomon claims that, although group work has been nominated by personal constructivists as an important learning activity, there seems to be little appreciation of the need for a pedagogy to shape the dynamics of groups.

Socialization: A Missing Dimension

In this study, the teacher maintained a well-established belief in a rationalist teaching approach. Ray's interpretation of the influential role of student socialization largely emphasized its apparently negative aspects. Task-related social communications were a central feature of the classroom, as indicated by the many informal small groupings of students who frequently consulted each other; some admitted to a preference for peer assistance, rather than teacher assistance. At times, student interactions that were directly related to their making sense of learning tasks conflicted with the teacher's whole class focus, and the perpetrators were admonished for interrupting the teacher. Ray demonstrated a strong interest only in students' logical reasoning and rarely addressed the social nature of students' learning processes, apart from discipline-related interventions.

Related to the teacher's rationalist perspective was his negative view of the prospects of his role for motivating students to engage in learning activities (both in class and homework). Although Ray was sure that motivation to learn was a necessary condition for academic success, he sustained an explicit belief that students' academic motivation was a fixed personality trait (especially amongst high school students) and that he was not able, therefore, to influence students in this regard. He expressed a strong reluctance to "coerce" students to engage in his prescribed learning activities. This belief enabled Ray to indulge in blaming the victims for their failure to learn. Ray believed that inadequate self-motivation among students was the major reason that students were unable to keep up with the teacher-determined rate of delivery of the curriculum. Although he admitted that it was not possible for him to determine a common curriculum pace that was suitable for all students, he countered this pedagogical dilemma by recourse to a belief in the inviolate nature of students' academic self-motivation.

The constructivist perspective that shaped the pedagogical reform process in this study seems to have reinforced a rationalist approach to constructivist-oriented teaching. The main focus of the reform process was on the relationship between students' prior knowledge and their newly constructed knowledge. There was little emphasis on the collaborative aspects of sense making, which continued to occur informally amongst students, by pedagogical default. Thus, the perspectives of students that shaped their beliefs about the nature and purposes of learning mathematics were largely ignored. The main subject of Ray's teaching appeared to be the logical structures of the discipline, rather than students. A constructivist perspective locates the structures of the discipline in the mind of the knower. Thus, a rationalist approach to teaching appears to sustain the pedagogical problem of teacher centralism.

Overcoming constraints: A cultural perspective

A sociological perspective suggests that teaching approaches should account for the cultural nature of learning, especially in relation to the emotionally laden social influences that shape students' cognitive constructive processes. The social consensual process of determining the culture of the classroom is as much a student activity as it is a teacher activity, and it occurs continuously in the social transactions amongst participants. It seems only reasonable that such an important process be the subject of determined and critical reflection by all participants and that a constructivist pedagogy take careful account of both the rational and social aspects of cognition.

A pedagogical reform process that incorporates a sociological focus will be required to overcome the constraining nature of teachers' rationalist perspectives. It is important to determine the extent to which rationalist beliefs are culturally derived and continuously reaffirmed in social exchanges between teachers. Tobin (1990a) differentiated between personal constraints and cultural constraints as classes of beliefs that prevent a teacher from implementing preferred changes. If a teacher's personally constraining beliefs result largely from the individual teacher's idiosyncratic cognitive constructive processes of sense making, then they might be more susceptible to change than if the beliefs are associated with the general culture of teaching.

Although the attribution of motivational personality traits to students might have currency amongst teachers, it seems reasonable to assume that it is likely to be a largely personal constraint that is susceptible to rational discourse and empirical verification. Nevertheless, a sociological perspective on knowledge highlights the intersubjective nature of beliefs and suggests that all beliefs have cultural roots. It seems necessary, therefore, to investigate the extent to which rationalist pedagogical perspectives result from teachers' enculturation into the teaching profession and are sustained by daily consensual discourse amongst teachers.

The Problem of Pedagogical Realism

The constructivist perspective brought to this study was based on the epistemological principle that learners actively construct their knowledge in relation to their preexisting conceptual structures, rather than passively receive knowledge from an external source. Knowledge is not a commodity for exchange and is not inherent in either the language of discourse or the visual field, but it is constructed in the mind as preexisting cognitive structures are restructured in attempts to make sense of new conceptual or sensory-motor experiences. Put simply, knowledge is a sense-making process and cannot be transmitted from teacher to student, or text to student, or diagram to student. The results of the study suggest, however, that the relatively well-known prior knowledge principle, associated with the personal constructivist position, is not sufficient to resolve the problem of pedagogical realism which constrained the teacher's constructivist-related pedagogical reform process.

Overcoming realism: A radical constructivist perspective

Pedagogical realism can be understood from a radical constructivist perspective by considering the metaphysical principle "that the function of cognition is adaptive and serves the organization of the experiential world, not the discovery of ontological reality" (von Glasersfeld 1987). In other words, although there might be an objectively real world with which we continuously interact throughout our lives, we have no direct access to that world other than through our senses and conceptually laden perspectives. We can know the world only by constructing interpretive models whose viability depends on their success in enabling us to deal, physically or cognitively, with the world. The model makers include the communities of scientists and mathematicians, as well as the individual student in the classroom.

The metaphysical principle of radical constructivism denies the realist perspective that science and mathematics comprise transcendental truths about an objective universe and that their adherents are able to discover and articulate ahistorical and depersonalized characteristics of an objective reality. Pedagogies based on a radical constructivist perspective identify knowledge as a subjective sense-making activity located in learners' minds and focus on developing the experiential fitness of learners' concepts for making sense of their intersubjective experiences. By contrast, realist pedagogies equate knowledge with discovering an objective reality, locate knowledge in sources external to the learner, and focus on developing accurate representations of that knowledge in learners' minds.

Information: A deceptive synonym

In this study, the teacher maintained his well-established realist pedagogy, albeit in a covert form. As Ray's pedagogy became more responsive and focused in relation to students' prior mathematical knowledge, I encouraged him to attend to the cognitive context within which students' mathematical knowledge was being constructed. At about that time, we attended a seminar on information processing psychology and became aware of the information processing models of mathematical problem solving (Resnick 1980). The language of our subsequent discourse became replete with references to key elements of an information processing model of learning. We shared a perspective about the need to ensure that the information conveyed to students in Ray's worksheets contained procedural instructions that would assist students to become adept at problem solving (especially with higher cognitive-level problems). The pedagogical challenge became one of facilitating students' recognition of the typology of mathematical problems that they encountered by retrieving appropriate knowledge from their long-term memories and combining it with newly acquired knowledge located in their short-term working memories. Procedural knowledge appeared to be the catalyst for this cognitive process.

The development of students' procedural knowledge became an explicit goal of the pedagogical reform process and seemed to be consistent with the constructivist focus on meaningfully integrating and utilizing newly constructed mathematical knowledge in relation to prior knowledge. Thus, the teacher's role was extended to provide students with models of problem-solving strategies that they could adapt to suit their individual needs and circumstances. Ray's models were presented in his worksheets and comprised procedural instructions and heuristic strategies.

From a radical constructivist perspective, an analysis of the influence of the information processing model of learning on the pedagogical reform process suggests that the model served as a two-edged sword. On the one hand, it provided an important pedagogical focus on the cognitive context in which students were constructing their knowledge. On the other hand, the language associated with the information processing model served as a subtle means of sustaining a realist perspective and a centralist classroom role for the teacher. Rather than transmitting his knowledge to the class, Ray came to believe that his prime instructional role was to transmit information.

In Ray's instructional model (see Fig. 1), his knowledge was transformed into "public information" in a process that involved reduction of complex knowledge structures into elemental components. Ray's model characterized students' learning roles as transforming (or constructing) the public information into their own private knowledge. In practice, Ray's questioning strategies were designed to "expose the explanation bit by bit." His worksheets were designed to "lead them through it in smaller steps," and his interventions with individual students aimed to "interrupt at a point where they've gone wrong or got some false method or impression . . . and correct them" (Taylor 1990a, b).

In a review of information processing psychology and mathematics education from a radical constructivist perspective, Cobb (1987) criticizes the realist assumptions that underpin production systems models that view learning as a process of encoding information deposited into students' working memories. This view of learning is consonant with the role of the teacher as

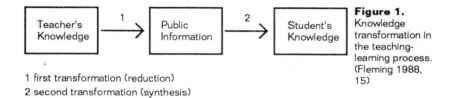

Figure 1. Knowledge transformation in the teaching-learning process. (Fleming 1988, 15)

1 first transformation (reduction)
2 second transformation (synthesis)

direct source of student knowledge, and it invokes a Platonic view of mathematical objects as elements of a mind-independent and preexisting mathematical reality which the teacher helps students to see. Cobb argues that production systems models promote a formalist view of mathematics as a system of syntactical rules for manipulating symbols which are regarded as mathematical objects. Cobb claims that this view of mathematics underpins the much-criticized contemporary mathematics education pedagogy of teacher explanations of definitions, theorems, proofs, and subsequent student practice. He associates this type of pedagogy with Skemp's instrumental learning and the empiricist and behaviorist traditions of psychology.

The concept as object: An alternative perspective

The main outcome of Cobb's epistemological criticisms is a call for a pedagogy that addresses the student's (or actor's) mathematical perspective, rather than the teacher's (or observer's) perspective. This requirement seems to be the main stumbling block for the schema theorists, the Piagetian-oriented information theorists, and their constructivist-related information processing models. Although they profess the prior knowledge principle of constructivism and proffer student-centered pedagogies that use concrete learning aids such as blocks and rods (physical) and analogies (cognitive), Cobb accuses them of being metaphysical realists. They seem to assume that students are capable of seeing the abstract mathematical representations that the teacher points out, a process which, Cobb argues, requires the student to already have made the construction in order to be able to see it (observation is theory laden). Thus, a realist assumption underpins a constructivist-oriented pedagogy that aims to provide students with clearer representations of a teacher's knowledge; the pedagogy is based on the teacher's mathematical perspective, rather than on students' perspectives. This realist pedagogy invokes an external authority (the teacher) to evaluate the correctness of students' attempts to construct replicas of the teacher's knowledge.

By contrast, a radical constructivist pedagogy promotes students as inquiry-oriented problem solvers who select and solve problems that they regard as problematic. The mathematical objects of reflection are identified as the objectified concepts that comprise students' own cognitive environments. Learning is a process of reflecting on the adequacy of existing knowledge structures, restructuring in order to neutralize cognitive perturbations, and determining the viability of the new knowledge structures. A radical constructivist perspective recognizes the social dimension of knowledge and the important role of language in its expression, and promotes a socially interactive approach to creating problems and to determining the viability of students' solution strategies. The pedagogical emphasis is on students' sense-making activities, rather than on the correctness of their answers in relation to the teacher's knowledge. Rampant relativism is avoided, however, by the pedagogical challenge of designing learning tasks that "embody the central ideas of the discipline and that will be problematic for students" (Wheatley 1991). The teacher has a facilitative role in promoting cooperative learning in small groups, sharing of alternative ideas amongst students, and encouraging students to develop intellectual integrity and autonomy.

The problem of pedagogical realism might be avoided by a pedagogy based on the metaphysical principle of constructivism, which counters the realist perspective that a truth correspondence exists between scientific/mathematical knowledge and an objective universe. The metaphysical principle highlights the adaptive function of cognition in relation to the experiential world, underpins a pedagogy that emphasizes the personal and social sense-making activities of students, and highlights the inappropriateness of pedagogies in which teachers try to transmit their knowledge and evaluate students' attempts at replication.

Reconstructing realist metaphors and images

This study indicated the pervasive nature of realist beliefs and their constraining influence on the development of a constructivist-related peda-gogy for overcoming the problem of teacher centralism. The study revealed how realist beliefs were unknowingly shared by Ray and me and unwittingly reinforced by our subscription to an information processing theory of learning. Therefore, a pedagogical reform process that wishes to overcome the problem of pedagogical realism must address the issue of transforming teachers' well-established realist perspectives and their underpinning beliefs.

A promising approach to this problem has been described by Tobin's (1990a, 1990b) investigations of the nature of teachers' beliefs from a *cultural anthropological perspective*. Tobin argues that the beliefs that shape teachers' classroom roles are organized by powerful metaphors, images, and myths associated with the culture of teaching. These high-level referents for thoughts and actions provide teachers with normative world-views of the teaching-learning process but are not necessarily verbalized or readily accessible for critical assessment. Tobin argues that constructivist-related pedagogical re-forms should involve teachers in identifying and changing the metaphors, images, and myths that shape their conceptions of teaching and learning and constrain their classroom roles. In this study, the teacher's reformed pedagogy was underpinned by realist beliefs associated with his image of the teaching-learning process as an information delivery service (Fig. 1), and an interactive informer metaphor (Taylor 1990b).

The problem of radically transforming pedagogical realism should not be underestimated, however. To the extent that teachers' realist metaphors and images are culturally derived and continuously reaffirmed in the daily social discourse between teachers, they are likely to maintain a constraining influence on teachers' well-established centralist classroom roles. In view of this ongoing social process of enculturation, it seems that the focus of constructivist-related pedagogical reform processes should be broadened to include groups of teachers, especially the influential members (or "reality definers") whose daily interactions serve to reaffirm the pedagogical beliefs and classroom practices of other teachers. This strategy suggests that a radical pedagogical reform process that focuses only on the individual teacher might achieve little more than an isolated pocket of pedagogical disturbance in the overall cultural fabric of the school. From an ethical viewpoint, undesirable social consequences might be experienced by the reformed teacher as she attempts to renegotiate a viable social role with colleagues who continue to subscribe to the predominant culture of teaching and socially interact in accordance with its unquestioned prevailing myths.

Overcoming the pedagogical problem of realism requires a concern with more than the activities of teachers and students in individual classrooms. There are system-wide implications of constructivist-related attempts to redefine the culture of teaching, especially in centralized education systems that hold teachers and schools accountable for the delivery of externally designed curricula and for students' performance on state-mandated, externally controlled examinations. A constructivist-related pedagogical reform process needs to address the broader problem of control of students' learning.

The Problem of Pedagogical Control

At the heart of the constructivist perspective brought to this study is the notion that the *locus of control* of the cognitive processes of knowledge construction lies with the individual learner, rather than with the teacher. Although these processes are mediated by social interactions with others (teacher, students), the objects of reflective abstraction are personal constructs within the learner's internal cognitive environment, and their restructuring results from attempts by the individual learner to make sense of problematic experiences. The success of cognitive restructuring is experienced by the individual learner as a sense of the viability of her new knowledge structures for resolving cognitive perturbations. Although a teacher can organize social interactions in the classroom for the purpose of stimulating a greater range of experiences for reflective abstraction, the ultimate responsibility for making sense of those experiences lies with the individual student. In other words, the sense-making criteria and the processes of evaluation that utilize those criteria are internally located and controlled by each individual learner.

A constructivist pedagogy, therefore, should assist students to become aware of their central roles in the learning process, especially in relation to assuming responsibility for determining the viability of their knowledge. Focusing only on the "correctness" of an answer to a problem ignores the cognitive processes that gave rise to the answer and locates evaluative judgments in external sources such as the teacher, the textbook, or other students. A prime concern with meeting internally generated critical demands for making sense of problematic experiences is the hallmark of an intellectually autonomous learner. A constructivist pedagogy should strive to assist students to develop their intellectual autonomy by providing opportunities for exercising control over their learning activities and assuming responsibility for their cognitive processes.

From a constructivist perspective, the centralist role of the classroom teacher seems to be antithetical to the development of self-determined students. To the extent that teachers continue to assert their control of students' learning activities and assume responsibility for determining the success of students' cognitive processes, centralist pedagogies are highly likely to be sustained. But to what extent can the individual teacher relinquish control, especially in a centralized education system that prescribes curriculum content and sets systemwide examinations for determining successful learning outcomes? This question has important implications for any constructivist-related theory of pedagogical change.

A question of interests

The ideological interests that constitute the theoretical underpinnings of many traditional curriculum structures have been the subject of criticism by critical social theorists for some time (Giroux 1981; Apple 1979). Recently, an alternative curriculum theory, based on Habermas' (1972) theory of knowledge-constitutive interests and grounded in the work of teachers, has been articulated by Grundy (1987), whose perspective was adopted in this study for the purpose of analyzing the curriculum-related constraints that governed the pedagogical reform process. Grundy discusses three main types of curriculum interests and their pedagogical implications.

A *technical curriculum interest* is primarily an interest in controlling the educational environment so that an educational product which accords with prespecified objectives may result. Students become objects whose behavior and learning are closely managed. The teacher's main concern is to control student behavior so that she can deliver the curriculum. A culture of positivism is promoted: knowledge is viewed as a set of rules and procedures that are impersonal and objective. The teacher is under pressure to be productive, and quality is judged by the products of her work (i.e., students' grades). A technical curriculum interest disempowers both the teacher and students because the power to determine and judge what teachers and students must do is vested elsewhere. This study revealed that technical curriculum interests were major determinants of the teacher's pedagogical reforms.

By contrast, at the center of a curriculum informed by *practical curriculum interests* is a concern with the intersubjective processes of interpretation, personal judgement, and reflective deliberation, rather than with the objective products of learning. This curriculum is concerned with human interactions in which teacher and students are regarded as subjects engaged in sense-making activities. The teacher is interested in developing the judgement-making skills of her students, as part of an overall goal of personal development and improvement. The curriculum content is justified in terms of moral criteria associated with the "good" of the students and is selected on the basis of facilitating interpretation and meaning making, rather than on rote learning of prespecified skills. The teacher is an interpreter of curriculum documents, rather than a technical implementer, and is an active participant in curriculum decision making.

The practical curriculum interests of Grundy seem to be consonant with the constructivist-related perspective that shaped the pedagogical reform process in this study. Both have a pedagogical orientation towards the cognitive state of the individual learner (whether teacher or student), both are concerned with enabling the learner to make sense of her learning experiences, and both promote learner autonomy and responsibility. In this study, however, the full realization of these goals was frustrated by the prevalence of the teacher's technical curriculum interests that underpinned his reformed pedagogical beliefs.

External curriculum control: Teacher as agent

Ray complemented his interactive informer teaching role with a *managerial role* that served to strongly control the curriculum, particularly students' engagement in prescribed learning activities. Rather than controlling the class from the center of the classroom, Ray vicariously exerted control through the agency of his worksheets which delineated a single "learning route" along which all students traveled. Although the casual observer would perceive students to be mostly working individually or in small groups at self-determined rates, Ray's (approximately) weekly regrouping of the whole class for the purpose of reviewing the topic of study was evidence of his expectations that the class should cover common curriculum content. Although Ray espoused a responsive and focused teaching approach that accounted for individual student learning needs, he believed that he should concurrently provide both a universally appropriate and highly prescribed common curriculum.

Ray's sustained technical curriculum interests were also obvious in the production and travel metaphors that he used to express his reformed pedagogical priorities. He often referred to the ideal of students' industriousness and his desire to have a productive classroom environment in which students were efficiently working on prescribed tasks. Ray was also concerned with his tour leader role, which he expressed in terms of "we must press on", "we haven't got anywhere", "kept up with the top class", "progress through the syllabus", and "leading them through it" (Taylor 1990b). Although Ray's transformed pedagogy was more student-centered, in the sense that he attempted to focus his attention on the individual student and his interactions with students were more responsive, Ray's overall curriculum concerns were shaped by prevailing technical interests. Although Ray's constructivist-related pedagogy seemed to be both theoretically feasible and practically feasible (two key conditions of the pedagogical change theory) in relation to students' short-term learning needs, it was not globally feasible in relation to the teacher's longer term curriculum priorities associated with the state-mandated curriculum requirements.

Compromise: A cultural constraint

The inconsistency in the feasibility of the teacher's constructivist beliefs was a source of cognitive conflict for Ray. It is interesting to consider the dominant role of the teacher's technical interests in resolving this conflict and, in effect, drawing to a close the prospects of further constructivist-related pedagogical reform. Ray acknowledged the dilemma he faced in trying to reconcile the conflict between his practical curriculum interest of facilitating learning with understanding and his technical curriculum interest in covering the syllabus in the available time. His sense of professional accountability for preparing all students for the frequent school-mandated topic tests and end-of-year state-mandated external examinations underpinned the latter interest.

Ray rationally reconciled his conflict of interests by appealing to a principle of compromise. Although he believed that his forcing of a common curriculum pace unduly compromised his constructivist-related belief in facilitating students' meaningful learning, Ray asserted that being compromised was a normal experience amongst teachers that "you learn to live with, and most teachers learn to live with it without probably even realizing it" (Taylor 1990b). Clearly, the teacher perceived himself to be a hapless victim of a largely external locus of curriculum control. Ironically, Ray's students are likely to have a similar perspective in relation to the teacher's control of their classroom learning activities.

From the perspective of Grundy's theory of curriculum interests, the pedagogical reform process of this study was curtailed by the teacher's prevailing technical curriculum interests, which denied a global feasibility for his pedagogical reforms. The coexistence of two conflicting curriculum interests set up pedagogical demands that remained in dialectical tension until resolved by the teacher's recourse to a belief in the inevitability of being disempowered by the prevailing positivist culture of teaching. Assisting teachers to overcome their technical curriculum interests poses a great challenge. A theory of pedagogical change is required that enables teachers to sustain a critical focus on the nature of their curriculum interests and to resist the temptation to acquiesce in the face of potentially disempowering, culturally derived constraints.

Empowerment: An alternative perspective

Grundy's (1987) third curriculum interest provides a promising perspective from which to consider the nature of this challenge. *Emancipatory curriculum interests* transcend the dialectical relationship between the technical and practical curriculum interests. At the center of a curriculum informed by an emancipatory interest is a concern for liberating teachers and students from the disempowering constraints of the ideology (or dominant ideas and values) that underpin the curriculum and define the social norms of the classroom. This study found that a principle constraint to be overcome by a constructivist-oriented pedagogical reform process is the ideological distortion of the pedagogical process by the technical interests of the external curriculum designers with whom the locus of control of knowledge ultimately resides. An emancipatory interest requires this locus of control to be relocated into the classroom and shared between teachers and students in order that they collaborate to become empowered as autonomous intellectuals.

However, the *hegemonic* nature of ideology makes this a very difficult task. Ideology constitutes teachers' and students' subjectivities and serves as an invisible interpretive framework for making (restricted) sense of the teaching-learning process and the prospects of changing the natural order of the classroom. Ideology is subtly but effectively propagated in the hidden curriculum of rituals that socialize teachers and students into the dominant order of schooling (McLaren 1986). To the extent that the prevailing ideology remains unrecognized, teachers and students will be unable to discern its socially constructed nature and will remain disempowered agents of its propagation. This study found that the ideology of technical curriculum interests prevailed and that the teacher remained an unwitting agent of his own ideological oppression by continuing to act out a natural classroom role, a role that promoted an external locus of knowledge control and that compromised the emancipatory interests of both his students and himself.

To what extent is it possible for a pedagogical reform process to liberate teachers and students from the pedagogically distorting influence of a pervasive ideology that promotes a culture of schooling based on technical curriculum interests? Grundy (1987) advocates *critical pedagogical practice*, which she describes as a form of action research operating in an emancipatory mode. In critical pedagogical practice, teachers and students not only actively engage in processes of reflective deliberation, personal judgement, and interpretation (in accord with practical curriculum interests), but they also bring a critical focus to bear on the social interactions and contexts of learning. With recourse to critical social theories, teachers and students participate in ideology critique for the purpose of identifying "the constraints imposed upon their practices by social structures and interactions which are informed by interests in domination and control" (146).

Emancipation requires the development of a critical consciousness that recognizes the culturally constructed nature of the education enterprise and fuels debate about the fundamental assumptions and interests taken for granted, which underpin the culture of teaching and learning and give rise to social inequalities and inequities. In critical pedagogical practice, teachers and students share the locus of control of their mutually constructed and intersubjective knowledge and collaboratively struggle to make sense of both the perceived world and the ideology that constrains their perceptions and actions. Thus, emancipation commences with enlightenment about the nature of ideological distortion and continues with political action aimed at reforming the social structures that constrain the emancipatory interests of teachers and students.

Critical consciousness: A feasible pedagogical goal

Grundy acknowledges that emancipation might not be an immediately attainable goal for pedagogical reform and that the development of a critical consciousness amongst teachers and students is the most feasible goal in the shorter term. Liberating students and teachers from the "false consciousness" that distorts current communications that constitute the teaching-learning process is a collaborative enterprise in which the teacher-as-student and students-as-teachers share authority and control in relation to the social construction of their knowledge.

In the classroom, the teacher could facilitate a critical focus on the sociocultural influences that shape teachers' and students' perceptions of their roles and that govern classroom social relations and the predominant discourse of learning. A possible example of this strategy is the classroom-based research program of Cobb et al. (in press) in which teachers maintain a pedagogical focus on the nature of classroom discourse and prompt students to "talk about talking about mathematics" in order to reveal and reshape the social norms that underpin the negotiations that constitute students' knowledge development. Teachers can draw on critical social theories related to issues such as gender equity, democracy, or ethnicity, to focus these meta-analyses of classroom discourse.

A pedagogical focus on the history and philosophy of the discipline also could reveal its socio-cultural genesis and provide a perspective for examining the current socio-cultural context of the discipline (particularly as it is represented in texts). Comparing current interpretations of the world with their historical antecedents (especially in relation to the names behind the formulas) might serve to humanize the discipline and highlight the tentative nature of modern knowledge.

Students could be encouraged to reflect on their perceptions of the locus of control of knowledge and examine the assumptions and actions that sustain its externality and disempower them from becoming autonomous intellectuals. Critical examination of learning theories underpinning students' life-worlds could be related to the actions of students and teachers in the classroom. The coexistence of conflicting curriculum interests could be an open agenda item that teachers and students discuss in order to frequently determine appropriate learning activities and classroom priorities and to evaluate the effectiveness of previously negotiated decisions. These discussions could include a critical examination of the role of the school in sanctioning specific behaviors and social values, particularly in relation to the prevailing ideological interests of the government as they are expressed in curriculum policies that shape students' future prospects. In these ways, a meaningful political education could proceed as an integral part of any discipline.

Critical constructivist pedagogical reform

From the perspective of critical curriculum theory, enabling teachers to develop a critical consciousness is an important goal of a constructivist-related pedagogical reform process. Strategies for working towards this goal might be adapted from strategies employed in this study. It is important that teachers become critically aware of the nature of the prevailing ideology that shapes their pedagogical beliefs and practices and of their social roles in propagating the ideology amongst colleagues and students. A critical focus on the metaphors, images, and myths embedded in the discourse between teachers and students is likely to reveal the socio-cultural underpinnings of established teaching and learning practices. However, there is a danger, as this study demonstrated, that teachers might be numbed into cynical conformity by the seemingly overwhelming constraints of "the system" that appears to be beyond their control.

It is important, therefore, that the reform process assists teachers to develop a viable alternative pedagogical vision and a realization of its longer-term feasibility. From the perspective of critical curriculum theory, an alternative vision would focus on the development of both teachers and students as autonomous intellectuals who are capable of both critical reflection and socially responsible action in relation to the democratization of their life-worlds. It seems that, initially, the reform process should be grounded in the daily actions of teachers and students so that its theoretical and practical viability is established in the short term, and the constraints of the prevailing ideology are thrown into sharp relief for critical examination, both within and outside the classroom. An appropriate analogy might be drawn between reform-minded teachers and environmental conservationists whose philosophy is encapsulated in the entreaty "think globally and act locally."

It is also important that the reform process involves teachers in initiating reform of the prevailing culture of schooling by engaging in critical discourse at all levels of their profession, with an aim of becoming instrumental in renegotiating policies that determine the nature of their professional actions and responsibilities. In other words, teachers need to become active participants in the political forums that extend from their classes and staffrooms into their local communities and beyond. The reform process, therefore, should assist reformist teachers to develop an authoritative voice. As researchers of their own pedagogies, reformist teachers need to effectively communicate about the processes and outcomes of their endeavors and to persuade disinterested, complacent, or recalcitrant colleagues to participate, at least, in a critical examination of the claims and evidence for the need to radically reform the prevailing culture of teaching.

CONCLUSION

A constructivist-related theory of pedagogical change was employed to facilitate a radical transformation in the pedagogical beliefs and classroom practice of a high school mathematics teacher. The theory of pedagogical change was based on theories of conceptual change in science education and theories of cognition associated with teacher education literature. Earlier papers assessed the influence of the teacher's beliefs on his pedagogical reforms and concluded that, although the teacher's pedagogy had become much more student-centered, it sustained an essentially centralist classroom role for the teacher and maintained a high degree of student compliance and dependence on the teacher.

This chapter shifts the analytical focus away from the teacher and onto the theory of pedagogical change. The key issue addressed in the chapter is the efficacy of the theory for overcoming teacher centralism and facilitating radical pedagogical reform. Multiple, constructivist-related perspectives were employed to determine whether the theory contained the seeds of its own delimitation and, if so, what a more effective theory might comprise.

A sociological perspective indicated that the "personal constructivist" perspective underpinning the theory of pedagogical change had promoted a rationalist view of knowledge and a pedagogical focus on assisting students to individually replicate the conceptual structures of the teacher in a pedagogical context of social isolation. A rationalist view of knowledge seems to have sustained the teacher's disregard for the central role of social influences in shaping students' interpretations of their experiences.

A radical constructivist perspective highlighted the influence of the theory of pedagogical change in contributing to a covertly realist view of knowledge and in sustaining a pedagogical focus on transmitting the teacher's knowledge to the students. A lack of concern with the ontological aspects of knowledge seems to have sustained the teacher's authority as an expert knower of an objective and formalist mathematical world of symbols and sustained students' roles as compliant replicators of the teacher's knowledge and apprentice manipulators of mathematical symbols.

The perspective of critical curriculum theory indicated that the short-term concern with establishing the theoretical and practical feasibilities of the teacher's pedagogical reforms ignored the teacher's global curriculum priorities, which continued to be closely associated with his technical interests. It seems, therefore, that the theory of pedagogical change sustained the teacher as an agent of the locus of control of knowledge that resides with the externally located curriculum and examination authorities. Thus, the theory failed to address an important cultural constraint that continued to disempower both students and teacher.

This chapter has revealed the limitations of a theory of pedagogical change based on a constructivist theory of knowledge currently popular amongst many science and mathematics educators. The multiple perspectives reported in the chapter have revealed the tenacious and multifaceted character of the problem of teacher centralism and have helped to provide a better understanding of the nature of the constraints that restricted the efficacy of the pedagogical reform process. A cultural anthropological perspective has proved useful for identifying the apparently culturally derived nature of major constraints and suggests that a pedagogical reform process should aim to assist teachers to change the metaphors, images, and myths that sustain their centralist classroom roles. A sociological perspective on knowledge suggests that these key beliefs are established and sustained in the daily discourse between teachers and that radical pedagogical reform might require teachers to engage in a renegotiation of the culture of teaching, rather than going it alone.

Finally, and most important, a critical curriculum perspective suggests that the pedagogical reforms achieved in this study are consonant with establishing the ascendancy of practical curriculum interests that recognize the central importance of students' subjectivities. The establishment of practical interests might be a useful first step towards identifying the socio-cultural influences that tend to deny the centrality of students' subjectivities and disempower them from developing as autonomous intellectuals. Having established these fertile conditions, the next step of a critical constructivist pedagogical reform process is to germinate a critical focus among teacher and students on the constraining influences and to promote collaborative efforts aimed at overcoming them.

REFERENCES

Apple, M. *Ideology and Curriculum*. London: Routledge and Kegan Paul, 1979.

Ausubel, D.P. *Educational Psychology: A Cognitive View*. New York: Holt, Reinhart, and Winston, 1968.

Burger, P.L., and T. Luckman. *The Social Construction of Reality: A Treatise in the Sociology of Knowledge*. London: Penguin Books, 1966.

Cobb, P. "Information-processing Psychology and Mathematics Education: A Constructivist Perspective." *Journal of Mathematical Behavior*, 6, (1987): 3–40.

Cobb, P., T. Wood, and E. Yackel (in press). "Coping with the Complexity of Classroom Life." *Harvard Educational Review*.

Driver, R., and J. Easley. "Pupils and Paradigms: A Review of Literature Related to Concept Development in Adolescent Science Students." *Studies in Science Education*, 5, (1978): 61–84.

Driver, R., and G. Erickson. "Theories-in-Action: Some Theoretical and Empirical Issues in the Study of Students' Conceptual Frameworks in Science." *Studies in Science Education*, 10, (1983): 37–60.

Driver, R., and V. Oldham. "A Constructivist Approach to Curriculum Development in Science." *Studies in Science Education*, 13, (1986): 105–122.

Eisenhart, M.A., J.L. Shrum, J.R. Harding, and A.M. Cuthbert. "Teachers' Beliefs: Definitions, Findings, and Directions." *Educational Policy*, 2, (1988): 51–70.

Erickson, F. "Qualitative Methods in Research on Teaching." In M.C. Wittrock (Ed.), *Handbook of Research on Teaching* (3d ed.). New York: Macmillan, 1986. 119–159.

Fenstermacher, G.D. "A Philosophical Consideration of Recent Research on Teacher Effectiveness." *Review of Research in Education*, 6, (1979): 157–185.

Fenstermacher, G.D. "Philosophy of Research on Teaching: Three Aspects." In M.C. Wittrock (Ed.), *Handbook of Research on Teaching* (3d ed.). New York: Macmillan, 1986. 37–49.

Fleming, B. *Enhancing the Teaching-learning Process in Mathematics through the Development of a Constructivist Teaching Style.* Unpublished postgraduate diploma project report, Curtin University of Technology, Perth, Western Australia, 1988.

Floden, R.E. "The Role of Rhetoric in Changing Teachers' Beliefs." *Teaching and Teacher Education*, 1(1), (1985): 19–32.

Fraser, B.J., and D.L. Fisher. "Use of Actual and Preferred Classroom Environment Scales in Person-environment For Research." *Journal of Educational Psychology*, 75, (1983): 303–313.

Gilbert, J.K., and D.M. Watts. "Concepts, Misconceptions and Alternative Conceptions: Changing Perspectives in Science Education." *Studies in Science Education*, 10, (1983): 61–98.

Gilbert, J., R. Osborne, and P. Fensham. "Children's Science and Its Consequences for Teaching." *Science Education*, 66(4), (1982): 623–633.

Giordan, A. *Bibliographie Concernant les Recherches sur les Conceptions des Apprenants en Biologie. Annales De Didactiques Des Sciences,* No. 2, 1987.

Giroux, H.A. *Ideology, Culture and the Processes of Schooling.* Philadelphia: Temple University Press, 1981.

Grundy, S. *Curriculum: Product or Praxis?* London: Falmer Press, 1987.

Habermas, Y. *Knowledge and Human Interest* (2d ed.). London: Heinemann, 1972.

Kelly. G.A. *The Psychology of Personal Constructs.* Vols. 1, 2. New York: Norton, 1955.

Kuhn, T.S. *The Structure of Scientific Revolutions* (3d ed.). Chicago: University of Chicago Press, 1962.

Kyle, D.W., and G. McCutcheon. "Collaborative Research: Development and Issues." *Journal of Curriculum Studies*, 16(2), (1984): 173–179.

Lakatos, I. "Falsification and the Methodology of Scientific Research Programmes." In I. Lakatos and A. Musgrave (Eds.), *Criticism and the Growth of Knowledge.* New York: Cambridge University Press, 1970.

McLaren, P. "Making Catholics: The Ritual Production of Conformity in a Catholic Junior High School." *Boston University Journal of Education*, 168(2), (1986): 55–77.

Munby, H. "A Qualitative Approach to the Study of a Teacher's Beliefs." *Journal of Research in Science Teaching*, 21(1), (1984): 27–38.

Nespor, J. "The Role of Beliefs in the Practice of Teaching." *Journal of Curriculum Studies*, 19(4), (1987): 317–328.

Novak, J. "The Importance of Emerging Constructivist Epistemology for Mathematics Teaching." *Journal of Mathematical Behavior*, 5, (1986): 181–184.

Nussbaum, J., and S. Novick. *Brainstorming in the Classroom to Invent a*

Model: A Case Study. Israel Science Teaching Centre, The Hebrew University, Jerusalem, 1980.

Olson, J. "Teacher Constructs and Curriculum Change." *Journal of Curriculum Studies*, 12(1), (1980): 1–11.

Olson, J. "Teacher Influence in the Classroom: A Context for Understanding Curriculum Translation." *Instructional Science*, 10, (1981): 259–275.

Osborne, R.J. *The Framework: Towards Action Research.* (Learning in Science Project Paper No. 28.) Hamilton, New Zealand: University of Waikato, 1981.

Piaget, J. *The Child's Conception of the World.* New York: Harcourt Brace, 1929.

Pines, L., and L. West. "Conceptual Understanding and Science Learning: An Interpretation of Research Within a Sources-of-Knowledge Framework." *Science Education*, 70(5), (1986): 583–604.

Pfundt, H., and R. Duit. *Bibliography: Students' Alternative Frameworks and Science Education* (2d ed.). Kiel, FRG: University of Kiel, 1987.

Pope, M., and J. Gilbert. "Personal Experience and the Construction of Knowledge in Science." *Science Education*, 67(2), (1983): 193–203.

Posner, G.J., K.A. Strike, P.W. Hewson, and W.A. Gertzog. "Accommodation of a Scientific Conception: Toward a Theory of Conceptual Change." *Science Education*, 66(2), (1982): 211–227.

Putnam, R., M. Lampert, and P. Peterson. "Alternative Perspectives on Knowing Mathematics in Elementary Schools." *Review of Research in Education*, 16, (1990): 57–150.

Resnick, L.B. "The Role of Invention in the Development of Mathematical Competence." In R. Kluwe and H. Spada (Eds.), *Developmental Models of Thinking.* New York: Academic Press, 1980. 213–244.

Resnick, L.B. "Toward a Cognitive Theory of Instruction." In S.G. Paris, G.M. Olson, and W.H. Stevenson (Eds.), *Learning and Motivation in the Classroom.* Hillsdale, NJ: Erlbaum, 1983.

Ruthven, K. "Ability Stereotyping in Mathematics." *Educational Studies in Mathematics*, 18(3), (1987): 243–253.

Saphier, J. "The Knowledge Base on Teaching: It's Here, Now!" In T.M. Amabile and M.L. Stubbs (Eds.), *Psychological Research in the Classroom.* Brandeis University: Pergamon Press, 1982. 76–95.

Solomon, J. "Social Influences on the Construction of Pupils' Understanding of Science." *Studies in Science Education*, 14, (1987): 63–82.

Steffe, L. "Principles of Mathematics Curricular Design: A Constructivist Perspective." In J. Malone, H. Burkhardt and C. Keitel (Eds.), *The Mathematics Curriculum: Towards the Year 2000.* Perth, Western Australia: Curtin University of Technology, Science and Mathematics Education Centre, 1989.

Taylor, P.C. "The Influence of Teacher Beliefs on Teaching Practices." In D. E. Herget (Ed.), *More History and Philosophy of Science in Science Teaching.* Proceedings of the History and Philosophy of Science in Science Teaching First International Conference, Florida State University, Tallahassee, 1990a.

Taylor, P.C. *The Influence of Teacher Beliefs on Teaching Practices.* Paper, annual meeting of the American Educational Research Association, Boston, 1990b.

Taylor, P.C., and B. Fleming. *Improving Teacher Effectiveness: A Collaborative Approach.* Paper, First Joint Conference of the Australian and New Zealand Associations for Research in Education, University of Canterbury, Christchurch, New Zealand, December 1987.

Tobin, K. "Social Constructivist Perspectives on the Reform of Science Education." *Australian Science Teachers Journal*, 36(4), (1990a): 29–35.

Tobin, K. "Metaphors and Images in Teaching." *What Research Says to the Science and Mathematics Teacher*, 5. Perth, Western Australia: Curtin University of Technology, The Key Centre for School Science and Mathematics, 1990b.

Tobin, K., and M. Espinet. "Impediments to Change: An Application of Peer Coaching in High School Science." *Journal of Research in Science Teaching*, 26, (1989): 105–120.

von Glasersfeld, E. "The Concepts of Adaption and Viability in a Radical Theory of Knowledge." In I.E. Siegel, D.M. Brodinsky, R.M. Golinkoff (Eds.), *New Directions in Piagetian Theory and Practice*. Hillsdale, NJ: Erlbaum, 1981.

von Glasersfeld, E. "Learning as Constructive Activity." In E. von Glasersfeld (Ed.), *The Construction of Knowledge: Contributions to Conceptual Semantics*. Salinas, CA: Intersystems Publications, 1987. 307–333.

Watt, D.H., and H. Watt. "Design Criteria for Collaborative Classroom Research." In T.M. Amabile and M.L. Stubbs (Eds.), *Psychological Research in the Classroom*. Brandeis University: Pergamon Press, 1982. 134–143.

West, L.H. "The Researchers and their Work." In C. Sutton and L. West (Eds.), *Investigating Children's Existing Ideas about Science*. Leicester, GB: University of Leicester, School of Education, 1982.

Wheatley, G. "Constructivist Perspectives on Science and Mathematics Learning." *Science Education*, 75(1), (1991): 9–21.

18

Learning to See Children's Mathematics: Crucial Challenges in Constructivist Reform

Jere Confrey

The United States of America has committed itself to educational reform. Nearly every state has enacted new legislation aimed at school improvement. Reports decrying the poor state of American education are released regularly. Our weak performance on international assessments has dealt our national pride a serious bout of shamefacedness, and our politicians leap to declare that school improvement must be a national priority. The calls for reform are easy to hear; what is critically less clear is what that vision for reform might be and how to achieve it in the second largest service profession, next to that of health. The changes must be systemwide, forceful, convincing, and rapid. However, like a swimmer stranded offshore without land in sight, choosing the wrong direction in which to move can lead to results equally as devastating as staying put.

REFORM IN MATHEMATICS EDUCATION

Direction has been provided by the National Council of Teachers of Mathematics (NCTM), which has responded gradually and persistently over the last five years to the warnings and has documented evidence within its community that serious change must be undertaken. In 1986, it established a Commission on Standards for School Mathematics that spent more than two years preparing the *Curriculum and Evaluation Standards for School Mathematics* (1991) for publication and gaining the endorsement and/or support of every major organization of mathematics educators. These *Standards* expressly articulate their social goals: to create literate workers, to encourage lifelong learning, to provide opportunity for all, and to ensure an informed electorate. The *Standards* were submitted to an intensive review process by teachers, parents, researchers, teacher educators, and administrators for a period of two years. What the *Standards* offer is a vision of improvement in mathematics education. Coupled with such documents as *Reshaping School Mathematics* (1990) and *Counting on You* (1991) from the Mathematics Sciences Education Board, the *Standards* offer a definitive picture of mathematics education into the new century. Although none of these documents explicitly declares itself as representing a *constructivist perspective*, most educators use the term to describe the program that is envisioned.

The reform program advocated in these documents, which have received such widespread support, can be summarized as calling for students who (1) value mathematics, (2) are confident in their own abilities, (3) are problem solvers, (4) can communicate mathematically, and (5) can reason mathematically. A few excerpts from the Introduction to the Standards communicates the flavor of the philosophy:

> First, "knowing" mathematics is "doing" mathematics. A person gathers, discovers, or creates knowledge in the course of some activity having a purpose. The active process is different from mastering concepts and procedures. We do not assert that informational knowledge has no value, only that its value lies in the extent to which it is useful in the course of some purposeful activity. . . . Instruction should persistently emphasize "doing" rather than "knowing that." (7)

> . . . some proficiency with paper-and-pencil computational algorithms is important, but such knowledge should grow out of the problem situations that have given rise to the need for such algorithms. (8)

> We recognize that students exhibit different talents, abilities, achievements, needs, and interests in relationship to the mathematics. . . . If all students do not have the opportunity to learn this mathematics, we face the danger of creating an intellectual elite and a polarized society. (9)

> Traditional teaching emphases on practice in manipulating expressions and practicing algorithms as a precursor to solving problems ignore the fact that knowledge often emerges from the problems. This suggests that instead of the expectation that skill in computation should precede word problems, experience with problems helps develop the ability to compute. Thus, present strategies for teaching may need to be reversed, knowledge often should emerge from experience with problems. Furthermore, students need to experience genuine problems. (9-10)

> . . .in many situations individuals approach a new task with prior knowledge, assimilate new information and construct their own meanings. . . . As instruction proceeds, children often continue to use these routines in spite of being taught more formal problem-solving procedures. They will accept new ideas only when their old ideas do not work or are inefficient. Furthermore, ideas are not isolated in memory but are organized and associated with the natural language that one uses and the situations one has encountered in the past. The constructive, active view of the learning process must be reflected in the way much of mathematics is taught. (10)

This view of mathematics education requires a significant change in the role of the teacher. In addressing this question, the NCTM produced the *Professional Standards for Teaching Mathematics* (1991). Subject to a similar process as the *Curriculum and Evaluation Standards*, these *Teaching Standards* identify the need for a teacher to change from an orator exposing information into an "intellectual coach" who can be a role model (modeling problem solving), a consultant (helping a student to achieve and monitor progress), a moderator (raising questions and encouraging group decisions), an interlocutor (encouraging reflection), and a questioner (ensuring there are defensible positions held).

If all of these forces are then devoted to fundamental educational reform, what could possibly prevent its occurrence or lead to unintended and unfortunate results? In my opinion, we are poised on a cusp; the outcome of reform can be either very positive and exciting or fragmented, superficial, and ineffective. The political astuteness of the activities of NCTM, the director of the *Standards* project, Dr. Thomas Romberg, and the other board members is indisputable. They have created a carefully articulated agenda for change and made it available at the time that the cries for reform abound. They provide what the politicians need: a clear articulation of an appropriate direction for reform. Admittedly, however, the *Standards* can only set the agenda. Achieving those goals will depend on the actions taken to enact that agenda.

The success of the reform movement will hinge on the answer to the question: What do teachers need to know and need to be able to do in order to implement these reform changes, and how can they gain the knowledge and dispositions to act if they do not already have them? In this chapter, I will examine this question and try to show where, from a constructivist perspective, the key issues lie. I will make the claim that the *Standards* will only be successful in reforming practice if teachers learn how to achieve what I will call a "close listening" to students. It is close listening that provides the impetus for a serious and deep reconsideration of alternative viewpoints about the subject matter. Furthermore, learning to listen closely demands that teachers use student methods as opportunities to examine, strengthen, and revise their own understanding of mathematics. What this amounts to is an epistemological revolution about what mathematics is, its structure, and its basis for authority. This call for close listening is not an explicit part of the *Standards*, nor is it a major reconceptualization of mathematical knowledge; however, both are consistent with the *Standards*. Before I discuss the implications of close listening, I want to share a true story of an elementary school in which the excitement of reform has ignited the teachers to address these issues and has prepared them to be ready to embrace changes. The question addressed specifically in this story, and more broadly in the rest of the paper, will be: How can we assist them in this change process?

A STORY

Parents in this elementary school were becoming aware of changes in the mathematics program. A primary reason for this was that less mathematics homework was being assigned, especially in the form of practice on sheets of exercises. The children came home and reported less about computational practice and gave vague descriptions of the mathematics they were doing, often questioning in their own minds if it was indeed mathematics. Parents were unintentionally left in the dark, and they were concerned. In response to parental concerns, a group of teachers, the school principal, and the curriculum committee chairperson in mathematics for the district organized a meeting to present an explanation of the school's mathematics program. The meeting was well attended. The curriculum coordinator started the meeting. She asked the parents for their first reactions to recalling their school mathematics experiences. The answers were "drilling," "enjoyment," "frustration in a black box," "times tables," "word problems," "flash cards," and so on. A majority had negative memories; a minority recalled an ease of learning and an excitement for it.

Then, the teachers presented their reasons for changing the curriculum. The discussion was placed into the context of widespread educational reform of the *Standards,* which were distributed along with the state syllabus at grade level. The teachers' presentations were filled with examples of students' work, videotapes of their discussions, and examples of materials. The methods of the new approaches to mathematics were connected to whole language approaches, interdisciplinary inquiry, writing across the curriculum, and project and group-oriented activities, all examples of other innovative approaches used successfully in the district. Some of the teachers' told stories of inventions by students. A second grade teacher reported on an activity of sorting fictional animal characters called "Zogs." Zogs had a variety of attributes: color, number of antennae, number of feet, and straight or curly hair. The teacher had given the students a rectangular array to sort Zogs. On the second day, he watched as the children realized that they needed a way to place Zogs in multiple categories. They needed intersections. To his astonishment, they invented a version of "Venn diagrams."

Another teacher told of how she and her team-teaching group had decided to begin long division by giving the fourth grade children a problem: "It took 18 scoops to fill the jar with 426 jelly beans. How many jelly beans were in each scoop?" The children figured out a variety of methods. One girl wrote, "18+18 = 36. Then 36+36 which was 72 and I put 4 tallies. Then 72+72 which was 144. Then I put 8 tallies. Then I added 144+144 which gave me 288. Then I put 16 tallies. Then I put 288+288 and got 476 and I had 32 tallies. But the number 476 was too big, so I subtracted 476-18 and crossed out 1 tally which gave me 458 and I subtracted 458-18 which gave me 440 and crossed out a tally and then I subtracted 440-18 and got 422 and crossed out a tally and then that was less than 426 but at that point I had 29 tallies with a remainder of four jelly beans!"

At the end of the presentations, parents were invited to ask questions. A parent from another country expressed her concerns that her children would not be able to return home and keep up with the skills required there. Another parent skeptically commented that the pendulum had already swung twice in mathematics, and this was just another swing and should be undertaken more moderately, and thus it should include traditional methods to "hedge one's bets." There were comments about "not throwing the baby out with the bath water," referring to computations and drill and practice as the baby. The principal responded by saying that it was not a matter of "rejecting traditional methods," since the computational practice was included in these new models, not as isolated drill and practice, but rather as skills and tools to solve problems. A teacher explained that the computation in this approach included explicit discussion of students' methods, such as solving 8+9 as 8+8+1 or 9+9-1, or

8+10-1, a systems-based knowing. A parent reported that, during the previous summer, her nieces and nephews brought over a traditional workbook. Her daughter, having learned mathematics the "new way," enthusiastically sat down with them, and, under her tutelage, they did the entire workbook in one afternoon, "for fun." Another reported that her son had corresponded with four students who had returned to their home countries. All had first asked about the kick-ball team at the school and then went on to comment that their studies in mathematics were proceeding well. By the end of the meeting, the majority of the parents were endorsing the program, and some who remained skeptical were simply asking to be kept informed of the children's progress by means of some form of assessment. One parent concluded, "It's the parents that need to be reeducated in this."

Watching this meeting as a professional mathematics educator who had worked in the school and as a parent, I felt a variety of reactions. I was impressed by the power of the videotapes and presentations of student work in affecting the parental audience. The sincerity, excitement, and commitment of the teachers was convincing. The potential for further development of mathematics in the presentations was intriguing. The honesty and genuine concern for the welfare of children was readily apparent. I knew that the presenters were the strongest advocates for the new methods and that other teachers would be less articulate and knowledgeable, a fact that would be expected in any reform. And I knew that these teachers who were leading the reform, as committed to the methods as they were, were also aware that they did not know mathematics as well as they would like, in order to be well prepared to take on this challenge. My own contributions to the meeting were to (1) advocate parents to encourage the district to support more professional development resources, (2) argue against the use of good and bad, stronger and weaker students as singular dimensional descriptions of children, and (3) let parents who were worried about the students' scores on standardized tests know that by the time these children reach high school, they are likely to be assessed on a variety of measures, such as demonstrations, portfolios, and standardized tests.

One week later, as I was reflecting on the meetings, one of the teachers approached me and said, "I am aware of my own deficiencies in mathematics. I learned it as a closed and constrained subject. I would like you to help me. Not so much in my mathematics lessons, but in *learning to see mathematics*. I can see science, social studies, writing in the kids' other activities, *but I don't know how to see their mathematics*." This indeed was the piece that I needed to hear. The question that needs to be addressed is how to assist teachers in seeing and acting in concert with children's mathematics.

The question is how can we help these teachers to effect the changes called for in the *Standards*. The willingness to experiment and the devotion to the children are there. The question posed by that teacher "How do I learn to see children's mathematics?" demonstrates her clear understanding of the challenges facing her and the changes they will require. These teachers have learned to appreciate the diversity and cleverness of students' methods. Learning to promote the development of these untutored ideas places a new set of demands on teachers. Promoting a diversity of response is a major first step. Using it to achieve developmentally strong conceptual understanding is the completion of the dance. It is this completion of the dance that will require teachers to learn to hear the possibilities in student methods and to use them to recast their own understanding of mathematics and to chart potential paths to deeper understanding. Learning to listen deeply to student methods requires a particular kind of knowledge of the mathematics and the development of mathematical concepts. How can we assist teachers in their quest to build such knowledge?

PASSAGES THROUGH PAST INNOVATIONS IN MATHEMATICS EDUCATION

In mathematics education, the need for dramatic change has been recognized by the educational research community for years. Responding to the Piagetian research program, researchers began conducting clinical interview studies in the mid-seventies and documented two significant themes: children were learning by memorization algorithmic approaches that they did not understand; and children demonstrated the ability to invent creative solutions to problems when the problems related well to the interests and intentions of the students (Davis 1985; Ginsburg 1977; Duckworth 1987; Erlwanger 1975). Serious questions were raised about the validity of standardized tests for assessing understanding (Easley 1977), and doubts were cast on the usefulness of a procedural, algorithmic approach inherent in the "Back to Basics" movement.

Both movements in mathematics education, the "Back to Basics" of the seventies and the "New Math" of the sixties, had failed. "Back to Basics" did not promote the diligent, determined accumulation of skills and procedures, for the children of the seventies were not really motivated to learn from pages of workbook practice. This approach was further undermined by the ready access to calculators, and automatic cash register calculation of change and sales tax, and so on. Questions were asked whether the procedural skills of the fifties were really the most appropriate content for an information age where the display of data, production of figures, and other forms of representation swamped the media with quantitative objects that needed comprehension and critique.

Of course, the "Back to Basics" movement was itself a response to the "New Math." "New Math," which largely represented a mathematician's view of mathematics, failed for a variety of reasons. Where teachers did not understand its content themselves, it never entered the door of many classrooms. Then, in the cases where it was well taught, some students profited from it but also carried with them its weaknesses. They learned the algebraic structures, worked in a variety of bases, and could complete truth tables and Venn diagrams, but their ability to apply that knowledge to the contexts of their lives remained limited and was largely irrelevant.

Alongside of these two movements was a Piagetian tradition which has taken longer to mature and is only now coming of age. Fundamental to this tradition was the recognition that a child's perspective may not be that of an inadequate or incomplete adult. Piaget's "genetic epistemology" was a view of knowledge as developing over time in response to the tasks and perturbances encountered by the child. These breakthroughs regarding the psychology of the child were complemented by developments in the philosophy of science, emphasizing the importance of theory in generating and gathering data, and of paradigm shifts wherein the basic assumptions, commitments, methodologies, and standards of evidence were revised. This yielded a socio-psychological perspective on the growth of knowledge which rejected knowledge development as an accumulation of facts independent of theory and community. An individual learning mathematics or science was likened to a member of a scientific field— theorizing, relating, and evaluating—and hence she was subject to similar upheaval and revisions in her own thinking. Her learning also involved theories (Karmiloff-Inhelder 1975), methods, and negotiation (Sinclair 1987).

Rather than asserting an absolutist, external view of mathematical knowledge, the growth of knowledge was seen to occur most effectively when one saw a need for an idea, considered a variety of possible solutions, tried out the solutions through a series of conjectures and refutations, and finally settled on a solution through a period of reflection and negotiation with peers (Lakatos 1977; Balacheff 1991). It is this view of knowledge that accounts for the revised focus in the *Standards*, for it is within such a view that issues such as the communication of mathematical ideas, the use of multiple methods of representation, and the need for dispute and resolution evolve naturally.

TYPICALLY SECONDARY CLASSROOMS

The school and the classrooms described earlier reflect a progressive approach to education. Having observed classrooms across the country, I can report a most alarming situation that exists in most mathematics classrooms in the United States, especially at the secondary level; the silence of the students is deafening. Breaking this silence in our classrooms and making use of the richness of the students' resourceful ideas are the first steps in dramatically changing the character of mathematics instruction. Doing so is, however, no easy task, for it is silence that is reinforced explicitly or tacitly by nearly all the groups that participate in the educational process. After a few years, students themselves wish to avoid the risk and potential embarrassment of speaking. They seem to have figured out that silence is an effective way to limit their personal involvement in an instructional system which devalues or ignores their contributions. Many teachers, feeling the press of mandated curricula (such as the Regents in New York) and standardized testing, come to prefer or, at least, grow accustomed to the silence; it allows them to "cover" the required content and to uphold the appearance of control in the classroom. When students do speak, it is startling how seldom their meanings seem to be understood. Twice in the past year I have seen first year teachers fail to be renewed by administrators who felt that their classrooms were too noisy—an evaluation made without any evidence of attention to the quality of the interactions that generated the noise within.

Instruction involves far more actors than are actually in the classroom. Textbook publishers have been presenting the material with as low a reading level as can be tolerated, providing step-by-step examples as models. These allow students to complete the exercises by imitation, avoiding the need for verbal discussion. The majority of individualized instructional programs, while tailored to individual needs, promote isolation and the avoidance of student confusion. Classroom periods in secondary schools are carved into tiny chunks by administrations and supervisors, 42-minute bits in which minimal serious intellectual inquiry can possibly take place. Cultural norms dictate who can and cannot talk, lead, or be seen in classrooms, as there is a tendency for economically poorer children and for girls to learn to accept silence as their interactional pattern. Often, the high achievers provide the only voices in the classrooms, diligently giving the right answers, so that confusion, disagreement, and alternative opinion have no hold over the conduct of the mathematics classroom. Charitably, one might assume that in none of these cases do the individuals involved intentionally promote silence—it is one of those tacit contractual agreements that one makes as one crosses the threshold of the mathematics classroom, and its effects are devastating.

This grip of silence is choking the lifeblood from the development of genuine and creative quantitative competence by our youngsters. It seems that the solution to the silence in classrooms ought to be simple: encourage more talk. However, after working extensively with teachers to teach constructivist approaches, we recognize now that encouraging talk is not enough. We must find ways to assist teachers in recognizing and making effective use of students' ideas. This is not simply a matter of having teachers ask more questions, not even if these questions are process and not product questions—that is, questions of *how* and *why*, and not *what*. Nor is it a matter of simply generating student opinion—a mere expression of opinion is a good way to get things started but it will not sustain an intellectual exchange unless there is a genuine kernel of an idea around which these opinions can be focused and negotiated. There is a need for purposeful activity in the mathematics classroom. We all talk best when (1) we care what we are talking about, (2) we are heard, (3) we believe that doing so will prove worthwhile. And we listen best when we believe what we hear will help us. These are tenets of good conversation, so the question is, "How do purposeful listening and talking turn into useful knowledge?" For this we turn to a theory of knowledge.

How Does One's View of Epistemology of Mathematics Influence Classroom Knowledge?

Let us suppose that we are in a classroom in which the teacher has given the students a problem or situation to work on that is interesting and compelling to the students. The materials invite and depend on students' participation; they support a diversity of approaches. How does the teacher's view of knowledge critically influence the course of events? We have repeatedly witnessed situations in which teachers have overlooked opportunities to explore rich student ideas. At other times, they have unintentionally distorted students' statements to fit with their own understanding of the content. To illustrate the potential for this kind of silencing, consider the following story. In discussions about pyramids, children debated intensely whether the "point" on the top of the pyramid should be counted as a corner in a chart of sides, edges, and corners. Over time, many alternative views of corners and points emerged: a corner is where three faces meet; a corner is on the inside, and a point is on the outside; a point is not a corner because it can have more than three sides meeting; a point is not a corner because it does not touch the base. The teacher who guided this discussion, "attempted to follow the students' thinking, asking questions designed to challenge and clarify their definitions" (Russell and Corwin 1991, 180). Now suppose instead that she had quickly introduced the term *vertex* to include both corners and points. The discussion would have been prematurely terminated, and an opportunity for children to understand the role of definition and distinction would have been lost.

Teachers who silence students may not want to remain in an authoritative teaching stance. They may believe in the value of students' input, but their preparation in mathematics does not prepare them to cultivate these opportunities. Teaching teachers to cultivate student invention, to make use of opportunities when they occur, and to challenge and revise their own views of subject matter is a critical matter. We must learn more about how to do it and provide the types of materials and activities that assist teachers in this area of professional development.

Because classrooms are for learning, classroom discourse is unique; it does not simply involve discourse between two voluntary participants. *Discourse* is conversations between and among a person who is more expert and a group of novices. Teaching is not simply imparting knowledge to students— it is structuring learning tasks and opportunities that invite students to enter into the intellectual exchange. Doing this is not a simple question of motivating students (von Glasersfeld, this volume). Teaching constructively involves building models of student thinking (an enterprise that demands evidence from multiple data sources), solving problems, working in groups, participating in class discussions, and performing on assessments. Doing this effectively requires that the teacher is able to view the students' activity in certain ways. The teacher must know how to promote fruitful mathematical exchange informed by constructivist pedagogy.

I wish to focus on what kinds of communicative exchange can come about within a constructivist perspective and what they mean for the mathematical and educational preparation of our teachers. I will do so by making four claims and by providing an example from actual practice that illustrates each claim.

Claim One: Traditional preparation in mathematics is an inadequate basis for recognizing students' inventive methods.

The contributions of Piaget and the constructivist program have documented time and time again that the formal presentation of mathematics by mathematicians forms an *inadequate basis for engaging students in the process of learning mathematics.* Knowing formal mathematics may be necessary, but, certainly, it is not a sufficient condition for teaching students (i.e., promoting student learning). Why? Because, as Piaget demonstrated so aptly, children see the world and experience objects *qualitatively different than adults.* He and his colleagues demonstrated that the world of mathematics is not simply placed into the experiential field of the child but that distinctions must be built by the child over time as she confronts challenges and novel situations. Piaget demonstrated how the roots of many of our mental operations involved in doing mathematics can be found in our actions on physical objects. From these repeated actions come mental operations, operations that become coded into the operations and concepts of the disciplines. Adding evolves from combining, affixing, joining, etc.; dividing from sharing, splitting, comparing, etc. The paths to mature ideas are not uniform across children; they do not consist of straightforward progress but of fits, starts, twists, and turns. A traditional preparation in mathematics will not be sufficient to allow a teacher to recognize these circuitous routes.

The epistemology of mathematics that dominates textbooks and classrooms at this time is a formalist one. It is one that seeks to display proof in a notational form (that tends to hide the questions and struggles) and becomes a manipulation of form rather than meaning; it strives for consistency as a demonstration of rigor. The epistemology of mathematics in a constructivist classroom is far more like that described in Lakatos' (1976) *Proofs and Refutations*—full of conjectures. These conjectures are followed by the creation of counter examples that challenge those conjectures. Efforts to explain the counter examples and resulting shifts in the definitions then lead to new conjectures and tentative proofs. The standards for proof evolve along with the ideas themselves.

Piaget's approach to the development of concepts was labeled "genetic epistemology" (Piaget 1970) and, by that, he referred to the microgenetic analysis of how concepts can develop in the minds of children through their activities and their attempts to resolve outstanding problematical situations. The processes of accommodation and assimilation were used to explain adaptation; these were followed by the process of reflective abstraction, which allowed for the stabilization and internalization of the changes. Relieving educational psychologists of the handicaps inherent in viewing children as inadequate adults (i.e., as lacking pieces of mature conceptual development), Piaget sought to demonstrate that children's methods are *rational, and they exhibited coherence, explanatory value, and consistency across age groups.* Believing that children's approaches almost invariably have these characteristics is the fundamental tenet of a constructivist teacher. Finding that rationality and finding a way to express it and discuss it with students is a most worthwhile and difficult goal of instruction. Disturbingly little classroom research and practice take these two principles seriously.

Should we call the results of our efforts to decipher and explore the methods (that differ markedly from conventional methods) of children and novices mathematical knowledge? This is a serious question, for the trade offs in making this decision either way demonstrate how politically risky constructivist practice can be. Many label the knowledge of how to teach particular content topics "pedagogical content knowledge"; however, such a label includes a wide variety of educational decisions and includes more than just issues of student conceptions. Other issues implied by the use of the phrase include the scope and sequence of a mathematics curriculum, a potpourri of techniques for teaching, an awareness of forms of assessment and proficiency in available materials, and so forth. For the most part, in educational research, researchers who have used such a label have not vigorously challenged epistemological aspects of the ideas of mathematics. Although they recognize the importance of content-specific approaches to pedagogy, their treatment of the mathematics itself is highly traditional.

If we call this knowledge of the view of mathematics from a child's vantage point "mathematics," two other outcomes may lead to the dismissal of student conceptions and methods. First, mathematicians tend to rename and recast the children's statements into the formal definitions and theorems that are familiar to them. In doing so, they miss opportunities to hear how students might conceive of the problem differently. Second, mathematicians are often scornful that inquiry into pedagogy can produce genuine mathematical insights. By demonstrating how a student's method is compatible with the extant mathematics, they again dismiss the possibility that representing or describing mathematics in a new way is legitimate mathematics. This bias is frequently witnessed not only in mathematicians' treatment of mathematics but also in their treatment of the history of mathematics, a tremendous resource for all those engaged in unearthing alternative perspectives (Unguru 1976). It seems reasonable to suggest that the bias is both the result of training that indoctrinates mathematicians narrowly to their fields of specialization and to a Platonist view of knowledge that reduces mathematics to a single body of knowledge, necessarily determined and simply awaiting human discovery.

An example might be helpful—my six-year-old daughter Kate was working on a problem; we had eight cinnamon rolls for breakfast, and only three of us in the family wanted to eat them. How should we share them fairly? After considering the possibilities of unequal sharing, favoring her and her brother over their parent, she realized that she could give each of the people two rolls and "split the other two into three parts." When she set about doing this, she used the following sequence of moves. Drawing her intentions for me, she first tried cutting one muffin in half and then dividing one of the halves into two parts (see Fig. 1). She saw that this did not produce equal amounts but produced as she put it, "two littles and one big." To adjust this, she slid her knife across the larger piece, keeping it parallel to the original cut until she judged that she had created equal amounts.

When I shared this result with a mathematician friend of mine, his reaction was to attempt to solve the problem of how far to displace that line to create equal-sized pieces. He reported back to me that it could not be done without

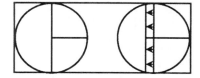

Figure 1.

certain techniques in calculus. For him, the need for calculus in her solution challenged the validity and usefulness of constructivism, for he asked, "If a first grader can generate a procedure that requires calculus to solve it precisely, then how valuable can such insights be for elementary teachers?" Having to reach a solution that had a precise, calculational answer (even though successive weighing could have worked empirically) was essential.

I question how helpful it is to recast her efforts into this sophisticated mathematical approach. To me, this episode had a different significance. I will first relate the end of the story and then offer a different interpretation of the significance of the events. When Kate showed her solution to her nine-year-old brother, he thought, at first, her solution was in error, not recognizing that she had offset the cut. He admitted that her method would work, and then he showed her the method that he had learned, which was to cut in the shape of a Y to get three congruent pieces (See Figure 2). Finally, her solution led him to consider making two vertical cuts, dividing the diameter into three equal parts, but resulting in unequal parts of the roll. He quickly realized this and dismissed this solution as unworkable.

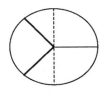

Figure 2.

Kate's reaction was to prefer her big brother's method. I do not know whether this was due to the symmetry of it or in deference to his age, experience, and confidence. However, with my reassurance that her method was valid, she tried it out, cutting a real roll into three parts. Hers produced a large piece and two small ones. His produced a small piece and two large ones. By mixing and matching, they gave each person a large piece and a small piece and ended up quite satisfied. In the car on the way to school, Brian commented that he preferred his method because he could check to see if it was accurate by placing the pieces on top of each other. We decided we could check Kate's method by weighing the pieces.

I am not claiming that coding this activity into a formal mathematical operation, such as calculus, would be an uninteresting pursuit. However, as a mathematics educator, its primary interest lay in the window it provided into a child's thinking. It provided a genuine opportunity for engaging in thought-experiment about the mathematical insights that it potentially represented. It demonstrated the power of the primitive binary structures of halving and quartering, etc., which were her starting points. Then she assessed the outcome in terms of "two littles and a big" and recognizing her goal was parity, she adjusted her outcomes. Her form of adjustment, sliding the cut to the side, preserved the equity (and congruence) of the two little pieces. However, it did not bring the third piece into this congruence. It leaves one pondering two questions: How strongly do children want to preserve this congruence in shape in equal sharing? What might have led her to consider folding her diameter into two parts at its center and rotating the radii to not only maintain the congruence of the two parts but create a congruence with the third. This is the type of analysis that is practiced in delving into "genetic epistemology." It is this type of analysis that a teacher needs to be able to engage in, in order to respond effectively to students' ideas.

In my curricular and research work, I try to ask myself the question, "What problem, activity, or situation might 'create a need' for a certain idea?" The problem just discussed, sharing eight rolls among three people, leads to a situation in which Kate needs to coordinate her circle concept with her understanding of equal shares. The concept of a circle has more than one possible scheme connected to it. A *scheme* is a framework used by people to anticipate, recognize, approach, and evaluate a mathematical concept, a kind of cognitive habit (see Confrey 1991; Steffe 1988; Thompson, in press). Formed through a process of experiencing a perturbation, acting to resolve the perturbation, and reflecting on the action to create operations, schemes underlie the development of concepts. A particular scheme will allow a student to work effectively with some aspects of a concept but to ignore others.

Thus, two possible schemes that underlie circles are to see the circle as bilaterally symmetric for the set of all possible diameters, based in a folding action, or to see it as rotationally symmetric, formed from a sweep of a radius about a given point. The first circle scheme is compatible with most schemes for equal sharing among two or four people, in that vertical and horizontal cuts intuitively connect up circle schemes and equal sharing schemes. Kate uses this construction to make her first approach to the problem; she creates a large and two smalls and then acts to equalize her outcomes by sliding the diameter to the left, creating a chord.

The second circle scheme, a sweep, creates a correlation between the angle traversed during the sweep and the area enclosed by the sweep. Considering how to create equal-sized sweeps might lead to "create a need" for the Y-shaped cut described by Kate's brother. This scheme is also more likely to be generative than if the specific Y cut is taught directly because it leads to a method for making five, six, or seven cuts. Pursuing this idea just a bit further, one can see that when an adult is challenged to cut a pie for seven, she will probably coordinate these schemes—knowing that the sweep scheme for circles means that it can be cut into any number of equal pieces, including seven, and then using the symmetric cuts to imagine either six or eight pieces and adjusting.

These two activities of genetic epistemology, "creating a need" and "identifying schemes," describe fundamental teaching competencies that are consistent with constructivist theory.

Returning to the question about whether to call this mathematics, I would be inclined to call Katie's method "mathematics," for it allowed her to solve a legitimate mathematical problem of equal sharing. It also leads a more sophisticated observer to consider the questions, "What assumptions have I accepted when I place the circle on a Cartesian plane and attempt to determine algebraically the displacement of the line from the diameter?" "What other method might I have used to examine this problem?" These types of questions lead us as teachers not only to try to explain a child's thinking but to reexamine and challenge our own mathematical knowledge. When we find alternative methods through historical analysis, they are called mathematics (though they may not be the mathematics practiced today). Should not the mathematics of the child be labeled as such?

Claim Two: "Close Listening" to students requires teachers to revitalize their understanding of mathematical ideas.

In literature, there is a concept of a *close reading*. Close reading requires very careful attention, sentence by sentence, indeed, word by word, to the writing in the text. I have invented the term *close listening* as an analogical form to literature for listening to students. Close listening involves an act of decentering by an adult or possibly a peer, in order to imagine what the view of the child might be like. It includes repeated requests for a child to explain what the problem is that she is addressing, what she sees herself doing, and how she feels about her progress. It requires one to ask for elaboration from the child about what, where, how, and why. The questioner should repeat her language as much as possible and invent tests of whether one's own interpretation is consistent with the statements and actions of the child. New tasks should be devised based on one's own understanding of one's position and knowledge of the subject matter. A willingness to assume a rationality in the child's responses and to relinquish or modify one's own expected responses is a necessary part of the process.

Close listening is something that one gets better at over time but only if one practices its methods. An experienced teacher may or may not take easily to learning close listening. It requires a deep and flexible knowledge of the subject matter, a high degree of awareness of one's own beliefs and knowledge, and excellent interviewing skills. Familiarity with student methods through previous exchanges or research reviews is an important part of it.

For instance, I see a variety of issues in Kate's method for the division problem described earlier. Firstly, the fourth grade student uses a quotitive approach to the division. *Quotition* is typically identified with repeated subtraction, but she uses repeated addition through "building up." There is research evidence that this is a widespread phenomena (Olivier et al. 1991). What is most notable is that the problem as written would typically be cast as a partitive, not a quotitive, problem. That is, if the number of beans per scoop had been given, the quotitive approach of seeking the number of scoops through repeated subtraction or addition would have been expected. Since the number of scoops was given, the expectation would be that partitioning, or "dealing" the beans to each scoop, would have been assumed to be used to find out the number of beans per scoop. It is possible that this student saw her tallies as keeping track of the beans mounting up per scoop, but an alternative interpretation would be that she identified the operation as division and used a method she had already developed for division, that of repeated addition. In close listening, a teacher might have asked her before she finished, "What are the tallies keeping track of?" If she could not answer, then the second interpretation gains more credence. The fact that her answer is left in the form of tallies and remainders also supports the second interpretation. What was her interpretation for the remainder? In the widely discussed *National Assessment of Education Progress* (1988), students who were given a problem to find the number of buses needed to transport soldiers answered 34 remainder 5, never considering that they needed another bus to make sense of the problem.

Secondly, in this student's method, I see an example of what our research group has coined "splitting." The girl used the method of doubling to reach her solution quickly. To do this, she "reunitized" (Confrey 1991), making the size of the unit of measure twice as large each time. As she did this, she kept track of the number of original units by doing a "double count." This is reminiscent of the Egyptian method of multiplication found in the Rhind papyrus. In early textbooks, it would be found under chapters entitled "duplication" and "mediation" (Bunt et al. 1976).

Finally, this problem was designed to introduce long division; how now might the teacher build a bridge to this type of division? There are undoubtedly a number of other possible interpretations and observations about this story of children's methods. One can ask, among other things, Who participated? and How did they resolve differences?

Close listening demands a variety of skills, but it is not a skill mastered instantaneously. It is a habit of listening, and it requires a commitment to rethink the mathematical ideas in the context of student learning.

> *Claim Three: To recognize the inventive and alternative student ideas, our understanding of mathematics must include less reliance on single representations, and more recognition of the role of multiple representations.*

In an "Apple Classroom of Tomorrow" in Columbus, Ohio, an inner-city school, a teacher had worked with a software version of "Guess My Rule" called *Function Finder* (Confrey 1988). The functions under examination were linear, unbeknownst to the student who only saw the notation as $y = \boxed{?} x + \boxed{?}$. The student selected x values and the computer generated y values. The student was challenged to guess the rule for the function. Discussing this example in class, the question was raised of how to identify the constant value that is added to the mx in the expression y=mx+b. (Note: A reader may comment that the most

312

obvious way to solve this problem is to put in the value for zero. However, the software is designed in levels. At any level higher than the first, the student is not allowed to use a zero value. The method described by this student overcomes this constraint.) Actually, since the task did not involve graphing, the two coefficients of m and b in the linear equation in this form were called the multiplier and the adder. One student proposed the following method illustrated in Figure 3:

He said, multiply the 3x9 and the 4x7. You get 27 and 28. Subtract them. You get 1. This is the adder. The multiplier is 2 because the right column goes up by two.

This description of the conjecture was all that the teacher had to work with. It worked, producing the equation y=2x+1, and it was unexpected. Luckily for us as researchers and for the teacher, the fire bell went off. Recognizing that there was a fair amount of face validity to the method, we all struggled to remake our knowledge of linear functions to allow us to assess the sense of his method. To see his method, consider the table for a linear function with b=0. For a consecutive set of x values, the y values proceed as an arithmetic progression, producing again the m value. Now trying his method, two identical values are produced. Since this is a direct variation, the ratio of x/y is a constant and his method produces the cross multiplication of $x_1 * y_2 = y_1 * x_2$. For example, 3*8= 4*6.

$3 * 9 = 27$
$4 * 7 = 28$
$28 - 27 = 1$
Therefore the equation is:
$y = 2x + 1$

Figure 3.

In the table represented by Figure 4, one can see the rationality in his method. The adder translates the values in the y column by b. Thus, for consecutive x values, the cross-multiplication produced by the proportion is off by b amount. Whether it is added or subtracted from the multiplied quantity was probably a matter of trial and error for this student, although he could quite possibly have figured out a rule to predict it.

$3 * 8 = 24$
$2 * 4 = 24$
$24 - 24 = 0$
Therefore the equation is:
$y = 2x$

Figure 4.

Seeing how his method works, one can think of a way to challenge it. How would this student have responded if the x values were not given in consecutive order? It turned out that he responded beautifully. As demonstrated in Figure 5 where the x values are given in steps of size two, he used the same method of multiplying diagonally but then divided the difference by the size of the step. Thus, for example, he multiplied 3x11 and 5x7 and compared 33 to 35. The difference of 2 over the two step from x=3 to x=5 meant the adder for a step of 1 was 1.

x	y
1	3
3	7
5	11
7	13

Figure 5.

A couple of points can be illustrated by this example. The first is that it was important for this teacher to go beyond the recognition of the validity of the method to produce the correct outcome. This method illustrates a different notion of a linear function than is often held by classroom teachers. The most common ways to approach these problems from a teacher/textbook perspective are to use a two-point form of an equation or to use two points to calculate slopes and substitute into y=mx+b to find the b value. To assume that students will automatically use a two-point approach is to assume that students recognize that the pattern and the form of the answer, shown on the screen as y=☐☐☐ x+☐, are linear. Few do, even as college freshman. For students engaged in building their concept of a linear function from a pattern of x and y values, what seems to emerge first is the recognition (through trial and error at the simple levels) that a combination of multiplication and addition is required. Then, students' methods vary. Some see zero as a useful value for x since it isolates the y value. Others discover that by using large values for x, the adder becomes conspicuous. This student may have discovered that finding the multiplier can be achieved by using consecutive values, and he could have been left to discover how to find the adder. He seems to have seen the adder as creating a deviance between the proportional equity of consecutive values.

In exploring this further, a constructivist teacher might try a switch of representation. Is there a way to view this students' method graphically (or algebraically) that could further enrich his concept of a linear function? Doing this type of exploration requires a person to change representational forms but also to attempt to carry out the same sequence of actions in the new environment.

These types of examples of inventive methods are not rare in classrooms that allow at least a minimal amount of student discussion. It turns out that as one learns to observe classrooms through the eyes of students, mathematical ideas suggested by students are often potentially productive routes of action. The apparent homogeneity of mathematical methods seen from instruction is not a natural occurrence; it is the product of a very conservative publishing policy that is admittedly resistant to the risks of producing something very different. And given the lack of preparation of many teachers to be able to adapt to new approaches and given the stability of the testing systems, the publishers are themselves captors of a self-reinforcing system.

A second point to be noted is that this student was by no means a high achiever in this class. He was an average student who would tune in and out and perform unevenly across the school year. It is not uncommon to see these ideas coming from students who do not typically excel. In fact, as one watches these students, one suspects that their innovations are due to their stubborn nature in insisting on pursuing their own methods. They are frustrated, and so they do drop out or tune in sporadically to see if anything has changed. We often dissuade the high achievers out of risking a proposal of original and inventive methods, rewarding them more for quickly accepting teacher-produced methods.

Another important point is that by promoting diversity in student methods, we also promote diversity in who is seen as capable in mathematics. Such an outcome is extremely attractive, given the demographic predictions for the next few years. If we do not develop classroom methods that attract and promote the growth of a more diverse group of students, we will witness massive alienation in the mathematics classrooms of this decade.

The primary issue to be discussed from this example is what kind of knowledge of mathematics and of students must a teacher possess to be able to take advantage of such opportunities. Two characteristics of teachers must be developed for this to happen. The first is that teachers must come to believe that student methods are reasoned and need serious consideration. This is a belief that must be "bred in the bone" of good mathematics teachers. Believing it as a merely rhetoric will not be effective. We observe many good teachers who try these methods and begin to elicit provocative student responses but then quickly transform their originality into forms that correspond to the forms that were originally intended by the teacher. "Bred in the bone" believers in students' inventiveness must have a combination of good interviewing skills and good improvisational acting skills. They must be willing to suspend judgment and to allow themselves and the students to reside in confusion. They must be willing to "think differently."

Understanding student methods also requires that teachers recognize that other forms of representation (differing from the traditional algebraic "general forms" that dominate the secondary curriculum) can prove useful. This particular solution relied heavily on a pattern in a data table. A provocative follow-up to such a proposal might be of how we might see on the graph that the product of x_n and y_{n+a} minus the product of y_n and x_{n+a} divided by a is equal to b.

> *Claim Four: When students present new ideas, these may lead to new representational systems. We must be willing to change the mathematics to allow students to explore new alternatives.*

When a computer scientist is working with a programming language and encounters repeated difficulties in using the language to accomplish a certain task, she is likely to consider changing the language, creating new metaphorical structures to accomplish her goals. Constructivism is a pragmatic philosophy; the constructs do not mirror a stable, known reality and thus possess an authority secured by their isomorphic qualities. The constructs must be effective in solving the problems faced by the individuals working with the ideas as perceived by these individuals. By giving students tasks that present genuine challenges to them, we witness the invention of original approaches to mathematical problems. These allow us to reconsider the mathematical ideas involved.

For instance, in a teaching experiment, we asked a college freshman to place a series of 15 dates of significant events on a time line. The events went from the Big Bang to the present. A number of the dates were given in scientific notation. One student, Suzanne, constructed a series of time lines. At first, she equally spaced the events and then dismissed that as inappropriate. Then she divided the time line and marked it in equal intervals of 10^9, 10^8, 10^7, etc. When challenged to locate the years within an interval, she first created 10 divisions and then adjusted this to nine divisions, and starting at 100,000, 10^5, counted off and labeled them as, for example, 200,000, 300,000, 400,000 . . . 900,000. The next mark, the beginning of the next interval was 1,000,000 or 10^6 (see Fig. 6).

The question that I would pose is how legitimate is Suzanne's unconventional representation. It is the case that the additive unit remains constant within the intervals from $10n$ to 10^{n+1}, while the multiplicative unit remains constant across intervals. This is not typical of most forms of scaling. However, when she was asked to compare the length of two intervals when the longer length interval covered a shorter period of time (its units were smaller), she was

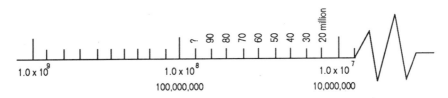

Figure 6.

able to compare them correctly. She also demonstrated that she realized that if she wanted to measure the time spanned across two intervals, she needed to calculate the time span on one interval and then add it to the time span on the other. Thus, it seems fair to say that she became fairly competent in the use of her representation. The only issue that she did not resolve in a relatively satisfactory way was where to place the point representing now. Since her intervals would have gone from 10, 10, 10, 10, or 10, 1, 1/10, 1/100, . . . , she would not be able to reach the current time. She decided to place "now" at the tic mark labeled one.

What was most enlightening about her representation was its fidelity to the form of scientific notation. Numbers put in scientific notation can be viewed as numbers which are divided into two parts, with two different systems of comparison. That is, the power of 10 alerts one to whether the numbers are in the same realm for comparison. Once they are, we can then compare them in the most familiar way, thinking of them as numbers on a conventional arithmetic number line. Her representation suggests the same kind of distinction. The large intervals set the "realm" of the number. Between large intervals, the smaller comparisons are made in a familiar fashion. (See Confrey 1992, for a fuller discussion of this interview study.)

This example is meant to illustrate, especially for scientists, how the students can make suggestions that lead provocatively to a revision in notational or representational form. Other examples in our work are not unusual. We found that, for students, the notation for linear functions of $y = x \, (6) + 3$ conveys more obviously to them that as x increases by 1, the y value increases by 6, as they read the statement as y equals x sixes plus three. This led us to design a piece of software, *Function Probe* (Confrey 1991 and 1992), that allows, among other features, a student to build a button of the form:

$$j \, 1 : \pmb{\theta} \, {}^{*} \, 6 + 3 =$$

In a final example, we recently watched a class in which the teacher had the students build a series of open-topped boxes, each built from a flat sheet of paper with set dimensions of 9 by 13 inches, with the corners of various lengths cut out to form different-sized boxes. They were seeking the one with the maximum volume. Students in the class spontaneously piled the boxes into the configuration shown in Figure 7. Using this configuration and considering how much area is lost and gained between the levels leads us to another approach to the problem, which is quite distinct from the conventional comparison of volumes by finding the maximum of a third-degree equation for volume. This is leading us in our current work to approach rate of change as a fundamental construct and, in cases such as these maximum-minimum problems, to develop the notion of a "trade" as having a significant role in the development of functional thinking. (See Confrey 1992, for a discussion of quadratic equations.)

Figure 7.

What Kind of Preparation Do Teachers Need to be Able to Make Use of these Opportunities?

The four claims listed above suggest that a traditional preparation in mathematics will not prove sufficient for educating teachers to promote this type of constructivist learning. Moreover, they offer a specific interpretative framework for implementing the *Standards* from within a constructivist perspective. Putting forth a series of specific competencies and ways of viewing the teaching-learning enterprise, I have argued for going beyond the *Standards* to undertake explicit epistemological reform.

These competencies can be summarized as follows:

- examining the "genetic epistemology" of a concept

- creating a need for a concept

- identifying schemes

- learning close listening

- coordinating multiple representations

- exploring new representational forms

- changing our mathematical knowledge

The question that remains to be addressed is what kind of education will assist teachers in promoting the development of students' ideas effectively. We have been examining this question in light of our work (Confrey et al. 1991) with experienced teachers and offer a few of the insights we have gained from that experience:

- Inservice education is probably the most appropriate time of intervention. Preservice teachers need to be taught mathematics in ways that are compatible with a constructivist epistemology, but they lack adequate experience in the classroom to anticipate student needs and ideas. They are more concerned with establishing their own legitimacy in the classroom. However, the difficulty in intervening in in-service education lies in the absence of incentives for professional development. Once teachers have received permanent certification, there is little economic incentive for them to continue their education. The changes we seek in instruction are demanding for classroom teachers, and they take a considerable amount of time. To engage in them, teachers need release time and significant support for sum-mer salary. Professional development days should be fully scheduled, with the expectation that they be used for devel-opment and not for completion of outstanding work assign-ments. The National Science Foundation played a signifi-cant role in educating teachers in mathematics during the "New Math" movement. Many of the best teachers now assert that this training was invaluable to them. Perhaps NSF should play a similar role in these initiatives; however, to do so effectively will require a modification in the current NSF practices, which tend to look to professional mathemati-cians to establish or endorse the programs run for teachers. The change to a Piagetian constructivist framework will require the assistance of professional mathematicians but should not depend solely on their existing expertise.

- It is preferable to involve teams of teachers in the activity of change. It is difficult for any single teacher to sustain the dedication, enthusiasm, and determination necessary to change within systems that remain unsupportive of such attempts in their forms of testing, grading, and rewarding teachers. Teams of teachers within and across grade levels and schools need ample time to discuss and debate these initiatives.

- Interventions must be focused around the curricular topics taught in the schools. One may pursue these topics though advanced mathematical courses, but, for the most part, many fundamental issues of relevance to classroom dis-course come out of careful examination of topics already within the scope of the secondary curriculum.

- Tasks and materials should be designed to "create the need" for an idea. Genetic epistemology has a commitment that ideas do not spring up full-blown; they evolve in response to a need for the idea. Designing tasks that create a need for a concept is a powerful method to build up and test one's understanding of the idea.

- Topics should be examined thoroughly, in contextual situ-ations and cross-representationally. Teachers need oppor-tunities to experience the content themselves, as learners, in a different way. This should include the use of new technologies, computers, graphing calculators, and video-discs. Larger themes should be selected as topic organizers to avoid the fragmentation so apparent in most textbooks and state-mandated curricula.

- Starting with student work is a key place to begin. It allows teachers to draw on and build on their wealth of knowledge of student's methods, allowing them to see viable alternatives in student work that they previously would have cast as simply erroneous. It allows all of us to begin with our ultimate clients, the students, and cast our efforts towards an improvement in their welfare. It allows reexamination of the mathematics in a less threatening and highly motivated context. Actual examples of classroom events should be used frequently in the materials. This means having teachers conduct actual interviews with students, analyze segments of high-quality video tape, and instruct in small group teaching experiments. Opportunities for teachers to be observed and critiqued should be provided, and opportunities to reflect on their own practices should be offered regularly.

- We need more research that documents the power of students' methods. This work must be content-specific. Its products need to be (1) the identification of schemes of student approaches to particular topics, (2) the preparation of case studies of students' development of these concepts, (3) the development of curricular materials designed to elicit student ideas, and (4) the examination of the process of teacher's development towards implementing new approaches to constructivist teaching.

It is an exciting time for mathematics education. The opportunity to reconsider mathematics in light of changing societal needs, building connections to everyday activity, responding to the new technologies, and promoting accessibility by more diverse groups has created a promising time for genuine reform. Constructivist interpretation and extension of the *Standards* provide the possibility that reform will be well-grounded in current developmental and epistemological perspective. Instead of seeing reform become hopelessly lost in a morass of political rhetoric, where we switch our descriptive language but leave the stifling atmosphere of mathematics learning in place, we have an opportunity to revitalize the teaching and learning of mathematics.

The success of these changes will hinge on how well we support our professional staff in their efforts to implement these changes. Teachers, mathematicians, scientists, users of mathematical and quantitative reasoning, and mathematics educators must join together in producing serious and informed change.

REFERENCES

Balacheff, N. "The Treatment of Refutation: Aspects of the Complexity of a Constructivist Approaches in Mathematics Learning." In Ernst von Glasersfeld (Ed.), *Radical Constructivism in Mathematics Education.* Dordrecht, The Netherlands: Kluwer Academic Publishers, 1991. 111–138.

Bunt, L., P. Jones, and J. Bedient. *The Historical Roots of Elementary Mathematics.* Englewood Cliffs, NJ: Prentice Hall, 1976.

Confrey, J. "Learning to Listen: A Student's Understanding of Powers of Ten." In Ernst von Glasersfeld (Ed.), *Radical Constructivism in Mathematics Education.* Dordrecht, The Netherlands: Kluwer Academic Publishers, 1990. 111–138.

Confrey, J. "The Concept of Exponential Functions: A Student's Perspective." In Leslie Steffe (Ed.), *Epistemological Foundations of Mathematical Experience.* New York: Springer-Verlag, 1991. 124–159.

Confrey, J. *Function Probe* [Computer Program]. Santa Barbara, CA: Intellimation Library for the Macintosh, 1991.

Confrey, J. *Function Finder* [Computer Program]. (3.0). Santa Barbara, CA: Intellimation Library for the Macintosh, 1991 and 1992.

Confrey, J. "Using Computers to Promote Students' Inventions on the Function Concept." In S. Malcolm, L. Roberts, and K. Sheingold (Eds.), *This Year in School Science 1991: Technology for Teaching and Learning.* Washington, DC: American Association for the Advancement of Science, 1992. 141–174.

Confrey, J., E. Smith, S. Piliero, and J. Rizzuti. *The Use of Contextual Problems and Multi-representational Software to Teach the Concept of Functions.* Final Project Report to the National Science Foundation (MDR-8652160) and Apple Computer, Inc., 1991.

Davis, R.B. "Solving the 'Three Switch' Problem: A Case Study." *Journal of Mathematical Behavior*, No. 4, (1985): 281–291.

Duckworth, E. *"The Having of Wonderful Ideas" and Other Essays on Teaching and Learning.* New York: Teacher's College Press, 1987.

Easley, J. *On Clinical Studies in Mathematics Education.* Columbus, OH: Ohio State University Information Reference Center for Science, Mathematics, and Environmental Education, 1977.

Erlwanger, S. "Case Studies of Children's Conceptions of Mathematics—Part I." *Journal of Children's Mathematical Behavior*, 1(3), (1975): 157–268.

Ginsburg, H. *Children's Arithmetic: How They Learn It and How You Teach It.* Austin, TX: Pro-Ed., 1977.

Karmiloff-Smith, A., and B. Inhelder. "If You Want to Get Ahead, Get a Theory." *Cognition* (3), (1975): 193–212.

Lakatos, I. *Proofs and Refutations: The Logic of Mathematical Discovery.* J. Worrall and E. Zahar (Eds.), Cambridge: Cambridge University Press, 1976.

Mathematical Sciences Education Board, National Research Council. *Reshaping School Mathematics: A Philosophy and Framework for Curriculum.* Washington, DC: National Academy Press, 1990.

Mathematical Sciences Education Board, National Research Council. *Counting on You: Actions Supporting Mathematics Teaching Standards.* Washington, DC: National Academy Press, 1991.

National Assessment of Educational Progress. *The Mathematics Report Card.* Princeton, NJ: Educational Testing Service, 1988.

National Council of Teachers of Mathematics. *Curriculum and Evaluation Standards for School Mathematics.* Reston, VA: NCTM, 1991a.

National Council of Teachers of Mathematics. *Professional Standards for Teaching Mathematics.* Reston, VA: NCTM, 1991b.

Olivier, A., H. Murray, and P. Hunan. "Children's Solution Strategies for Division Problems." *Proceedings of the Thirteenth Annual Meeting of the North American Chapter for the Psychology of Mathematics Education, 2,* (1991): 15–21.

Piaget, J. *Genetic Epistemology.* New York: Norton, 1970.

Russell, S.J., and R. Corwin. "Talking Mathematics: 'Going Slow' and 'Letting Go.'" *Proceedings of the Thirteenth Annual Meeting of the North American Chapter for the Psychology of Mathematics Education, 2,* (1991): 175–181.

Sinclair, H. "Constructivism and the Psychology of Mathematics." *Proceedings of the Eleventh International Conference for the Psychology of Mathematics Education,* Vol. 1, (1987): 28–41.

Steffe, L. "Children's Construction of Number Sequences and Multiplying Schemes." In M. Behr and J. Hiebert (Eds.), *Number Concepts and Operations in the Middle Grades.* Reston, VA: National Council of Teachers of Mathematics, 1988. 119–140.

Thompson, P. (in press). "The Development of the Concept of Speed and Its Relationship to Concepts of Rate." In G. Harel and J. Confrey (Eds.), *The Development of Multiplicative Reasoning in the Learning of Mathematics.* Albany, NY: SUNY Press.

Unguru, S. "On the Need to Rewrite the History of Greek Mathematics." *Archive of Exact Sciences,* 15(2), (1976): 67–114.

von Glasersfeld, E. "Constructivist Notes on Teaching." In Kenneth Tobin (Ed.), *The Practice of Constructivism in Science Education.* Washington, DC: American Association for the Advancement of Science, 1992.

PART 4

Conclusions

19

On Considering Constructivism for Improving Mathematics and Science Teaching and Learning

Thomas M. Dana and Nancy T. Davis

The concept of constructivism has kindled a great deal of interest in many educational circles. Within the fields of mathematics and science education, many scholars from a diversity of backgrounds and interests are beginning to use constructivism as a way to make sense of educational phenomena. The purpose of this book is to explore the nature of constructivism as an epistemology and the role constructivist thinking can have in teaching and learning contexts. The idea of constructivism has stimulated debate recently in educational circles, and the chapters of this book add important contributions to the ongoing discussions about knowledge, knowing, teaching, learning, and research in mathematics and science education.

This book is likely to serve as a starting point to understanding constructivism for many readers while challenging those already familiar with constructivism to reconceptualize educational situations—particularly those associated with efforts to vastly improve the current state of mathematics and science education. As such, the primary purpose of this concluding chapter is to examine the arguments and cases presented by the authors in order to highlight some of the common notions interwoven throughout their various stories. While the authors have guided the reader through many philosophic arguments and practical examples, we will examine the question of how embracing the notions of constructivism must also require a collective rephrasing and rethinking of many taken-for-granted educational practices. To that end, we have organized our comments in this final chapter around five themes woven throughout this book: knowledge and knowing, teaching and learning, educating teachers, educational research, and future directions.

KNOWLEDGE AND KNOWING

Consider, for a moment, an image of a teacher with an epistemological perspective that posits the process of coming to know as the search for truth. Learners are like discoverers who have the task of finding knowledge, perhaps with the purpose of coming to understand the world the way it "really" is. Several authors in this volume have labeled this perspective *objectivism* or *positivism* and, accurately, so we believe, pigeon-holed this perspective as the foundation of the traditional model of education. In this model it is assumed that an already developed body of knowledge, developed, proven, and accepted by society, can easily be transmitted to students through generally passive instructional means. From this perspective, coming to know and making progress toward knowing the truth is a matter no more complicated than precisely following a set plan or scientific method. Furthermore, success in teaching and learning contexts can be defined by what the students learn. The assessment of student knowledge, traditionally, is limited to a focus on the recall of factual information and labels. No wonder the process of teaching is defined to be telling, and the process of learning is often considered to be memorization and recall. The myths of the teacher as "great communicator" and student as "quiet listener" and "effective memorizer" make much sense if one accepts this epistemological perspective. However, if another epistemological perspective becomes the referent, these all-too-familiar conceptions of knowledge, knowing, curriculum, and instruction all come into question.

Ernst von Glasersfeld stated that "constructivism is an attempt to cut loose from the philosophical tradition . . . that knowledge has to be a representation of reality." This apparently simple statement can have the effect of a cadre of backhoes and landmovers: once they get going, the landscape will never be the same. The educational implications of constructivism are earthshaking in the sense that knowledge is not viewed as a set of truths that exist outside the learner and that learning is not passively receiving knowledge from sources such as a teacher or textbook. All of the authors of this volume have stated, in one form or another, that we cannot transmit meaning but that others must construct it for themselves. How this might be done and in what contexts is at the heart of some of the differences among the authors. Von Glasersfeld, Bettencourt, Cobern, and Good et al. lay out for the reader some differing perspectives on constructivism with discussions of *radical constructivism, trivial constructivism, social constructivism, personal constructivsm,* and *contextual constructivism.* These perspectives, while espousing some fundamental differences, support a coherent point of view that knowledge is built by individuals as a result of interactions with prior knowledge, events, and other people.

Cobern's chapter helps the reader appreciate the notion that culture impacts teaching and learning. By defining learning as only occurring within a context and linking this idea to anthropological world view theory, Cobern places into question conceptions of "how students make sense of the world." Race, language, economic and educational levels, occupation, geographic location, gender, religion, and philosophy are all mentioned by Cobern as sources of considerable cultural variation that impact knowledge construction. As Cobern notes, "A student constructs knowledge so that the knowledge is meaningful in the student's life situation." Another important contribution of this chapter is the historical look at the evolution of constructivism. Cobern's chart of the "genealogy" of constructivist educational research is a useful tool for understanding the origins and relationships between differing constructivist orientations, such as the influence of personal constructivism on the conceptual change movement and the sociology of knowledge on contextual constructivism.

The chapter by Good, Wandersee, and St. Julien serves as a strong reminder that knowledge must stand the test of, in von Glasersfeld's words, viability. Constructivism, just as other models of knowing, should not be uncritically embraced as the "right" or "real" way people come to know. Each of us must attempt to make sense of the ideas of constructivism in as many contexts as possible, determining where the ideas work and where they do not. The arguments presented by Good et al. challenge us to reflect on differing conceptions of constructivism and to test the viability of our understandings with their sometimes critical notions of the promises, possibilities, problems, and pitfalls constructivism offers for understanding educational practices. This kind of activity helps each of us to enrich our understandings, and in turn, our research and practice in mathematics and science education contexts.

TEACHING AND LEARNING

The premise that students are knowing beings and that knowledge is personally meaningful and a result of a constructive activity has promoted renewed efforts to understand what it means to learn and to teach. As individuals begin to use constructivism as a referent for making sense of teaching and learning, they must reexamine beliefs and practices in a new light. The activity of reflection on beliefs and practices is a continual process as the individual struggles to make sense of her new understandings of teaching and learning. Educational practices are developed by teachers over their entire experiences as teachers and learners, and unless they are reflected on, teachers may continue to use them without understanding why they do so.

It is without much argument that objectivist perspectives have been used to make sense of schooling in the past. The structure of the educational process in most schools reflects that perspective. Whether we consciously recognize it or not, we all hold mental images of what schools that are grounded in objectivism are like. The process of critically examining our beliefs, images, and practices takes time and constant attention. The chapter by Taylor points out the difficulties each of us has when we attempt to reconceptualize ingrained practices. Taylor found in his work with a teacher named Ray that many of his own beliefs used objectivism as a referent and needed reconceptualizing for understanding the implications of constructivism. The interactions he had with Ray assisted him in his own reflections in this area.

As with the case of Ray, teachers who begin to embrace constructivist notions become disturbed with the learning in their classrooms. Although readily accepting constructivism as a viable alternative, they are unsure of what their roles might be in classrooms based on this newly considered epistemology. Oftentimes, teachers request that others provide images of "constructivist classrooms." This can be dangerous because it leads to the impression that constructivism is a method of teaching rather than a theory of knowing and learning. We must realize that every learning environment where learning is occurring can be called a constructivist classroom. Take, for example, a large lecture class in introductory biology or college algebra in the university setting. The typical course consists of three one-hour lectures each week, problem sets, and textbook readings. Students do learn from those lectures and from the readings in the textbook; but what students learn may not be exactly what the instructor or author intended. Students are sense makers. They interpret what has been said or read in light of what they already know, constructing knowledge about biological or algebraic concepts that fits their understandings of the world. The challenge for teachers, especially university teachers, becomes the question of how to design learning opportunities that result in maximal learning. Several chapters in this book present ways in which this might be achieved by focusing on group learning and negotiation, problem-centered activities, laboratory activities, and learning through the use of technology.

As one considers the pedagogical implications of constructivism, the focus of teaching moves from proficiency at content delivery to assisting individuals in their interpretations of concepts. Wheatley, Jakubowski, and Linn and Burbules each present ideas about involving students in the activity of learning. Cooperative learning offers much promise in moving away from teacher-centered to student-centered instruction. Wheatley discusses the role of both the individual and the group in negotiation of meaning. He emphasizes the importance of language in developing understanding of others and in communication of ideas. Wheatley also delineates aspects of establishing meaningful tasks for the learners to undertake. Jakubowski addresses the need to establish learning environments that are conducive to nurturing students' confidence in their own ideas and in their ability to learn. Likewise, Linn and Burbules' review of cooperative learning reminds us that it is not the method as much as the motivation for learning that is important and that cooperative learning, just as any other method of instruction, does not offer an educational panacea. All three of these chapters caution about unjustified implementation of cooperative learning just because it may be fashionable and in vogue. Linn and Burbules stress the need to understand the goal behind the learning activity. If a learning environment is based on constructivist ideals, then one must recognize that cooperative group activities can both foster and hinder knowledge construction. Groups engaged in healthy debate or problem-solving activities are likely to construct and test many ideas while groups without motivation or direction will meet with limited success.

As with groups, learning can take place in a variety of contexts. Roth's analysis of student learning in classrooms and laboratory situations provides insights into the value of providing a variety of learning activities and how individual students make sense of science from those activities. Roth's chapter serves, again, as a reminder that it is not the method that is of importance, it is the focus on learning that matters most. Roth provides the reader with illustrative vignettes of himself in role as a teacher—the facilitator of the learning through the design of meaningful activities. He shows us that he must constantly be aware of individual student's experiences and have a repertoire of methodologies so that he can use his professional judgment in facilitating student learning. Roth's discussion of the notion of scaffolding is particularly noteworthy as it is his justification (professional judgment) for promoting collaborative group learning. Furthermore, in admitting that, as a teacher, he values the opportunity for students to interact with individuals who can challenge each other and that, as a teacher, he cannot be that individual to every student all of the time, Roth reveals that teachers as the designers of learning situations often face dilemmas that influence the choices they make in their classroom.

As educators focus on the individual learning rather than on content to be taught, new issues arise. As constructivism becomes the referent for making sense of teaching and learning, ethical, social, and gender issues need to be revisited. Just as compelling is Gallard's powerful, personal story of his learning experiences in traditional schools. He reminds us what it means to sacrifice our individuality in order to fit the accepted way of thinking and acting. Each of us can think of situations where we have doubted our own ways of making sense and conformed to authoritarian pronouncements as we learned to play the game of education. Gallard raises issues of ethics that might stem from reconsidering traditional notions of schooling. The chapter helps us to realize that teaching and learning are moral acts involving the self-esteem and self-concept of humans.

Throughout the chapters in this book, we have repeatedly read the idea that teachers who begin to consider constructivism call into question certain educational practices. But without images of alternatives, the questionable practices often continue. Educators with alternatives in their teaching repertoires have a choice in what learning opportunities they provide for students. Considering instructional alternatives is one way in which teachers reflect on, and learn from, their practices. In Rieber's description of instructional technology programs, alternative models of curricula and curriculum design are presented. As students make sense based on their individual interpretations of reality, the teacher who has a large repertoire of possible curriculum designs can be more flexible in adapting classroom activities to individual student's needs.

TEACHING TEACHERS: VIEWING THE TEACHER AS A LEARNER

The various chapters of this book challenge readers to consider constructivist views about the nature of knowing, teaching, and learning. By far, the most challenging issue that comes to mind when considering the notion of constructivism is the same challenge that has faced many other educational scholars who have attempted to improve teaching and learning practices. This challenge occurs at the level of practice—what goes on in classrooms once the bell has rung and the doors are closed. A number of authors in this volume embrace the challenge. From a constructivist perspective, it might be argued that student learning is at the heart of teaching. In an ideal situation, teachers might, with a goal of assisting students to make sense of the world, actively involve the learner in tasks which promote reflection on their learning, inference making, cognitive conflict, and negotiation of meaning. But not all situations are ideal . . . and not all teachers consider these activities to be teaching. As Tobin notes, teachers who adhere to an objectivist epistemology will see little value in classroom activities that are consistent with the notion that learning is something the learners do and that mathematical and scientific knowledge are something that individuals construct from interactive settings with materials and other people. A potential key to helping teachers adopt teaching strategies that promote active learning might come in the form of assisting teachers to understand learning from a constructivist perspective. But how might this be done?

All of the efforts at teacher education reported are based on the premise that prospective and practicing teacher education activities, traditionally one-day, one-shot, tell-them-what-to-do phenomena, have often failed because they advocated that teachers change teaching approaches without fully engaging these teachers in the activity of examining beliefs about knowing and learning. Conversely, the models of teacher education presented in this book are built on the epistemological foundation of constructivism. Using the best of what is known about learning, the authors regard teachers as learners—inquirers into what it means to know, learn, and teach.

Several of the authors provide models that are consistent with the notion of teachers as learners. Basing their model on a case study of Jessica, an experienced grade five teacher, Shaw and Etchberger focus our attention on the processes teachers like Jessica might encounter as they wrestle with reconceptualizing teaching and learning activities. They devote their chapter to highlighting the events that Jessica faced as she began to embrace a constructivist epistemology. They argue that "for successful and positive change to occur teachers need to be perturbed, they need to be committed to do something about the perturbation, they need to establish a vision of what they would like to see in their classrooms, they need to develop a plan to implement this vision, and they need support to carry out their vision."

Likewise, other authors approach practicing teacher education as the context for assisting teachers to understand the nuances and subtleties of a constructivist-orientation to knowing. Peterman advocated that staff development opportunities should entice teachers to make explicit their implicit "core" beliefs and personal theories about classroom practices and assist them in reconstructing beliefs about learning and teaching. One of the major issues that Peterman addresses is the preponderance of staff development institutes that focus solely on the implementation of classroom procedures, but not on the underlying principles of the procedures. In short, she justifies her ideas that staff development opportunities must engage teachers in iterations of considering something new, reflecting, and enacting by indicating that "as constructivist notions about students' sense making in classroom settings impacts instruction, similarly, teachers' sense making must be considered in the design of staff development."

Gallagher, too, recognizes that a mode of schooling based largely on behaviorist traditions has been the only educational model most teachers have ever encountered and that notions consistent with constructivism are often considered "strange" and "unwelcome." His plan to assist middle school mathematics and science teachers to make the necessary transformations to a constructivist epistemology begins with where the teachers are, both in a logistic and a cognitive sense. First, the teachers' union and administration endorsed the model that Gallagher and his cohorts at Michigan State University designed. Gallagher's program directly involves teachers in the sense-making process at various school sites. Through building a model of teacher enhancement that respects teachers' current practices and engages teachers in challenging readings and classroom research, Gallagher reaches out to teachers in a way that narrows the gap between practitioners and ivory tower scholars.

One thread that is interwoven throughout the chapters on teacher education is that of the notion of reflection. By engaging teachers in purposeful inquiry about constructivism and what it means to know and learn, the authors have perturbed teachers' notions of what it means to teach and learn. As indicated in several of the chapters, coming to understand constructivism for many teachers has involved iterations of examining what is happening in their classrooms, developing a vision of what mathematics and/or science lessons could be like, implementing new ideas, having successes, and making mistakes as they put their beliefs into action, and reflecting on what has occurred in their classrooms. As the cycle continues, teachers gain knowledge and understanding of what it means to be a teacher and a learner.

EDUCATIONAL RESEARCH

What counts as viable research for educators changes as they begin to use constructivism as a referent for thinking and practice. A paramount question that needs to be asked is, "What is the research for?" From an objectivist perspective, a reality outside of the individual is sought. Objectivists seek to divorce human subjectivity from research and eliminate values and consideration of contextual factors from educational discourse. Educational research reflects an interest in discovering cause-and-effect relationships in teaching and learning in order to suggest the most efficient manner in which instruction can be designed to transmit knowledge determined by experts to be important to the learner. Generalization across populations is considered a strength of good educational research. Much of the process-product research based on student achievement scores reflects the objectivist interest.

In contrast, educational research considered valid by constructivists is grounded in individual interpretations and constructions of meaning. Researchers are learners, involved in a problem-solving endeavor in a specific context. As such, research is not a search for a truth external to the cognizing individual. Rather, the focus of classroom research, for example, might be on developing understandings of the key players, say, teachers and learners, within a particular context. Educational research becomes a very personal and subjective endeavor. Emotions are an integral part of the sense-making process and should not be considered to be objectively put aside. Each participant, including the researcher, who is also a participant in the process, must struggle to come to understand what is happening and why. Action research, critical autobiographies and biographies, case studies, and critical ethnographies can be valuable for researchers and other participants to provide alternative perspectives that can assist in interpreting events. Interpretative research, as Tobin explains it, has a goal of critical subjectivity. He explained, "To be critically subjective . . . is to be conscious of one's beliefs, values, and epistemologies and construct alternative explanations that go beyond those which are associated with the beliefs to which highest value is assigned."

Many ethical issues arise as researchers use constructivism as the referent for making sense of teaching and learning. Issues of power and control need to be constantly examined and reflected on as the research process continues. Collaborating with teachers as they attempt to make sense of their teaching and learning situations, rather than conducting research on teachers, should become a focus. This may tend to add a dimension to the research that gets a bit "sticky" as each participant is asked to come to the mirror and reflect on her beliefs and actions. We must remain cautious that we are dealing with emotive persons, not objects, when we do research. Clearly, there is a need to be sensitive to the understandings and self-esteem of others.

FUTURE DIRECTIONS: A CALL TO REVOLUTION?

With the state of mathematics and science education receiving closer attention by politicians and media and with the United State's goal of having its students be world leaders in mathematics and science achievement, we should expect to hear the clamor of reform movers and shakers well into the next decade. Mathematics and science curricula, teaching strategies, using technology, equity and gender, and student assessment are at the center of many of these reform movements. Interestingly, some of these movements are based on notions that might be considered consistent with some pedagogical implications of constructivism. In the case of the National Council of Teachers of Mathematics (NCTM), a set of innovative national standards is taking hold in many schools. Advocating "active student learning," the NCTM curriculum emphasizes problem solving, communicating, and making connections between mathematics topics. Students are expected to have hands-on experiences, use computers, and learn mathematics concepts in cooperative groups. Similar reform movements are beginning to take shape in science as well. In many ways, the ideas behind these movements are revolutionary, causing teachers, administrators, curriculum supervisors, parents, and policy makers to rethink the kinds of mathematics and science that are ingrained in the present educational system and seek out alternatives that will promote learning content and process with understanding. However, while these efforts seem promising, revolutions are a scary thing to many people, and "new ideas" are often skeptically shunned rather than critically embraced.

In addition to teaching and learning strategies, one issue that needs much more exploration is the role of assessment within a constructivist paradigm. Matching student achievement to predetermined objectives is based on an objectivist view that experts know correct answers, and student answers should reflect those of the experts. If we believe that learning occurs as meaning is given to experiences in light of existing knowledge, then assessment techniques must permit students to express their personal understandings of concepts in ways that are uniquely theirs. We need to find ways to determine a student's depth of understanding. Ideally, teachers might use these strategies not to solely determine what has been learned during a particular "unit" of instruction, but to understand students' prior knowledge and monitor their own instructional success. Alternatives to traditional testing programs such as performance assessments, cooperative problem solving, concept mapping, and portfolios have much promise in this area. However, assessment of student understanding continues to be a puzzle which needs continued attention.

Another issue that needs much more attention is what content knowledge should students learn in order to be considered competent. Knowledge within science and mathematical fields also continues to grow at exponential rates. The amount of knowledge needed to develop competence within even a narrow field of study is problematic as new technologies allow for further observations and refinements of scientific and mathematical models. The emergence of mathematics and science as new national education goals raises questions about not only the content of the curriculum, but also its depth and breadth. As reform efforts continue in their loops of refinement, we will continue to hear questions such as, "what are appropriate mathematical understandings for seven year olds?" and, "is there a 'core of knowledge' for high school chemistry?" The question of what content should be addressed cannot be divorced from the question of who determines that content and whether it is all students or only college-bound students that should know it. Within the objectivist paradigm, "experts" determine what should be taught. Constructivists must be able to answer these questions, too.

It is often said that resistance to change is a formidable barrier to be overcome. However, as we have read in the cases of several teachers, coming to understand constructivism was nothing short of a revolution in the way they approached their teaching. Perhaps the central lesson to be drawn from this book is that constructivism provides a set of referents that can be used to make sense of all situations where someone is learning. Some of those situations are easier to understand than others. Yet, many tough challenges remain.

Index